ADVANCED PLANT TAXONOMY

ADVANCED PLANT TAXONOMY

[For Undergraduate and Postgraduate Students]

Dr Amal Kumar Mondal MSc, BEd (Science), PhD

Postgraduate Department of Botany
Barasat Government College
Formerly Postgraduate Department of Botany
Darjeeling Government College, Darjeeling

New Central Book Agency (P) Ltd

Co-Published by

Bijay Publications

E-mail : bijaypublications@gmail.com

Exclusively Distributed by

Bijay Pustak Bhandar (P) Ltd

90/6A M. G. Road, Bhawani Dutta Lane, Kolkata-700007

✆ : (033) 4605-7856, 8777898576, 8274916104

E-mail : bijaypustakbhandar@gmail.com

NCBA

REGD OFFICE

8/1 Chintamoni Das Lane, Kolkata 700 009, India
email: ncbakolkata@gmail.com

OVERSEAS

NCBA (UK) Ltd, 149 Park Avenue North, Northampton, NN3 2HY, UK
email: ncbauk@yahoo.com

CORPORATE OFFICE

212 Shahpur Jat, New Delhi 110 049
email: ncbaexports@ncbapvtltd.com

BIJAY PUBLICATIONS
90/6A M. G. Road, Bhawani Dutta Lane, Kolkata-7
E-mail : bijaypublications@gmail.com

BIJAY PUSTAK BHANDAR (P) LTD
90/6A M. G. Road, Bhawani Dutta Lane, Kolkata-7
E-mail : bijaypustakbhandar@gmail.com

ADVANCED PLANT TAXONOMY • Dr Amal Kumar Mondal

First Published: September 2005
With Additions: March 2007
Reprinted: January 2009
Reprinted: July 2011
Reprinted: January 2014
Reprinted: July 2016
Reprinted: January 2018
Reprinted: 2023

PUBLISHER • TYPESETTER • PRINTER
New Central Book Agency (P) Ltd
8/1 Chintamoni Das Lane, Kolkata 700 009

CO-PUBLISHED BY
Bijay Publications
90/6A M G Road, Bhawani Dutta Lane, Kolkata 700007

COLOUR PHOTOGRAPHS
Dr Amal Kumar Mondal

PROJECT SUPERVISOR
Prabir Ghosh

ISBN: 978 81 7381 491 4

Price: ₹ 715.00

Dedicated to my beloved parents

Dedicated to my beloved parents

Preface

Taxonomy is an ever-changing field of biology and is a complex mixture of biology, philosophy and mathematics. Taxonomy and systematics, either directly or indirectly, encroach many other areas of science, including medicine, pharmacology, agriculture, horticulture, etc., as well as traditional areas of Botany, Zoology and Microbiology, and within each of these, taxonomy provides names for organisms and a framework within which the organisms can be classified. Systematics, in turn, provides an insight into the evolutionary relationships between organisms, which form the basis for comparative studies and drawing unbiased conclusions about the relationships among and between characters and environmental factors.

Unfortunately, in spite of its obvious importance, taxonomy has largely been a neglected branch of biological science in India, and taxonomists have long drawn a measure of scorn from other biologists, who consider taxonomic work as a rather imprecise area of research. The role of taxonomists has often been misunderstood and many biologists, who are not directly involved, hardly have any concept of the rigor involved in the proper execution of taxonomy. This widespread downturn in taxonomic research is largely due to the lack of adequately trained taxonomists, failure in revival of interest in the subject in the present generation, lack of adequate literature and text books to cater to the needs of the students and researchers working in this field, and the lacunae in the syllabi of most universities and colleges, which have largely remained marginalized just to the traditional aspects of plant taxonomy and more trendy areas, excluding the modern methods and the recently developed aspects and techniques in modern taxonomy. Ironically, this downturn has occurred at a time when circumstances have made the need for more taxonomic research urgent. In several countries, just as with other fields of science, taxonomy has continued to advance in leaps and bounds, both in procedure and in philosophy. Modern taxonomy has a potential to reshape a large part of our current biological thought.

This book aims to stem down this downtrend in taxonomy, by introducing the students not only to the traditional aspects of plant taxonomy, but also to the modern techniques and recent trends that forms the basis of good taxonomic practice and interpreting evolutionary relationships. The book includes wide range of consideration about characters, taxa, current rules of nomenclature, classification systems, range of techniques, including the modern techniques in molecular biology and methods used to collect taxonomic characters, as well as a comprehensive account of plant collection and botanical gardens.

A number of practical features within the textbook enhance its usefulness for students. It is richly illustrated with figures that include new information and provide fresh perspective.

- Each chapter starts with an outline.
- Important concepts and taxonomical terms are boldfaced.
- Cross-references direct students to other areas of the text where they can refresh their understanding or clarify an unfamiliar subject.
- Important relevant literature is cited at the end of each chapter to connect the readers to further references and resources.
- Interesting boxed essays are included in some chapters to present subjects or illustrations.

Chapter I: Introduction

This chapter introduces the discipline, defines central concepts and provides some principles, definitions of terms associated with taxonomy and phases of plant taxonomy.

Chapters II & III: Plant Identification & Plant Nomenclature

The main role of a taxonomist is to properly identify and name plants. These chapters explain the intricate rules of plant identification and nomenclature.

Chapters IV & V: Plant Herbarium & Botanical Garden

These chapters introduces the students to the proper techniques that should be followed for preparing an authentic herbarium and the role of Botanical Gardens as well as the up-to-date information on important herbaria and Botanical Gardens of the world.

Chapters VI & VII: Plant Classification & Systems of Plant Classification

Challenges to existing phylogeny are arriving at a very frequent rate with the extensive research being carried out in different parts of the world. Many contradictory aspects in comparison to the existing classifications are arising. Students need to be prepared for new phylogeny and classifications, rather than committing one to memory forever. These chapters include a historical overview, the earlier traditional systems of classification, as well as the current phylogenetic classification systems including the APG system.

Chapter VIII: Taxonomy — A Synthetic subject

It discusses the role of various botanical disciplines — external morphology, ultrastructural systematics, cytology, embryology, palynology, phytochemistry, serology, ecology and paleobotany in solving various taxonomic problems.

Chapter IX: Numerical Taxonomy

Science is a process of ongoing research. Nowhere, perhaps, is that more evident than in phylogenetics. Construction of phylogenetic hypothesis via numerical analysis and cladistics, is one of the major additions to functional and evolutionary morphology. Thus, the basics are included in this chapter.

Chapter X: Origin and Evolution of Angiosperms (Anthophyta)

This chapter presents the latest information on phylogeny of angiosperms, the various theories regarding their origin and evolution, along with exquisite illustrations of various fossil forms.

Chapter XI: Remote Sensing Technology and Vegetation Classification

This chapter expands on one of the latest developments in science and technology i.e. the use of remote sensing in the classification of vegetation. It enriches the reader with some of the basic principles of remote sensing, its pros and cons, and utility in classification and resource management.

Chapter XII: Future in Plant Taxonomy

Today there is a great dearth of botanists with broad-based knowledge on taxonomy. Most present-day students and researchers hardly care for taxonomic involvement in the advancement of plant sciences. Thus, this chapter critically analyzes the main causes for the down stem in plant taxonomy and explains the various prospects and future of plant taxonomy.

I hope this book will benefit not only the undergraduate students of universities, but also postgraduates, teachers, researchers and other professionals interested in this field. I welcome comments and suggestions from all the users of this book and would be grateful if the reader notifies my errors, which I am very much willing to rectify for the improvement of this book.

Kolkata, August 2005

DR. AMAL KUMAR MONDAL

Acknowledgements

During the preparation of this book, my colleagues and students shared their special knowledge, volunteered their co-operation, thoughtful suggestions and criticisms, which have greatly influenced this text and I extend my sincere thanks to all of them.

Writing a book with a scope as broad as this one requires the assistance of many people. In particular I recognize and am especially grateful to the following individuals for their generous responses to my requests for information and their comments and suggestions :

- Professor Amy McPherson
- Gerald D. Carr
- Professor Angela Fössl
- Rosita Sturm (Ms.)
- Professor Dr. Dietrich Gotze,
- Dr. Les Watson
- M. J. Dallwitz
- Prof. Craig Brough
- Professor Monique Dubrule Reed
- Professor Andrea Tarozzi
- Fun Science Gallery
- Mr. Tarozzi
- Giorgio Carboni
- Dr. Les Watson
- Professor Nicholas Plummer
  The Librarian, **Government College University Lahore**, Lahore, Pakistan
- Professor Dr. Khalid Aftab, Vice Chancellor, **Government College University Lahore**, Pakistan
- Professor Hugh D. Wilson, Curator
- The Editor, Annals of Missouri Botanical Garden
- Professor Dr. Dietrich Gotze
- Dennis W. M. Stevenson, The New York Botanical Garden Press
- Professor James R. Manhart, Associate Professor of Biology
- Professor Monique Dubrule Reed
- Professor John J. Leggett
- Professor Richard Furuta
- Professor Andrea Tarozzi, Fun Science Gallery
- Professor Dr. Toni Paine
- Professor Dr. Mike Dallwitz
- Professor Dr. Eric Zurcher
- Professor James Reveal, Northern Brown Herbarium, University of Maryland
- Professor R. Liesner, Missouri Botanical Garden

Their suggestions have shaped nearly every aspect of this book. I cannot sufficiently express my gratitude for their efforts.

I am indebted to the patient, able and good-humored people at New Central Book Agency, who were so important in seeing this text-book through, to the book you now hold in your hands. To friends and family, I remain grateful, especially to my parents who helped, encouraged, tolerated and supported me during the preparation of the manuscript of this book. I extend my thanks to my wife Dr. Sanjukta Mondal (Parui), Senior Lecturer in Zoology, Department of Zoology, Barasat Government College for embellishing this book with the beautiful illustrations drawn by her, which greatly enhance the value of the text. I also extend my full-hearted love and affection to my dear daughter Annanya, who, although just three while I wrote this book, was always understanding and sure never to disturb me while at work, but was always there to rejuvenate me with her mischievous pranks, which gave me immense joy after my day's work. Special thanks are due to my Ph.D. student Rajat Pramanila for his whole-hearted cooperation and kind help.

I want to express my deep appreciation to my former teachers of Suri Vidyasagar College, Suri, Birbhum, where I completed my Honours graduation; my teachers at the Post-Graduation level at the University of Burdwan (Department of Botany), and my other teachers of Visva-Bharati University (Department of Botany), specially my guide and supervisor Professor Sudhendu Mandal, Professor of Botany, presently the Director of National Library, Belvedere, Alipore, Kolkata 700 027, INDIA, for conveying me a realistic picture of what Botany and Plant Taxonomy are all about. I have pleasant memories of many stimulating discussions about Botany and Plant Taxonomy that took place during my days as a M.Sc and Ph.D. student at the Universities.

DR. AMAL KUMAR MONDAL

CONTENTS

Chapters	Pages

Chapters		Pages

Chapters		Pages

Chapters			Pages

Chapters	Pages

Chapters		Pages

Chapters		Pages

Chapters **Pages**

Chapters	Pages

Chapters	Pages

1 INTRODUCTION

AT A GLANCE

Taxonomy "the theory and practice of classification", is one of the oldest fields of biological sciences. The wealth of vegetation that adorns the earth includes, a conservative estimate of about 400,000 described species of plants, of which, 286,000 belong to the flowering plants. They show a wide diversity of form and mode of life. Some are gigantic trees attaining a height of about 100 meters with very complex structure, while some of them are minute floating herbs with a very simple construction. Since the early stages of civilization, man has been absolutely dependent on plants for food, shelter, clothing, fuel, medicine, paper, etc. Thus, the primitive man tried to group the plants according to their economic uses.

1.1 Need and Importance of Taxonomy

With the huge variety of plants surrounding us, it is extremely essential to pinpoint a particular plant of our interest by noting the similarities or differences with other plants. A biologist studying a particular plant needs to know his field of study. Thus it becomes extremely necessary that the plant is first identified, given a proper name so that he can communicate his ideas about it, and also know the group to which the plant belongs. These three activities are, in fact, the main functions of taxonomy.

Thus taxonomy, which is probably the oldest of sciences, is the science of classification, including its bases, principles, rules and procedures. Its general purpose is to arrange plants in such a way as to provide an information retrieval system, i.e. the name and rank or position in an ordered hierarchy, which tells us something about the plant. It comes from two Greek words, *taxis* meaning arrangement and *nomous* meaning law or rule. Thus, **taxonomy means classification**

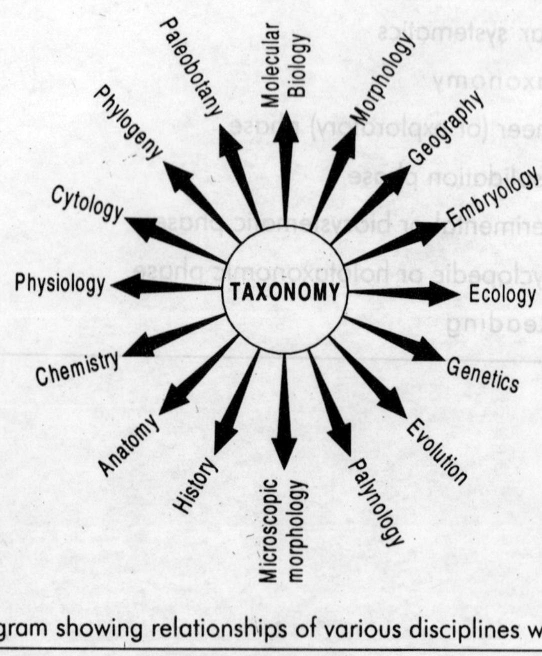

Fig. 1.1 : Taxogram showing relationships of various disciplines with plant taxonomy

following certain rules or, principles. But it is very important to note that a plant's name is the key to its literature, and grouping can only be possible when its identity is revealed and are named for the sake of convenience and communication of ideas about it.

The term taxonomy was first introduced to the plant science in 1813 by A. P. de Candolle, which meant the theory of plant classification. But later this term became more inclusive and at present it includes identification of plants, their nomenclature and classification. Traditionally taxonomy was based largely on gross morphological features of a plant. However now virtually all other scientific information is used from branches like anatomy, genetics, cytology, chemistry, reproductive biology, ecology, physiology, molecular biology, etc. (Fig. 1.1). Hence taxonomy is a very dynamic and synthetic science and is the basic to all other sciences. At the same time it is dependent upon them, i.e. the information used by taxonomists is gathered using information from other disciplines such as genetics, ecology, morphology, anatomy, physiology, etc.

1.2 Different Aspects of Taxonomy

The three functions of taxonomy include, identification, nomenclature and classification (Fig.1.2). Its main aim is to provide a convenient method of identification and communication about a taxa and provide a classification which is based on natural affinities of plants as far as possible. The word 'taxon' (taxa) was first used by a German Biologist Adolf Meyer in 1926 for animal groups. It was later proposed for the plant system in 1948 by Herman J. Lam. It is a taxonomic group of any rank, e.g. family, genus, species, subspecies, etc.

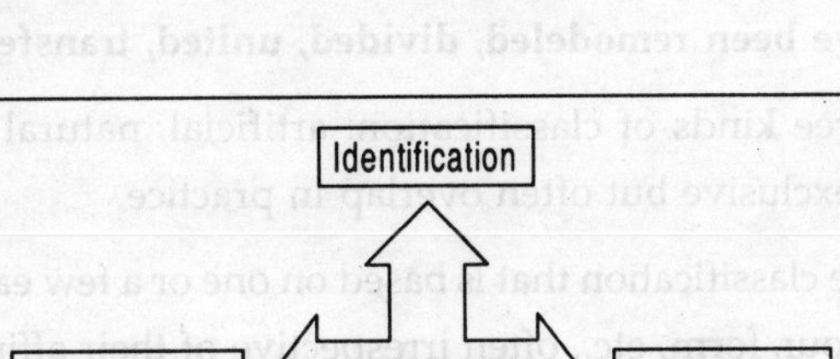

Fig. 1.2 : Taxonomic fundamentals

(a) Identification

Identification of a taxon is a prerequisite for any study based on it. It is the determination of a taxon based on overall similarities and differences with other taxa. Identification is generally done by comparing representative specimen of a given taxon with the help of key descriptions, illustrations, etc. Sometimes, the specimen may not agree or compare with the existing predetermined specimen. In such a case it is taken to be new to science. **Identification is thus assignment of additional unidentified plants to a correct rank once a classification has been established. It is the determination of a name for a specimen. This also implies its rank.**

(b) Nomenclature

Once the taxon has been identified, it becomes necessary to give it a scientific name. **Thus, nomenclature is the naming of a taxon correctly. It is a precise and universal system of rules used by all botanists of the world for naming newly discovered plants.** Often constant study of plants results in differences of opinion concerning past work. This often results in splitting, uniting, changing ranks etc., all of which may affect the plants name. Thus, universal rules are needed to determine the correct name for a plant and the rules for naming the plants are governed by the International Code of Botanical Nomenclature (ICBN). A new name is given to a new taxon.

(c) Classification

Keeping in view, the enormously high number of plant species, it is impossible for any one to study all plants individually. To overcome this problem, the plants can be placed into small or large groups based on their similarities and differences, and then arranged in a sequential manner into categories according to their levels, and each category given a name following the rules of nomenclature. For example, similar plants may be grouped under a "species", similar species under a "genus", and so on. **Classification is thus, the arrangement of groups of plants with particular characteristics by rank or position according to certain criteria, and placing them within the taxonomic hierarchy. It includes the determination of position or rank for new taxa as well as old taxa, which have been remodeled, divided, united, transferred or altered in rank.**

There are basically three kinds of classification: artificial, natural and phylogenetic. These however, are not mutually exclusive but often overlap in practice.

1. **Artificial :** It is the classification that is based on one or a few easily observable characters such as habit, colour, form, etc., often irrespective of their affinity.
2. **Natural :** It is the classification that is based on overall similarity, mostly on gross morphology.
3. **Phylogenetic :** It is the classification based on common evolutionary descent.

(d) Systematics

It is the part of classification that involves the arrangement of plants into related groups. It is the scientific study of the kinds and diversity of plants and the relationships among them if any, and includes not only the traditional activities of taxonomy but also the investigation of such things as evolution, speciation, natural variation, reproductive biology, and a variety of biological phenomena. According to Radford (1986) **systematics is the study of phenetic, genetic and phylogenetic relationship among taxa.**

(e) Molecular Systematics

Molecular systematics is that part of systematics, which utilizes data from nuclear, chloroplast or mitochondrial DNA, to elucidate phylogenetic relationships between plants. Gene sequences obtained from protein-coding genes, ribosomal RNA or non-coding spacer regions are often used in molecular systematic studies. Other DNA-based methods such as Restriction Fragment Length Polymorphism (RFLP) and Random Amplified DNA Polymorphisms (RAPD) "fingerprinting" are used to study problems or issues on genetic diversity (which has implications for conservation and management), the detection of hybrids, and taxonomic delimitation at the level of species.

1.3 Phases of Taxonomy

Classification of a taxon, is generally achieved in four successive phases, as formulated by Davis and Heywood (1963).

(a) **The pioneer (or exploratory) phase :** This is the discovery and collection phase and different members of a taxon are collected from different parts of the world (exploration) and identified, at least provisionally.

(b) **The consolidation phase :** This is the synthesis phase mostly based on morphology. This phase deals with the study of the plant material in the field as well as in the herbarium. All the new groups, if invented, are described. Thus a compilation of a flora or a monograph is done in this phase.

(c) **The experimental or biosystematic phase :** This deals with a wide variety of things and a more thorough knowledge of a taxon based on analysis of a wide variety of things such as the geographical variation, chemical studies, cytogenetic studies, physiological features, fertility relationships, evolutionary patterns and other population studies.

(d) **The encyclopedic or holotaxonomic phase :** This is a coordination of the three phases mentioned above and involves the analysis and synthesis of all information and types of data to develop classification systems based on phylogenetic relationships.

The first two phases corresponding to the "alpha" classification are mainly descriptive and based on gross morphological features. The last two phases correspond to the "omega" (ω) classification. The data for these phases are accumulated from researches in the field, laboratory, garden, herbarium and library, and may be analysed with the help of computers.

SUGGESTED READING

- Bell, C.R., 1969, *Plant Variation and Classification*, Macmillan, London.
- Davis, P. H. and V. H. Heywood, 1963, *Principles of Angiosperm Taxonomy*, Oliver & Boyd, Edinburgh.
- Heywood, V.H., 1967, *Plant Taxonomy*, Edward Arnold Publishers, London.
- Naik, V.N., 1984, *Taxonomy of Angiosperms*, Tata McGraw Hill, New Delhi.
- Radford, A.E., 1986, *Fundamentals of Plant Systematics*, Harper and Row, New York.
- Sivarajan, V.V., 1991, *Introduction to the Principles of Plant Taxonomy*, Oxford & IBH, New Delhi.

2

PLANT IDENTIFICATION

AT A GLANCE

We share our planet with millions of species of different plants and animals. In order to communicate, retrieve, store, and accumulate information about our co-inhabitants, it has been necessary for biologists and others to : (1) identify these organisms; (2) name them; and (3) place the organisms into groups that reflect our current knowledge of their evolutionary relationships. Collectively these activities - identification, nomenclature and classification, make up the discipline of taxonomy.

Identification is a basic activity and one of the primary objectives of systematics. **It is the recognition of characteristics of organisms and the application of a name to an organism with those particular characters** (Jones and Lucksinger, 1986). Thus, although identification is a separate activity or process, in practice it involves both classification and nomenclature. It is simply the determination of the similarities or differences between two elements, i.e., whether the two are the same or they are different. The comparison of an unknown plant with a named plant and the determination that the two plants are the same also involves classification, i.e., when one correctly decides that an unknown plant belongs to the same group (species, genus, family, etc.) as a known specimen. Both the processes of identification and classification involve comparison and judgment and require a definition of criteria of similarities. Identification is, therefore, a basic process in classification with nomenclature playing an essential role in the retrieval of information and as a means of communication.

2.1 Characters often Considered before Plant Identification .....

1. Whether a plant is herbaceous or woody, and annual or perennial in nature.
2. Whether or not milky or coloured sap is present in the leaf, stem or other plant parts.
3. The leaf type, phyllotaxy, and venation.
4. Presence or absence and type of stipule on young shoots.
5. The distribution and kinds of surface coverings (i.e. hairs, trichomes, spines, etc.).
6. The parts of the flower and the number of sepals and petals, whether separate or fused, and also their arrangement i.e. aestivation.
7. Whether perianth is present in one or more series, or absent.
8. Whether pappus (e.g. Asteraceae) or epicalyx (e.g. Malvaceae) or similar structures are present.
9. Whether a nectar-secreting disc is present in the flowers (e.g. Rutaceae).
10. Whether the flowers are actinomorphic or zygomorphic.
11. The number and attachment of stamens and if there is any fusion of anthers or filaments.
12. The number of pistils, styles and stigmas of the gynoecium, observation of a transverse section of the ovary, the number of locules, number of ovules per locule, and also the placentation.
13. The position of the ovary and fusion of the perianth by observing a longitudinally cut section of the entire flower through its center.

2.2 Identification Methods

The methods of identification include the following :

(a) **Expert determination :** The best method of identification is expert determination in terms of reliability or accuracy. In general the experts have prepared treatments (monographs, revisions, synopses) of the group in question, and it is probable that the more recent floras or manuals include the expert's concepts of taxa. Experts are typically found in botanical gardens, herbaria, museums, colleges, universities, etc. However, although of great reliability, this method presents problems of requiring the valuable time of experts and creating delays for identification.

(b) **Recognition :** It approaches expert determination in reliability. This is based on extensive, past experience of the identifier with the plant group in question. In some groups this is virtually impossible.

(c) **Comparison :** A third method is by comparison of an unknown with named specimens, photographs, illustrations or descriptions. Although this is a reliable method, it may be very time consuming or virtually impossible due to the lack of suitable materials for comparison. The reliability is, of course, dependent on the accuracy and authenticity of the specimens, illustrations, or descriptions used in the comparison.

(d) **The use of keys and similar devices (synopses, outlines, etc.) :** This is by far the most widely used method and does not require the time, materials, or experience involved in comparison and recognition.

2.3 Taxonomic Keys

A key is a device, which when properly constructed and used, enables a user to identify an organism. **Keys are devices consisting of a series of contrasting or contradictory statements or propositions requiring the identifier to make comparisons and decisions based on statements in the key as related to the material to be identified.** Thus, a taxonomic key is a device for quickly and easily identifying to which species an unknown plant belongs. The key consists of a series of choices, based on observed features of the plant specimen. It provides a choice between two contradictory statements resulting in the acceptance of one and the rejection of the other. **A single pair of contradictory statements is called a *couplet* and each statement of a couplet is termed a *lead*.** By making the correct choice at each level of the key, one can eventually arrive at the name of the unknown plant.

2.3.1 Suggestions for the use of Keys

(a) Appropriate keys should be selected for the materials to be identified. The keys may be in a flora, manual, guide handbook, monograph, or revision. If the locality of an unknown plant is known, a flora, guide, or manual treating the plants of that geographic area may be selected. If the family or genus is recognized, one may choose to use a monograph or revision. However, if the locality is unknown a general work may be selected. If materials to be identified are cultivated, one of the manuals treating such plants should be selected, since most floras do not include cultivated plants unless naturalized.

(b) The introductory comments on format details, abbreviations, etc. should be read before using the key.

(c) Both the leads of a couplet should be read before making a choice. Even though the first lead may seem to describe the unknown material, the second lead may be even more appropriate.

(d) A glossary should be used to check the meaning of terms, which one does not understand.

(e) Several similar structures should be measured, when measurements are used in the key, e.g. several leaves and not a single leaf should be measured. No decision should be made on a single observation. Rather it is often desirable to examine several specimens.

(f) The results should be verified by reading a description, comparing the specimen with an illustration or an authentically named herbarium specimen.

2.3.2 Suggestion for Construction of Keys

(a) Constant characteristics rather than variable ones should be used.

(b) Proper measurements rather than terms like "large" and "small" should be used.

(c) Characteristics that are generally available to the user of the key rather than seasonal characteristics or those seen only in the field should be used.

(d) A positive choice should be made. The term "is" instead of "is not" should be used.

(e) If possible one should start both choices of a pair with the same word.

(f) If possible, different pairs of choices should be started with different words.

(g) The descriptive terms should be preceded with the name of the part to which they apply.

2.3.3 Types of Keys

There are two types of keys: (a) Dichotomous and (b) Polyclave (also called Multiple Access or Synoptic).

2.3.3.1. Dichotomous Keys

Keys in which the choices allow only two (mutually exclusive) alternative couplets are known as dichotomous keys. In constructing a key, contrasting characters are chosen that divide the full set of possible species into smaller and smaller groups i.e. the statements typically begin with broad characteristics and become narrower as more choices are required. Each time a choice is made, a number of species are eliminated from consideration and the range of possible species to which the unknown specimen may belong is narrowed. Eventually, after sufficient choices have been made, their range reduces to a single species and the identity of the unknown plant is revealed. Dichotomous comes from the Greek root *dich* meaning "two" and *temnein* meaning "to cut".

Couplets can be organized in several forms. The couplets can be presented using numbers (numeric) or using letters (alphabetical). The couplets can be presented together or grouped by relationships. There is no apparent uniformity in presentation of dichotomous keys.

Example of a numerical key with couplets
1. Seeds round—soybeans
1. Seeds oblong go to—2
2. Seeds white—northern beans
2. Seeds black—black beans
Example of an alphabetical key with same couplets
A. Seeds oblong go to—B
B. Seeds white—northern beans
B. Seeds black—black beans
A. Seeds round—soybeans
(*Courtesy:* Constructing a Dichotomous Key, Theodore M. Sperry Herbarium, Department of Biology, Pittsburg State University, Pittsburg, Kansas 66762)

(a) **Types of Dichotomous Keys :** There are two types of dichotomous keys. They differ in the method by which the couplets are organized and how the user is directed to successive choices.

(i) **Indented Keys** (also called yoked) — indents the choices (leads) of the couplet an equal distance from the left margin. The two choices of the couplet are usually labeled 1 and 1' or 1a and 1b. It is not necessary that the choices are numbered, but it helps. The user goes to the next indented couplet following the lead that was selected.

(ii) **Bracketed Keys** — provides both choices side-by-side. The choices of the couplet must be numbered (or lettered). It is very helpful if the previous couplet is given. This key has exactly the same choices as the first example. The choices are separated, but it is easy to see the relationships. While this key might be more difficult to construct, it gives more information to the user.

Example of an Indented Key on *Rhododendron*	
1a. Flowers in shades of red	
2a. Flowers blood-red, leaves oblong-ovate, leathery and thick matty texture	*R. sikkimense*
2b. Flowers crimson-red, leaves broad, oval to elliptic oblong, shiny green above	*R. fulgens*
1b. Flowers in shades of rose-pink	
3a. Calyx 3-5 mm long, leaf under surface covered with tufts of brown hair	*R. wallichii*
3b. Calyx obscure, 1-2 mm long, leaf under surface covered with continuous indumentum	
4a. Corolla in shades of deep rose-pink flushed externally with red-purple, young leaves aeruginose, leaf margins inrolled	*R. aeruginosum*
4b. Corolla pale lavender blue, mauve or rose-purple, rarely white, young leaves not aeruginose, leaf margins not inrolled	*R. campanulatum*

Example of a Bracketed Key on *Rhododendron*		
1a.	Flowers in shades of red	go to 2
1b.	Flowers in shades of rose-pink	go to 3
2a.	Flowers blood-red, leaves oblong-ovate, leathery and thick matty texture	*R. sikkimense*
2b.	Flowers crimson red, leaves broad, oval to elliptic oblong, shiny green above	*R. fulgens*
3a.	Calyx 3-5 mm long, leaf under surface covered with tufts of brown hair	*R. wallichii*
3b.	Calyx obscure, 1-2 mm long, leaf under surface covered with continuous indumentum	go to 4
4a.	Corolla in shades of deep rose-pink flushed externally with red-purple, young leaves aeruginose, leaf margins inrolled	*R. aeruginosum*
4b.	Corolla pale lavender blue, mauve or rose-purple, rarely white, young leaves not aeruginose, leaf margins not inrolled	*R. campanulatum*

(b) **Problems using Dichotomous Keys :** A key may be difficult to use at times because :

- the key may not include all potential variations in the species;
- the key may rely on features not present in that season;
- the key may not include "all" species of interest;
- one may misinterpret a feature or make a mistake.

2.3.3.2 Polyclave / Random Access / Synoptic Keys

Another type of key, which is relatively a new alternative to dichotomous keys and becoming increasingly popular, especially because of the ease of computerizing them, is termed multiple access or polyclave or synoptic key. The advantage of these keys is that they allow the user to enter the key at any point.

This key is based on the identification of organisms by a process of elimination. In a written polyclave key there is a series of characters and character states. Each state is followed by a number or code for the species that possess that feature. The user needs to select any character and then copy down the list of species that possess the feature. Then the user has to select another character and eliminate any species that is not common to both lists. This process has to be continued until the specimen is identified.

It's easy to imagine how these keys are computerized. Consider a series of standard playing cards. Imagine each card has four holes punched into it along the top margin. If the card is a spade we cut the first hole through the margin; if a club, the second hole is notched to the margin; a heart the third hole is notched; and finally if it is a diamond the fourth hole is notched. Further imagine that along the bottom of the card we punch 14 holes (2 - 10, J, K, Q, A) and cut a notch for the appropriate number. Thus, the Queen of Hearts will have a notch cut into the third hole on the top of the card, and the Queen notch on the bottom of the card.

Now, let's use our punch card deck of cards to identify an unknown card. Shuffle another deck of cards and pick any card. Let's assume that this "unknown" card is the Ace of Spades. To identify this unknown, we analyze the characters and two are obvious, suit and number. Let's start with suit - take a long needle and stick it through the "spades hole". Since, only spades are notched, the other suits will remain on the needle and spades will drop out of the deck. Now, collect the spades cards and put a needle through the next character, the Ace and, viola, the Ace of Spades falls out. This is the general principle of how the computerized version of polyclave keys work. The main difference is that a computer allows for countless holes (characters) and notches (states) to be included and does the needlework for us.

An example of a Polyclave key : Polyclave Key to Pollination Type

Pollination is the process of transferring pollen from one flower to another. Since plants can't move, they utilize vectors such as wind, water and animals to accomplish this process for them. Flowers are specialized by shape, color, odor, nectar etc. in order to maximize the chance that a certain vector will accomplish pollination. These flower adaptations are collectively known as pollination syndromes or systems.

Plants differ in the degree of their specialization for a particular pollination system. For example, many orchids are pollinated by only a single type of bee. Other flowers are not as specialized and may be pollinated by a variety of bees or perhaps beetles. In other cases, insects may visit flowers without actually transferring pollen. These factors make it difficult to determine with absolute certainty the pollination system by the polyclave key.

To illustrate how to use a polyclave key, let's determine the pollination system of a dandelion

1. Select any one of the FLOWER CHARACTERS in the key.
   Note : you can choose the characters in any order(i.e., multi - entry) ...... Let's choose FLOWER COLOR
2. Choose the character state (description) that matches the flower you are observing .... Dandelions are Yellow
3. Write down the possible pollination systems for this feature .... BE,BU,BI
4. Select another feature .... NECTAR
5. Choose the character state description that matches your flower .... b. present...WI,FM,BE,BA
6. Eliminate from the first character state selected the pollination systems .... BE,BU,BI
   not found on both lists. Continue this process until the pollination system is identified. Note: in some cases it will not be possible to narrow the choices to just one .... Thus, dandelions are bee pollinated

[*Courtesy :* Plant Taxonomy – Stephen G. Saupe, Biology Department, College of St. Benedict/St. John's University, Collegeville, MN 56321]

(a) **Advantages of Polyclave Keys :** The advantages of a polyclave (multiple-access) key are :

- They are easy to use.
- They allow multi-entry i.e. the user can start anywhere. This is a significant advantage because the user can rely on characters that are most easy to observe, rather than having to deal with characters that may not be present in the specimen or are poorly developed.
- They are order-free i.e. the user can work in any direction with any character.
- They are faster.
- They are easily computerized. In fact, these keys are most commonly used in this form. Paper versions are typically large and unwieldy because each character needs to list all possible taxa.

(b) Suggestions to use a Polyclave Key

- **Read through the list of characters to become familiar with the possibilities.**
- **Scan the list to find a character with a state that you observe in your specimen. Start with a readily identifiable character that has only a few numbers (taxa) associated with it.**
- **Write a brief description of the character and state and the numbers of the taxa that can be described by this state.**
- **Choose another character and state that describes your taxa. Write a brief description of this state below the name of the first state chosen. Then, scratch off the original list of any taxon that doesn't appear in the second.**
- **Continue this process until just one taxon remains for all of the states. If there is no single taxon described by the states chosen, and two or more remain, go back and check for errors.**
- **Read the name of the taxon after its number in the list of taxa. Check your identification with a description in a manual or the herbarium.**

[*Courtesy* : Plant Taxonomy – Stephen G. Saupe, Biology Department, College of St. Benedict/ St. John's University, Collegeville, MN 56321]

SUGGESTED READING

- Blackwelder, D., 1967, *Quick-Key Guide to Trees : Trees of Northeastern and Central North America*, Doubleday, Garden City, New York.
- Bossert, W., 1962, "Computer techniques in systematics", In : *Systematic Biology*, National Academy of Science Publication, Washington.
- Harrington, H.D. and L.W. Durrel, 1957, *How to Identify Plants?* The Shallow Press, Chicago.
- Hutchinson, J., 1967, *Key to the Families of Flowering Plants of the World*, Claredon Press, Oxford.
- Morse, L.E., J.A. Peters and P.B. Hamel, 1971, A general data format for summarizing taxonomic information. *Bioscience*, 21:174-180.
- Osborne, D.V., 1963, Some aspects of the theory of dichotomous keys. *New Phytologist*, 62:144-160.
- Pankhurst, R.S., 1970, A computer programme for generating diagnostic keys. *Computer J.*, 13:145-151.
- Pankhurst, R.S., 1971, Botanical keys generated by computer. *Watsonia*, 8:357-368.
- Pankhurst, R.S., 1974, Automated identification in systematics. *Taxon*, 23:45-51.
- Peters, J.A., 1969, Discussion, In : *Systematic Biology*, National Academy of Science Publication, Washington, D.C.
- Voss, E.G., 1952, The history of keys and phylogenetic trees in systematic biology. *Journal of the Scientific Laboratories of Denison University*, 43:1-25.

3 PLANT NOMENCLATURE

AT A GLANCE

Classification and nomenclature are two different aspects of biological system. Any object that is once known to human beings has to be given a name in order to communicate ideas about it and describe it. This art of naming objects is what we call nomenclature. As any other object, plants also need to be given names to communicate about it and to indicate relationships. Thus, **nomenclature is a process of naming plants based on international rules agreed upon by botanists in order to ensure a stable and uniform system.**

3.1 Common Names

It is the name given to a plant by people in a particular area or locality. It is the vernacular name or local name. In a multilingual country like India, where there are several languages, almost all plants have local names, which differ from language to language, and even from dialect to dialect. Thus, a particular plant may have several names in a particular country. This makes things extremely difficult, when a person speaking a particular language wants to communicate about a plant to a person speaking a different language. Hence, common names have proved to be misleading and inadequate for wider use throughout the country or the whole world.

3.2 Scientific Names

Plants should be named in such a way so that it can be used internationally. In order to overcome the problems raised by common names, botanists have given scientific names to all the known plants, for international communication. The first problem, which they faced while giving scientific names to plants was to select the language which can give a clear cut meaning and which would not change. Although Latin is a dead language now, it was agreed by botanists of the world that scientific names should be in Latin particularly for the following three reasons:

- Firstly, this language is not being used by any country or nation at present.
- Secondly, most of the European languages are derived from Latin
- Thirdly, in the past European scholars learned their subjects in Latin. It was a widely used language throughout the European countries and a lot of botanical literature has been written in Latin.

Since Europe, and particularly Greece, has dominated the whole world in the field of science during 1600 to 1850 A.D., Latin is now a language of botany and other allied sciences. However, the script is Roman.

3.3 Binomial Nomenclature

In the early attempts of naming plants, plant names used to be long, as a single name used to be followed by one to many descriptive terms or adjectives, giving rise to multiple terms or **polynomials** which have been used by many taxonomists including Linnaeus (1753), in his *Species Plantarum*, and he used additional *trivial* names for these polynomials. After Linnaeus system of trivial names, later workers however found it more convenient to adopt the **binomial system** consisting of two latinized words, a generic name followed by a specific epithet (a trivial name). Thus, in the binomial system, a botanical name of a species is a combination of a generic name and a specific epithet. As for example, the botanical name of sunflower is *Helianthus annuus*. The first word (*Helianthus*) designates the genus of the plant, while the second word (*annuus*) designates the species of this genus. The generic name is usually a substantive (noun), while the second term or specific epithet is an adjective or noun.

3.3.1 Generic Name

The generic name is usually a noun and singular, which is spelled or written with a capital letter. Generic names have different types of origin:

(a) A genera may be named in honor of a botanist or horticulturist or may be a patron of botany. For e.g. *Linnaea* has been named in honor of the famous botanist Linnaeus.

(b) Some generic names are of poetic or mythological origin. For e.g. *Nymphaea* denotes the lovely water-nymphs.

(c) Many generic names denote some characteristic feature of the plants. For e.g. *Liniodendron* or lily tree is named after the shape of the flowers of the Tulip Tree.

(d) There is another category of generic names in which the name is aboriginal in origin i.e. the names existed in the lands where the plants were discovered but later they were given Latin names. For e.g. *Betula* is an old Greek name for Birch.

3.3.2 Specific Epithet

The specific epithet is often an adjective and it is written with a small initial letter unlike the generic name. However, if the specific epithet is derived from common names, or from a generic name, or from a person's name, the initial letter of the epithet may be a capital letter. In the typed or hand written manner, both the generic names and specific epithet should be underlined, while if printed it should be in *italics* or boldface.

Like a generic name, a specific epithet may also have several origin:

(a) It may be in honor of a person.

(b) It may be derived from some special characteristic of the plant.

(c) It may be derived from a geographical location where the plant grows.

(d) It may originate from an old common name.

(e) It may be named arbitrarily.

3.4 International Code of Botanical Nomenclature (ICBN) ....

3.4.1 History of ICBN

Before the middle of the 18th century, plant names were usually polynomials i.e. made up of several words in a series. It was superseded by the binomial system, which was first applied for the plant kingdom by Linnaeus in his *Species Plantarum*. It was Linnaeus who proposed the elementary rules of naming plants first in 1737 in his *Critica botanica* and then in 1751 in *Philosophia Botanica*.

Elementary rules were framed to serve as a guide to botanists. Later in 1813, A.P. de Candolle in his *Theorie elementaire de la botanique* gave a detailed set of rules regarding plant nomenclature. However discovery of new plants from later explorations caused concern over procedures for naming these species. Thus, with the passage of time, the need for an international system and rule for naming plants became increasing apparent. It was then that Alphonse de Candolle, son of A.P. de Candolle convened an assembly of botanists of several countries to present a new set of rules. Candolle convened the First International Botanical Congress held at Paris in 1867. Subsequent meetings of the International Botanical Congress were held in 1892 (Rochester Code), 1905 (Vienna Code), 1907 (American Code) and 1910, but a general agreement regarding the internationally acceptable rules of plant nomenclature was reached in 1930 at the IBC meeting at Cambridge where for the first time in botanical history, a code of nomenclature came into being that was international in function as well as in name. This code is called the International Code of Botanical Nomenclature (ICBN). The modifications or amendments as suggested by the International Botanical Congress at the subsequent meetings have been incorporated in the ICBN on a regular basis.

(i) **Paris Code (1867)** : The First International Botanical Congress was held at Paris in August, 1867, and was aimed at the standardization and legislation of proper nomenclature practices. About 150 American and European botanists were invited to attend the congress. Laws of botanical nomenclature *(Lois de la nomenclature botanique)* were discussed and they were adopted with slight revision. It was resolved that the Laws, as adopted by the Assembly, shall be recommended as the best Guide for Nomenclature in the Vegetable Kingdom. These Laws were called the Paris Code, as they were adopted at the French capital, or de Candolle rules, as they were prepared by Alphonse de Candolle. According to the Paris Code, the starting-point for all nomenclature was fixed with Linnaeus. However no date or any work was specified. The rule of priority was considered as basic with no provisions for exceptions. It was very important that the publication be valid and attention was given to author citation and terms applied to categories of plants. Although the Paris Code guided taxonomic activity in most countries to a considerable degree, but their application showed many inherent defects. As time went on, American and British botanists deviated from the rules and they started practicing the unwritten law named *Kew Rule* according to which if a species was transferred to another genus, the specific epithet need not be transferred to the new genus but the author was free to use a new epithet in the new combination.

(ii) **Rochester Code (1892)** : In 1892 a batch of botanists headed by N.L. Britton met at Rochester, New York, in the United States and developed a set of rules to govern nomenclature. These rules, which are based, on modifications of the Paris Code, are commonly known as the Rochester Code and include some new recommendations which are as follows:

(a) Establishment of the type concept to ascertain the correct application of names;

(b) Strict adherence to the principles of priority;

(c) Acceptance of alternate binomials resulting from employment of the principles of priority, even if the specific epithet repeats the generic name (referred to as a tautonym); and

(d) Interpretation of priority to apply to the precedence of a name in a publication in addition to the date of publication.

(iii) **Vienna Code (1905)** : The third International Botanical Congress was held at Vienna in June, 1905. The new changes included the

(a) Establishment of Linnaeus *Species Plantarum* (1753) as the starting-point for naming vascular plants.

(b) Nomina generica conservanda by which generic names having a wide use would be conserved over earlier but less well-known names.

(c) Banning of tautonyms and

(d) Requirement that names of new taxa be accompanied by a Latin diagnosis.

(iv) **American Code (1907)** : Being dissatisfied with the results of the Vienna Congress, most proponents of the Rochester Code refused to accept the new rules and in 1907, they put forth a slight modification of the Rochester Code under the heading of the American Code. The first provision was that they would not subscribe to the principle of *Nomina generica conservanda* or of the requirement of Latin diagnosis, but they accepted the type concept. Another provision of the revised code was that 'a binomial may not be used again for a plant in any way if it has been employed previously for another plant, even though the previous use may have been illegitimate'. Thus the Rochester Code and American Code created two opposing schools of thought in the United States. One was under the leadership of Britton and his students, who followed the provisions of the American Code and the other was under the direction of Asa Gray's pupils who followed the international rules.

(v) **Brussels Code (1912)** : The Fourth International Botanical Congress was held at Brussels, 1910. The most significant decision made at this Congress was the establishment of different starting points for priority of names of non vascular plants, the recognition of value of type concept and classification of phraseology of the Vienna rules.

(vi) **Cambridge Code (1935)** : The basic differences between the Vienna Code and American Code was finally reconciled at the Fifth International Botanical Congress held in 1930 at Cambridge, England and it brought about harmony among the major botanical factions. The new rules legislated at Cambridge constituted the code of nomenclature that was truly international in name and function. The approved provisions were the following:

(a) The type concept should be pursued;

(b) There should be a list of *Nomina generica conservanda;*

(c) Tautonyms are not admissible; and

(d) Latin diagnosis is required after January 1, 1932.

(vii) **Amsterdam Code (1947)** : At the Sixth International Botanical Congress, Amsterdam (1935), a few major changes in the rules was made. It was resolved that "from January 1, 1935, names of new groups of recent plants, except the Bacteria, are considered, as validly published only when they are accompanied by a Latin diagnosis". An attempt to select a list of *nomina specified conservanda* was thwarted by an overwhelming vote.

(viii) **Stockholm Code (1952)** : The Seventh International Botanical Congress met at Stockholm in 1950. It introduced certain number of definitions on types. The word *taxon* (Pl. *taxa*) was introduced for the first time to designate any taxonomic group or entity.

(ix) **Paris Code (1956) :** The Eighth International Botanical Congress was held in Paris in July, 1954 in which great emphasis was laid on types (Appendix IV), but the rule of Latin came under fire. It decided that the code should be published in English, French and German languages. The separation of the Preamble and Principles from the Rules and Recommendations was the major feature of this code. Appendix II (*nomina generica conservanda et rejicienda*) was amended and supplemented.

(x) **Montreal Code (1961) :** The Ninth International Botanical Congress held at Montreal in August, 1959 appointed a special committee to study the question of conservation of family names (*nomina familiarum conservanda* for Angiospermae as Appendix II was introduced). In this code, it was specified that the naming of fossil plants should follow the same lines as that of recent ones.

(xi) **Edinburgh Code (1966) :** The report of the special committee appointed during the ninth Congress was submitted at the Tenth International Botanical Congress held at Edinburgh in August, 1964. Some of the important points in the report are as follows:

(a) A.L. de Jussieu's *Genera Plantarum* (1789) is the starting-point for family names. Many of the names included in the list are not a matter of dispute, but their inclusion in the list forms a ready reference providing correct names of families together with a type genus for each family.

(b) In the list of *nomina familiarum conservanda*, a few names are found with a spelling somewhat different from the long-established names, e.g. Cannabaceae for Cannabinaceae, Capparaceae for Capparidaceae, Haloragaceae for Haloragidaceae, Melastomataceae for Melastomaceae, etc.

(c) The rule of Latin for new diagnoses was finally settled and no amendments were proposed at this congress.

At this congress, a committee was set up to prepare a glossary of the main technical terms occurring in the code which resulted in *An annotated glossary of botanical nomenclature.*

(xii) **Seattle Code (1972) :** The Eleventh International Botanical Congress was held at Seattle in August, 1969 which proposed the Seattle Code which was edited by F.A. Stafleu and published in 1972. Most of the proposals submitted to the Eleventh Congress were concerned with refinements and increased precision. The chief issues in Seattle Code include the tautonymous designations of taxa between genus and species and below species, the perennial question of superfluous names and a reorganization of the rules for hybrids. The word autonym (automatically established names) was introduced in this code.

(xiii) **Leningrad Code (1978) :** The Twelfth International Botanical Congress was held in July, 1975 at Leningrad, Russia and its recommendations came out in 1978. This Code indicates only small differences from the Seattle Code and includes the following changes:

(a) The concept of organ genera is eliminated for fossil plants.

(b) The Code does not apply to names of organisms treated as bacteria and does apply to all other organisms treated as plants.

(c) The principle of automatic typification is extended to those names of taxa above the family rank that are ultimately based on generic names and the application of the priority principle is recommended while selecting among names thus typified.

(d) A name or combination published before 1953 without indicating the rank is considered validly published but imperative in questions of priority except for homonymy and certain names to be accepted at the varietal rank.

(e) Art. 69 of the previous Code is modified on the basis of type method and Art. 70-71 dealing with discordant elements and monstrosities were deleted, but the Art numbers are retained to facilitate the use of the Code.

(f) The section on orthography is thoroughly rewritten.

(g) Individual paragraphs on all Articles and Recommendations are numbered in a decimal-like system, some being rearranged.

(xiv) **Sydney Code (1983) :** The Thirteenth International Botanical Congress was held in August, 1981 at Sydney, Australia.

(xv) **Berlin Code (1988) :** This includes the proposals made at the Fourteenth International Botanical Congress at Berlin, Germany in 1986. *Nomina specifica conservanda* was for the first time introduced at this Congress. Two species names, *Lycopersicon esculentum* P. Miller and *Triticum aestivum* Linn., have been conserved against the rule of priority, as these names are widely used and any change may create confusion. Articles 66 and 67 were removed.

(xvi) **Tokyo Code (1994) :** The Fifteenth International Botanical Congress met at Yokohama, Tokyo in 1993. It has been translated into Chinese, French, German, Italian, Japanese, Russian and Slovak. In this code, extensive renumbering had taken place and therefore its preface included a tabulation comparing the placement of its provisions with those of the Berlin code.

(xvii) **St. Louis Code (1999) :** This includes the proceedings of the Sixteenth International Botanical Congress, which was held at St. Louis, Missouri in July-August, 1999 and supersedes the Tokyo Code. Like the Tokyo Code, it is anticipated that the St Louis Code, too, will become available in several languages in due course. The detailed provision of the Code is divided into Rules, set out in Articles, and Recommendations. The object of the "Rules" is to put the nomenclature of the past into order and to provide for that of the future. Names contrary to a rule cannot be maintained. The Recommendations deal with subsidiary points, their object being to bring about great uniformity and clearness, especially in future nomenclature. Names contrary to a recommendation cannot, on that account, be rejected, but they are not examples to be followed. The Rules and Recommendations apply to all organisms whether fossil or non-fossil, including fungi but excluding bacteria. A separate code called International Code of Nomenclature of Bacteria (ICNB) governs the nomenclature of bacteria.

The Seventeenth International Botanical Congress is scheduled to meet at **Vienna in 2005.

Broad outline of the Saint Louis Code (Greuter et al. 1999)

Preamble : (Pre.1-Pre.11)

Division I : Principles (I-VI)

Division II : Rules and Recommendations (Art. 1-62)

- Chapter I : Taxa and their ranks (Art. 1,2,3,4,5)
- Chapter II : Status, typification, and priority of names (Art. 6-15)
  - Section 1 : Status definitions (Art. 6)
  - Section 2 : Typification (Art. 7,8,9,10)
  - Section 3 : Priority (Art. 11 and 12)
  - Section 4 : Limitation of the principle of priority (Art. 13, 14, 15)
- Chapter III : Nomenclature of taxa according to their rank (Art. 16-28)
  - Section 1 : Names of taxa above the rank of family (Art. 16 and 17)
  - Section 2 : Names of families and subfamilies, tribes and subtribes (Art. 18 and 19)
  - Section 3 : Names of genera and subdividions of genera (Art. 20, 21, 22)
  - Section 4 : Names of species (Art. 23)
  - Section 5 : Names of taxa below the rank of species (infraspecific taxa) (Art. 24, 25, 26, 27)
  - Section 6 : Names of plants in cultivation (Art. 28)
- Chapter IV : Effective and Valid publication (Art. 29-50)
  - Section 1 : Conditions and dates of effective publication (Art. 29, 30, 31)
  - Section 2 : Conditions and dates of valid publication of names (Art. 32, 33, 34, 35, 36, 37, 38, 39, 40, 41, 42, 43, 44, 45)
  - Section 3 : Author citations (Art. 46, 47, 48, 49, 50)
  - Section 4 : General recommendations on citation (Rec. 50A – F)
- Chapter V : Rejection of names (Art. 51, 52, 53, 54, 55, 56, 57, 58)
- Chapter VI : Names of fungi with a pleomorphic life cycle (Art. 59)
- Chapter VII : Orthography and gender of names (Art. 60 – 62)
  - Section 1 : Orthography (Art. 60 and 61)
  - Section 2 : Gender (Art. 62)

Division III : Provisions for the governance of the Code

Broad outline of the Saint Louis Code (Greuter et al. 1999)

Appendix I — Names of hybrids (Art. H. 1, H. 2, H. 3, H. 4, H. 5, H. 6, H. 7, H. 8, H. 9, H. 10, H. 11, H. 12)

Appendix IIA — NOMINA FAMILIARUM ALGARUM, FUNGORUM, PTERIDOPHYTORUM ET FOSSILIUM CONSERVANDA ET REJICIENDA

- A. Algae
- B. Fungi
- D. Pteridophyta
- F. Fossil plants (excl. diatoms)

Appendix IIB — NOMINA FAMILIARUM BRYOPHYTORUM ET SPERMATOPHYTORUM CONSERVANDA

- C. Bryophyta
- E. Spermatophyta

Appendix IIIA — NOMINA GENERICA CONSERVANDA ET REJICIENDA

- A. Algae
- A1. Bacillariophyceae (incl. fossil diatoms)
- A2. Bodonophyceae
- A3. Chlorophyceae
- A4. Chrysophyceae
- A5. Cyanophyceae
- A6. Dinophyceae
- A7. Euglenophyceae
- A8. Phaeophyceae
- A9. Rhodophyceae
- A10. Trichomonadophyceae
- A11. Xanthophyceae
- B. Fungi
- C. Bryophyta
- C1. Hepaticae
- C2. Musci
- D. Pteridophyta
- E. Spermatophyta
- E1. Gymnospermae
- E2. Monocotyledones
- E3. Dicotyledones
- F. Fossil plants (excl. diatoms)

Appendix IIIB — NOMINA SPECIFICA CONSERVANDA ET REJICIENDA

- A. Algae
- B. Fungi
- C. Bryophyta
- D. Pteridophyta
- E. Spermatophyta
- F. Fossil plants (excl. diatoms)

Contd

Broad outline of the Saint Louis Code (Greuter et al. 1999)	
Appendix IV	NOMINA UTIQUE REJICIENDA
	A. Algae
	B. Fungi
	C. Bryophyta
	D. Pteridophyta
	E. Spermatophyta
Appendix V	OPERA UTIQUE OPPRESSA

Principles of ICBN	
Principle I :	Botanical nomenclature is independent of zoological and bacteriological nomenclature. The Code applies equally to names of taxonomic groups treated as plants whether or not these groups were originally so treated.
Principle II :	The application of names of taxonomic groups is determined by means of nomenclatural types.
Principle III :	The nomenclature of a taxonomic group is based upon priority of publication.
Principle IV :	Each taxonomic group with a particular circumscription, position, and rank can bear only one correct name, the earliest that is in accordance with the Rules, except in specified cases.
Principle V :	Scientific names of taxonomic groups are treated as Latin regardless of their derivation.
Principle VI :	The Rules of nomenclature are retroactive unless expressly limited.

3.4.2 Rules of ICBN

(a) Rule of Priority

It has been found that the number of published names for the various plants exceeds the actual number of taxa. This is because several names have frequently been attributed to the same taxon. Hence, it is very essential to have a way to decide which of the several names should be selected for a particular taxon. This is where the Principle of Priority comes into play, which insists that each family or taxon of lower rank with particular circumscription, position and rank can bear only one correct name and each taxon is to be known by the earliest name validly published for the taxon, special exceptions being made for nine families for which alternative names are permitted. Exceptions are also permitted at the family and genus levels, where names are conserved. The correct name for a taxon below the rank of genus, should be a combination of the earliest available legitimate epithet in the same rank with the correct name of the genus or species to which it is assigned, except in certain specified cases. Although the code recommends (Recommendation 16B) that while selecting from among the typified names above the rank of family, the rules of priority should be followed by the taxonomist, it is not mandatory and will not affect the legitimacy of names selected otherwise. An example of the nomenclatural history of a genus is given below:

Polygala L. (1753), *Poligala* Neck. (1768), *Polygaloides* Agosti (1770). According to the principle of priority the earliest name, *Polygala* L., is the correct name for the genus.

(b) Ranks of Taxa

A taxon is referred to as a taxonomic group of any rank. The early part of the code deals with the ranks of taxa. Every individual plant is treated as belonging to a number of successively higher rank with the species as the basic unit. The principal ranks of taxa in ascending sequence are: species, genus, family, order, class, division and kingdom (Art 3). If a greater number of ranks of taxa is required, the terms for these are made either by adding the prefix sub (*sub-*) to the terms denoting the ranks or by the introduction of supplementary terms (Art 4).

(c) Typification

The code has greatly emphasized on typification of taxa in order to bring about stabilization of names. The naming of taxonomic groups is determined by means of nomenclatural types where a nomenclatural type is that element, to which the name of a taxon is permanently attached, either as a correct name or as a synonym. It is not necessarily the most typical or representative element of the taxon. Following are some of the important nomenclatural types :

(i) **Holotype :** It is the one specimen or other element used by the author or designated by him as the nomenclatural type (Art 7.3). The holotype is chosen by the original author from a single gathering made by a collector at one time, and expressed definitely at the time of original publication.

(ii) **Isotype :** An isotype is any duplicate (part of a single gathering made by a collector at a time) of the holotype. For example, if a new species is based on a single gathering that consist of four specimens of the same plant with the same field number placed on four separate herbarium sheets, the original author will designate one of them as a holotype and the remaining three as isotypes.

(iii) **Paratype :** A paratype is a specimen cited with the original description other than the holotype or isotype. For example, if a new taxon is collected in one season without flowers and fruits and the same taxon is collected in a different season with flowers and fruits, then the two collections will be given different field numbers. Out of these two collections, the author will select one collection as the holotype and the isotype, while the next gathering cited in the protologues will form the paratype.

(iv) **Syntype :** A syntype is one of the two or more specimens cited by an author of a species when no holotype was designated, or it is any one of the two or more specimens originally designated as types (Art. 7.7).

(v) **Lectotype :** It is a specimen or other element selected from the original material to serve as a nomenclatural type when no holotype was designated at the time of publication or as long as it is missing (Art. 7.5). If the holotype is lost or destroyed, the lectotype is selected from the isotypes, or, when, two or more specimens have been designated as types by the author of a species, the lectotype must be chosen from among these types.

(vi) **Neotype :** It is a specimen or other element selected to serve as nomenclatural type as long as all of the material on which the name of the taxon was based is missing (Art. 7.8).

(vii) **Topotype :** It refers to the specimen collected from the same locality from where the holotype was collected.

(d) Effective and valid publication

The new name of a taxon is considered valid or effective for publication, only when it is distributed in a printed form to the general public, or at least to ten well established botanical institutions with libraries accessible to botanists generally. The new name is not considered valid if communicated at a public meeting or/by placing of names in collections or gardens open to public or by the issue of microfilm made from manuscripts, typescripts or other unpublished material. Further, the name of a taxon when published has to fulfill certain conditions for validity. Those that have been published on or after 1.1.1935 must be accompanied by a Latin description or diagnosis or by a reference to a previously and effectively published Latin description or diagnosis of the taxon. The name of a new taxon of the rank of family or below that has been published on or after 1.1.1958, is valid only when its nomenclatural type is mentioned along with the name of the place where the type specimen is permanently conserved. Those, that have been published on or after 1.1.1953, are valid only if there is a clear indication of the rank of the taxon and whether it is a new genus (*gen. nov.*), new species (*sp. nov.*) or a new combination (*comb. nov.*) etc.

(e) Author citation

The name of a taxon (unitary, binary or ternary) is incomplete unless the name of the author or authors who first validly published the name, is cited along with it. This helps in verifying the dates of publication and in imparting precision in biological nomenclature. There are several rules for author citation:

(i) Usually the names are cited in abbreviated forms but never underlined or typed in Italics, e.g. (1) *Vitex* Linn., (2) *V. trifolia* Linn., (3) *V. trifolia*, var. *simplicifolia* Cham.

(ii) These citations can indicate bibliographic references, which are especially helpful in the recognition of homonyms. For e.g. *Utricularia caerulea* Linn., and *Utricularia caerulea* Clarice are two names referring to two different taxa, but it would have been impossible for us to recognize this, if the citation of author's names appended to the respective plant names were not given.

(iii) If the name of the plant is jointly published by two authors, their names should be linked by means of *et.* or an ampersand (&)

(iv) When more than three authors are involved, citation is normally restricted to the first author and is followed by *et al.*

(v) If an author validly publishes a name but ascribes it to another person, for example to the author who suggested the name but failed to publish it validly, or to an author who published the name before the starting point of the group, then the name of the latter should be connected to the name of the person who validly publishes the name by an *ex* as for e.g. *Acalypha racemosa* Wall. *ex* Baill.

(vi) If a genus or taxon of lower rank is altered in rank or position, but retains its name or epithet, the name of the author who first published the name or epithet (basionym) must be cited in parenthesis followed by the name of the author who effected the change. This is called double citation, e.g *Leucaena latisiliqua* (Linn.) Gillis (1974)..Basionym: *Mimosa latisiliqua* Linn. (1753).

(vii) If a taxon is of garden origin, then while citing the name it should be ascribed to *hort.* *(hortulanorum)* and connected to the name of the author who published it by an *ex*, e.g. *Geaneria dwklarii* hort. *ex* Hook.

(f) Choice of names

Following are some of the important criteria for choice of name of a taxon:

(i) When the taxon rank is changed, for e.g., a species becomes a genus, the earliest legitimate name in its new rank is its correct name.

(ii) When two or more taxa of the same rank are united into one, e.g. two or more genera are united, the oldest legitimate name of these taxa should be retained as the name of the united taxon.

(iii) When a genus or a species is divided into two or more genera or species, the original name of the genus or species must be retained.

(iv) When a species is transferred to another genus without the change of rank, the original name must be retained.

(g) Rejection of names

A legitimate name or epithet must not be rejected merely because it is inappropriate or disagreeable, or because another is preferable or better known, or because it has lost its original meaning. However, a name must be rejected if it was nomenclaturally superfluous when published. Similarly, a name or epithet rejected is replaced by the oldest legitimate name or in a combination by the oldest available epithet in the rank concerned. The following types of name can be considered to be illegitimate and unusable :

- **Synonyms** — They are the different names used for the same taxonomic group or taxon.
- **Tautonyms** — These are names where the specific epithet exactly repeats the generic name with or without transcribed symbol.
- **Typonyms** — A name is rejected if there is an older valid name based on the same type.
- **Metanyms** — A name is rejected when there is an older valid name based on another member of the same group.
- **Homonyms** — A name is rejected when preoccupied i.e. identical names cannot be applied to two different taxa.
- **Hyponyms** — A name is rejected when the natural group to which it applies is undetermined.
- **Autonym** — It is an automatically created tautonym for infrageneric or infraspecific taxa.
- **Nomen nudum** — It is a name that does not fulfill the criteria set by the International Code of Botanical Nomenclature as a legally described scientific name and therefore cannot be used unless it is subsequently proposed correctly. It is a Latin term (plural *nomina nuda*) referring to a name that has been published or mentioned without a proper and complete description.

3.5 Phylocode : A Phylogenetic Code of Biological Nomenclature

The current Linnaean system of nomenclature as embodied in the preexisting botanical, zoological and bacteriological codes, has been found to be unsuitable to govern the naming of clades and species. **A clade is a group for which all the descendants of the last common ancestor of the members of the group are included in the group**. It is a group in which every member shares a common ancestor that is not shared with any other group (a unique common ancestor). They are theoretically the most significant entities above the organism level as they compose the tree of life. The major drawback of the preexisting code is that it fails to provide names that explicitly and unambiguously refer to a particular clade or species. The name of a species often changes with time i.e. a species name changes whenever a species is referred to a different genus as a result of phylogenetic or phenetic consideration. Often under the preexisting code, supraspecific names are associated with clades and because these names are operationally defined in terms of ranks and types, they often fail to retain their associations with particular clades. Due to these reasons, an entity whose hypothesized composition has not changed may be given a different name under the preexisting codes based on considerations of rank in case of a clade oı genus in case of a species. The former is particularly objectionable given the wide recognition that rank assignment is subjective and biologically meaningless. Thus, to avoid this, and in order to promote clear communication and efficient storage and retrieval of biological information, the Phylocode or Phylogenetic Code of Biological Nomenclature has been proposed which is designed for naming species and clades. The starting date of the Phylocode has not been determined and is cited as 1st January 200n in the draft code. The present version of Phylocode only governs clade names. The rules governing species names will be added in future.

While developing the Phylocode, much thought has been given to minimize disruption of the existing nomenclature and it has been designed in such a way, so that it can be used concurrently with the preexisting codes or (after rules governing species names are added) as the sole code governing the names of taxa, if it is ultimately accepted by the scientific community. The main intention of this Code is not to replace existing names but to provide an alternate system by which both the existing and newly proposed names can be applied together. Thus, rules and recommendations have been included to ensure that most names will be ured in ways that approximate their current usage. The names of clades will however be redefined in terms of phylogenetic relationships rather than taxonomic rank as previously done under the Linnaean system, so that they will not be subjected to subsequent change with time due to changes in rank. As according to the Linnaean and phylogenetic system, the membership of a taxon associated with a name may sometimes differ, suggestions have been provided to indicate which code governs a name when there is a possibility of confusion. However, names that were provided with published phylogenetic definitions before 1st January 200n, will not be considered to be established under the Phylo Code.

3.5.1 History of Phylogenetic Nomenclature

The theory of Phylogenetic Nomenclature which was built on the suggestions that a taxon name could be defined by reference to a part of a phylogenetic tree, was in development since the early 1980's until they were finally published in a series of theoretical papers in the 1990's, the most noteworthy among them are a series of papers by de Queiroz and Gautheir (1990, 1992, 1994). Several other papers have also discussed or applied phylogenetic nomenclature. Three symposia were also conducted, which focused on phylogenetic nomenclature. These include the annual meeting of the American Institute of Biological Sciences in San Diego, California, U.S.A. in 1995 entitled *" Translating Phylogenetic Analysis into Classification"* organized by Richard G. Olmstead. The second symposia (Southwestern Botanical Systematics Symposium) was organized by J. Mark Porter in 1996 at the Rancho Santa Ana Botanic Garden in Claremont, California, U.S.A. and was entitled *"The Linnean Hierarchy: Past, Present and Future"*. The third symposium was organized at the XVI Internationational Botanical Congress in 1999 at St. Louis, Missouri, U.S.A. by Phlip Cantio and Tarsten Eriksson entitled *"Overview and Practical Implications of Phylogenetic Nomenclature"*.

The preparation of the Phylocode however began in the autumn of 1997. A workshop was later organized at the Harvard University Herbaria, Cambridge, Massachusetts, U.S.A. from 7th August to 9th August, 1998 and was attended by 27 people from five countries. An initial draft of the Phylocode prepared by Philip Cantino and Kervin de Queiroz was provided to the participants in advance, which was later revised on the basis of decisions made at this workshop. The draft derived some terminologies and rules from the BioCode and several rules from one or more of the preexisting codes, particularly the Botanical and Zoological Codes. However, many rules of the Phylocode are totally unique, and have no counterpart in any code based on the Linnaean Taxonomic categories because of fundamental differences in the definitional foundations of this system. The initial draft of Article 21 was written by F. Pleijel, A. Minelli and K. Kron, which was subsequently modified by M. Donoghue and P. Cantino. The initial drafts of Recommendations 10D and 11.8 B were written largely by T. Eriksson. The Latin terms in Article 9.3 were provided by W.M. Owens.

3.5.2 Properties of Phylogenetic Nomenclature

The phylogenetic system of nomenclature has the following properties:

(a) Inspite of the hierarchial relation between the taxa, the system is rankless, as assignment of rank is not part of the naming process, thus, having no bearing on the spelling or application of taxon names.

(b) This system provides rules for naming clades. The rules for naming species will be provided soon. Till then, the naming of the species will be governed by the preexisting codes.

(c) According to this system, the categories "species" and "clade" are not ranks, but are considered as different kinds of biological entities. A species is a segment of a population lineage, whereas, a clade is a monophyletic group of species. Both of them, which are discovered by systematics, rather than being created, are products of evolution, and even if not named have an objective existence.

(d) In contrast to the preexisting code, where supraspecific names have types, according to the Phylocode, a supraspecific name is given a phylogenetic definition and is applied to the clade which fits that definition, irrespective of its hypothesized composition. Species, specimens and synapomorphies cited within these definitions are called specifiers as they specify the clade to which the name applies and function, somewhat like types in providing reference points that determine the application of a name.

(e) This code allows taxonomists to restrict the application of names with respect to clade composition. If one wishes to ensure that a name refers to a clade that either excludes or includes a particular subtaxa, this code has the provision of including a qualifying clause in the definition specifying conditions under which the name cannot be used.

(f) Since there is a basic difference in the rules governing how supraspecific names are defined according to the Phylogenetic and earlier traditional systems, this has led to operational differences in the determination of synonymy and homonymy. As for example, according to the preexisting system, synonyms are names of the same rank based on types within the group of concern, regardless of prior associations with particular clades. On the other hand according to the phylogenetic system, synonyms are names whose phylogenetic definitions specify the same clade, regardless of prior associations with particular ranks. However precedence determines the correct name of a taxon when synonyms or homonyms exist as according to the preexisting codes. All established names, need to be registered which will reduce the frequency of accidental homonyms. Although precedence is based on the date of establishment under the Phylocode, under exceptional circumstances there is a provision for later established names to be conserved over earlier names for the same taxon, at the discretion of the International Committee on Phylogenetic Nomenclature.

3.5.3 Advantages of Phylogenetic Nomenclature

The Phylogenetic system of Nomenclature has the following advantages over the traditional system :

(a) According to the preexisting codes, it is often difficult to name a newly discovered clade, which either requires the use of an unconventional intermediate rank (e.g. superfamily) or shifting of less or more inclusive clades to lower or higher ranks. This causes a major source of instability leading to a cascade of name changes resulting in a delay in naming clades, until an entire classification is developed. As a result of this, taxonomy falls progressively farther behind knowledge of phylogeny, as well supported clades remain unnamed. With the latest burst of new information about phylogeny due to the recent advances in molecular biology and computer technology, this is a serious drawback as most of the phylogenetic information are not being translated into taxonomy, with the use of preexisting system. The use of the Phylocode will allow researches to overcome this disadvantage and name newly discovered clades much more easily, as they are discovered.

(b) The Phylocode will also improve nomenclatural stability, when it is extended to the naming of species, where it will remove instability in species names leading to revision of generic limits under the preexisting codes. However, when species will be named according to the principles of Phylocode, there will be absence of hierarchical information in species names and one will be unable to infer phylogenetic relationship from the species name in the way one can infer genus assignment from species names governed by the preexisting codes. This is because, according to the Phylocode, the primary purpose of the name of a taxon is to provide a means of referring unambiguously to a taxon, and not to indicate its relationship. However, the phylogenetic position can easily be indicated, by associating the species name with the names of one or more clades to which it belongs.

(c) The abandonment of ranks in Phylocode will also eliminate the error caused by many researchers who treat taxa at the same rank as if they are comparable in some biological meaningful way, due to which they commit the error of counting clades or species that possess properties relevant to the question of interest, or tracing the evolution of these properties on a phylogenetic tree, leading to faulty reasoning.

SUGGESTED READING

- Greuter, W. et al. (eds.), 1988, *International Code of Botanical Nomenclature*, Konigstein.
- Jeffrey, C., 1973, *Biological Nomenclature*, Edward Arnold Ltd., London.
- Johnson, A.T., 1971, *Plant Names Simplified*, W.H. & L. Collingeridge Ltd., London.
- Lawrence, G.H.M., 1951, *Taxonomy of Vascular Plants*, Macmillan, New York.
- McVaugh, R., R. Ross and F.A. Stafleu, 1968, *An Annotated Glossary of Botanical Nomenclature*, Regnum Vegetabile, Bohn.
- Radford, A.E., 1986, *Fundamentals of Plant Systematics*, Harper & Row, New York.
- Stearn, W.T., 1966, *Botanical Latin*, Great Britain.
- St. John, H., 1958, *Nomenclature of Plants*, Ronald, New York.

4 PLANT HERBARIUM

AT A GLANCE

A herbarium is a store-house of dried and pressed plant specimens collected from different places, which are mounted on appropriate sheets and arranged according to some known system of classification, and kept in pigeon-holes of steel or wooden cupboards specially prepared for the purpose. Some fleshy members, like those of Cactaceae, are however, preserved in liquid preservatives instead of being pressed and dried, while, bulky plant parts like dry fruits, cones of gymnosperms, etc. are dried without pressing and stored in special boxes. According to Fosberg and Sachet (1965) a modern herbarium is a great filing system for information about plants, both primary in the form of actual specimens, and secondary in the form of published information, pictures and recorded notes.

4.1 History of the Development of a Herbarium

The art of herbarium was initiated by an Italian taxonomist from Bologna, Italy, named Luca Ghini (1490-1556), who collected plants, dried and affixed them on paper with gum, in the form of herbarium specimens. He had a collection of about 300 specimens in 1551. Not much is currently known about his herbarium, which is now lost. Later, his student Gherards Cibo, continued with this art, and his herbarium is still preserved in Rome. In the middle years of the sixteenth century, three Ghini students, namely Aldrovandi and Cesalpino from Italy and Turner from England, also made their own herbaria. Cesalpino's herbarium, currently in Firenze, is very important, as it can be compared with his book *"De Plantis Libri XVI"*, which introduced a scientific approach to the study and classification of plants. An Englishman, John Falconer, who probably met Ghini in Italy, prepared a herbarium in 1553. There are currently more than twenty herbaria created before 1600, preserved in various European cities. The first recorded written publication is that of a native of Brussel, Adrian Spieghel, in *"Isagoges"*, a treatise of botany, which dates back to 1606, dealing with how to dry plants under a press and what kind of paper one should use, along with other accurate information.

The word ***herbarium*** (plural *herbaria*) was however, first applied by Pitton de Tournefort, in the book *"Elemens"*. Other herbaria were developed during the seventeenth century. A very good example is that of the Muséum National d'histoire Naturelle in Paris. During this period, many collections of exotic plants were created, as a result of the many geographical explorations. Some of these exotic herbaria have been of great importance for the advancement of scientific knowledge of some areas like Asia or Africa, and can be currently seen in a few European museums. It was Linnaeus who first started the current practice of mounting the plant specimens on separate sheets and storing them horizontally. Before Linnaeus, the actual practice was to sew dried plants on a sheet by thread and binding them into volumes. With the simple beginning by Linnaeus, herbaria today have developed into facilities for storing millions of specimens. In the earlier days the herbaria included plants of local or regional significance. But now, most of the herbaria include plants from different parts of the world and have developed into centers of advanced research in the field of taxonomy.

4.2 Functions of a Herbarium

A modern herbarium serves numerous valuable functions. Some of the important functions of herbaria are as follows:

(a) A herbarium serves as an invaluable conservatory of plant material of flora, collected from different parts of the world. Thus, they provide at one place, basic material for study of flora and vegetation of different places or regions. Since it serves as a permanent record of flora of those regions, collections in the herbarium provide evidence of the vegetation of a region, which may be destroyed due to some natural catastrophes.

(b) The specimens in the herbarium carry valuable data on their labels. These include data on habitat, habit, local names, color of flowers or other characters of the plant, native uses of the plant, abundance or frequency of the species, associated plants, etc. Such data provides valuable material for proper morphological description and range of variation of a similar plant collected from a different region, range of distribution and variation in its uses in different places. Thus, a herbaria provide data for botanical, ethno-botanical and phytogeographical studies.

(c) The herbarium serves as an aid in teaching botany to students in institutions where a herbarium is present, as it helps a teacher to show his students a plant specimen which may not be available fresh at the time of giving the course. It also helps students to identify local plants collected by them.

(d) Preserved specimens of herbaria are used in almost all types of taxonomic research. It is believed to be an essential requirement for biosystematics research today, for correct identification and nomenclature of the plant under study. Since in biosystematics studies, a work may need material of a taxon from far and wide in studies involving 'all world species', it may not be always possible for a researcher to visit different areas of occurrence. In such cases one has to largely rely on the resources of herbaria.

(e) The specimens in the herbaria are very often used as a source of material for anatomical, palynological and chemo-taxonomical studies.

(f) The herbaria provide important data on actual places of occurrence, time of flowering and fruiting, associated species and other data for researches in embryology, cytology and ecology.

(g) The herbaria have proved to be very valuable source of information for ethno-botanical researches as many native uses of plants are recorded on the herbarium sheets.

4.3 Different Kinds of Herbaria

Depending upon the interest of the organization or institution, the contents of holding and the labels and notes on the sheets in a herbarium vary accordingly.

(a) The herbaria of organizations like Botanical Survey of India contain all collections from any part of the world.

(b) Those institutions, which are interested in drugs and medicines, have herbaria, which include specimens of plants of known medicinal properties.

(c) The herbaria of the universities and colleges generally contain specimens only of interest for teaching or those included in the syllabus and research.

(d) The herbaria of Agricultural Colleges and Universities include specimens of crop plants and weeds of cultivated fields.

4.4 Important Herbaria of the World

A large number of herbaria have been established in different parts of the world over a period of the last four hundred years. According to a census conducted nearly ten years ago, there are about one thousand two hundred recognized herbaria in the world, excluding a large number of smaller unrecorded herbaria of various universities, colleges, pharmacies, etc. The first such herbarium was founded in the University of Padua, Italy, in 1545, along with the establishment of the first botanic garden in the same year.

The following tables (Tables 1-3) gives the name of the institution, location, approximate number of sheets, year of founding and the standard abbreviation of some important herbaria in different parts of the world :

Table 4.1 : Some important herbaria of the world

Sl. No.	Name of Herbarium	Place	Year of founding	Abbreviation	Total number of sheets (Approx.)
1.	Royal Botanic Garden, Kew	London, UK	1853	K	6,500,000
2.	V.L. Komarow Botanical Institute Meseum National d' Histoire	Leningrad, USSR	1823	LE	5,000,000
3.	Naturelle Laboratoire de Phanerogramme	Paris, France	1635	P	5,000,000
4.	Coservatoire Et Jardin Botaniques	Geneva, Switzerland	1817	G	4,000,000
5.	New York Botanic Garden	New York, USA	1891	NY	3,000,000
6.	U.S. National Museum	Washington, USA	1868	US	3,000,000
7.	Vienna Botanischer Gaertn	Vienna, Austria	1748	W	2,500,000
8.	National History Museum	Chicago, USA	1893	F	2,350,000
9.	Royal Botanic Garden	Edinburgh, UK	1893	E	2,500,000
10.	Missouri Botanic Garden	St. Louis, USA	1859	MO	1,700,000
11.	National Herbarium	Melbourne, Australia	1857	MEL	1,500,000
12.	Zurich Botanischer, Gaertn	Zurich, Germany	1834	Z	1,500,000
13.	Gray Herbarium, Harvard University	Cambridge, USA	1807	GH	1,485,000
14.	Philadelphia Academy of Sciences	Philadelphia, USA	1812	PH	1,000,000
15.	Arnold Arboretum	Boston, USA	1872	A	700,000
16.	Department of Agriculture	Peradeniya, Ceylon	1817	PDA	85,000
17.	Gordon College	Rawalpindi	1893	RAW	60,000
18.	Herbarium of the Rangoon University	Rangoon, Burma	1947	RANG	15,000
19.	Botanical Research Institute of Texas	Texas	1985	BRIT	500,000
20.	Fowler Herbarium – Queen's University, Canada	Canada	1987	—	—
21.	G.F. Leningham Herbarium – University	Regina	1989	—	50,000
22.	New Mexico State Range Science Herbarium	New Mexico		—	20,000
23.	University of Florida Herbarium (FLAS)	Florida	1990	FLAS	25,000
24.	University of Minnesota Herbarium	Venezuela	1992	—	830,000
25.	Washington State University — Marion Ownbey Herbarium	Marion	1995	—	350,000

Table 4.2 : Some important herbaria of India

Sl. No.	Name of Herbarium	Place	Year of founding	Abbrevia-tion	Total number of sheets (Approx.)
1.	Central National Herbarium	Shibpur, Howrah	1793	CAL	2,500,000
2.	Botanical Survey of India, Southern Circle	Coimbatore	1874	MH	200,000
3.	Botanical Survey of India, Eastern Circle	Shillong	1956	ASSAM	100,000
4.	Botanical Survey of India, Western Circle	Poona	1956	BSI	125,000
5.	Botanical Survey of India, Northern Circle	Dehradun	1956	BSD	60,000
6.	Botanical Survey of India, Industrial section	Kolkata	1887	BSIS	50,000
7.	Botanical Survey of India, Central Circle	Allahabad	1955	BSA	40,000
8.	National Botanic Garden, Herbarium	Lucknow	1948	NGB	100,000
9.	Forest Research Institute Herbarium	Dehradun	1816	DD	300,000

Table 4.3 : Some other minor herbaria of India

Sl. No.	Name of Herbarium	Abbrevia-tion	Year of founding	Total number of sheets (Approx.)
1.	Andaman & Nicobar Circle, Botanical Survey of India, Haddo, Port Blair	PBL	1972	20,000 (all groups worldwide specially vascular plants)
2.	Arid Zone Circle, Botanical Survey of India, 775/80, Subhash Nagar, Khema ka Kuan, Pal-Basni Canal Link Road, Jodhpur, Rajasthan	BSJO	1972	16,500 (all groups worldwide specially vascular plants of Arid and Semi-arid zones)
3.	Arunachal Field Station, Botanical Survey of India, Sankie View, Itanagar, Arunachal Pradesh	ARUN	1977	10,000 (all groups)
4.	Sikkim Himalayan Circle, Botanical Survey of India, P.O. Rajbhawan, Below Rajbhawan Campus, Gangtok, Sikkim	BSHC	1979	10,000 (all groups)
5.	Deccan Circle at Hyderabad	—	1984	—
6.	High Altitude Circle at Solan, Himachal Pradesh	—	1984	—
7.	Delhi University Herbarium, Delhi	—	—	15,000
8.	Rajasthan University Herbarium, Jaipur	—	—	30,000
9.	Herbarium of Jiwaji University, Gwalior	—	—	15,000
10.	School of Plant Morphology, Meerut College Herbarium	—	—	25,000
11.	Llyod Botanic Garden, Darjeeling	—	1878	12,000

4.5 Method for Preparing a Herbarium

4.5.1 Requirements for plant collection and its equipments

The main objective of collecting plants in the field and preserving them in the herbarium, are to preserve the specimens for all times and notes on the herbarium that will yield the maximum information about the plant concerned. The techniques employed for the collection and preservation of the flowering plants are discussed below.

4.5.1.1 Rules for plant collection

(a) Almost all natural environments are suitable for searching of plants for the herbarium. So, places, which could appear sterile and dry, must not be overlooked. Only native and naturalized plants may be collected. One may rely on the richness of the flowers and plants, which grow on grasslands and prairies, but one may also find an interesting specimen during trips to the mountains, swamps, coast, woods or wherever the climate and the temperature are not too extreme for plant survival. However, collecting on private property, or National parks, or private farms should be avoided. In such cases, it does require prior permission from the owner or local administrative office. Further, some species of plants are protected by the law. So it is not wise to pick them up from the environment. Some rare plants are protected in the whole territory of some region or districts. It is the responsibility of the plant collector to get a list of the protected species and one should get permission to collect these protected species too. One can also substitute color photographs for samples of the protected species.

(b) One should avoid collecting all of something and the last of anything i.e. in general, a good rule of thumb is to collect only a small set of samples from a large population. It is wise never to pick up an excessive amount of specimens of the same plant, especially if one does not see many of them around. It is advised to harvest the minimum number of plants one needs for classification and collection. In case of annual plants, a similar guideline is usually well to follow, but for trees and shrubs, it is best to selectively trim a few pieces from different individuals so as not to damage, in any way, the long-term survival of any individual.

(c) The most important part of every collected specimen is the flower. So the best seasons for picking up plants are spring and summer. However, certain species show their flowers in autumn or even winter. Hence, the colder months should not be neglected for excursions. At the same time a few plants have a very short blooming time, which one should keep an eye on to catch the flowering time. Some plants have short-lived flowers, which are deliquescent and will break up in pieces, unless picked early in the day and immediately pressed in the folder.

(d) The plants which one is interested in should be fresh and not wet (apart from plants which live in aquatic environments). The most convenient specimens to pick up are those, which seem quite dry, and lacking any trace of surface moisture. For this reason, it is better not to look for plants during rainy days, or early in the morning. Also, the hottest hours during summer days should be avoided, as plants will not show their freshest appearance.

(e) While collecting annuals, one should attempt to gather roots, flowers and fruits. Some species may also require fruits or mature seeds for identification, as for example, Brassicaceae require fruits for identification. One may bring along some small bags or envelopes to carry the seeds and other small separate components. Sometimes during some particular excursion, one may happen to pick up the plant when the seeds are not yet formed. In such cases it is advisable to return to the same place some weeks later, hoping to find some newly grown seeds. Similarly, while collecting a biennial, both the first and second year forms are useful, but at least flowers are required (again, Apiaceae requires fruits). As for herbaceous perennials it is best to avoid taking the root system. A small piece of the root, or a single bulb, may be necessary for proper identification in some groups (Poaceae, Liliaceae). When collecting members of Poaceae, Cyperaceae and Juncaceae, for example, it is advised to make observations regarding rhizome, the presence of which can be critical in the identification of the plant. Many other groups may have stolons or rhizome, and while a piece of the stolon can easily be gathered, the presence of a rhizome in some other groups might better be just simply noted rather than collected. All dirt should be removed from the plant after collection as well as all the insects, spider-webs and foreign bodies attached to the specimens.

(f) When collecting trees and shrubs, reproductive structures are vital. Cones, catkins and similar structures need not necessarily be new. However, every effort should be made to collect fresh flowering and fruiting parts if possible. Several species will actually flower before leaves are produced. Such individuals may be collected, but for proper identification in some groups (for example, Fagaceae), leaves and even mature fruits may be necessary.

(g) A herbarium with scientific merit usually relies on the plants, which grow naturally in a specific geographical area. So it is important to distinguish between the wild species from those, which have grown after human intervention, as for example some garden flowers or most cultivated plants. Occasionally it has been found that a plant species previously introduced by man may continue to spontaneously grow outside its original artificial environment. In such a case, the plant can actually become a new element of the spontaneous flora and can consequently become part of a herbarium.

(h) When one picks up plants from the natural environment, one should keep in mind that the plants have to be fixed up on returning home and they will have to be classified too, in order to achieve a scientific value for the herbarium. These operations require a substantial amount of time. Hence, one should try not to collect so many specimens that one cannot find the time to settle and study all of them. Further, during classification, one may have to section and basically tear apart some flowers. Hence, it is suggested not to pick up just one or two flowers of each species while looking for plants, otherwise one may be forced either to sacrifice every flower for the classification job or to simply give up the classification.

4.5.1.2 Tools and equipments

The following are the essential items or equipments, which are indispensable for plant collection:

(i) Tools

- Some tools are rather important while picking up plants. These include a small knife, scissors, thorn-proof gloves and a small handy spade could be of great help.

- **Bags :** The collected specimens should be put into a strong bag made of cloth or polythene or similar material (plastic, etc.). The function of these containers is to protect the plants from damage during the outing. When collecting in plastic bags, the specimens should be folded to the correct length for a herbarium sheet and placed firmly, but carefully, into the bag. They should not be just dropped in. This will prevent entangling of separate collections and there will be less damage. Later while emptying the bag, it should be turned upside down. The material should not be pulled out of the bag. This usually breaks up the specimens. It is preferable to use large bags rather than small ones, as there will be less damage to the plants. If one collects large, heavy plants, it is best to put them into a separate bag as they may damage other more delicate plants in the bag. This is especially true for palms and large aroids. To prevent significant wilting, plants may be wrapped in moist newspaper and placed in a plastic bag, which should be kept shaded. When mass collecting in tropical areas, this is not always feasible. Small delicate plants should be placed inside smaller bags.
- **Folder or Plant Press** : If the excursion takes place in summertime or lasts for two or more days, it is better to bring a folder of approximately 45×30 cm or more, or the plant may be pressed after reaching home, whichever may be convenient for the collector. However one should keep in mind that a good specimen is made in the field, not in the herbarium. It is therefore necessary to keep in mind what one is pressing in the field and how it will look on a sheet of herbarium paper. The folder must be made of cardboard or some other strong stuff, e.g. aluminum, and it must contain some old newspapers. The folder can be covered with cloth and it should be closed with straps or belts, and a handle or shoulder-belt should be added for easy carrying (Figs. 4.1a & 4.1b). The folder could be made in some different way too, for example a couple of plywood boards containing some newspapers and the boards can be kept together with some large rubber straps or ropes. Blotting paper can be used instead of newspapers, but the latter is much cheaper. The picked specimens must be arranged inside the folder between a few layers of paper, so that every plant has some paper on both sides. The closed folder does not have to press the specimens too firmly between the newspapers. If the plants are fleshy, more newspapers are needed, and one can also add some additional cardboard inside the folder. A large waterproof bag can be carried too to carry aquatic plants and can be used to cover the folder in case of rain.

(ii) Camera and tripod

In order to substantially enrich the quality of the herbarium, not only aesthetically, but also from the scientific point of view, one may take color pictures of each plant in its natural environment. In that way the dried specimen can be placed together with one or more photographs, which can be particularly very helpful for bulky plants like trees or bushes, which obviously can not be entirely included in a herbarium. Also the habitat of a plant can be well described with a photograph, taking care not to be too distant from the nearby bushes or trees. The suggested equipment is a 35mm single lens reflex camera, with a standard lens and a macro-lens, the latter very useful for close-ups of flowers and other specific features. Also a tripod can be very important if many close-ups have to be made, allowing the camera to remain steady. A tripod can also alleviate the need for a flash, which may be used when taking pictures in low light, but has the disadvantage of giving quite unnatural looking images. The speed of print films can range from 64 - 100 ISO to 200 or 400 if pictures in the woods are planned.

(iii) Field notebook

An important part of plant collection is recording of field data at the time of collection, which can actually increase the botanical knowledge of certain geographical areas. Hence, it is a good idea to write down, various information regarding the specimens one is picking up. These notes will not only help the identification of the specimens, but also can be compared and added to the current botanical knowledge. For this, special types of permanently bound, pocket-sized notebooks are used. Each book should have at least one hundred leaves, which are serially numbered, and six detachable tags should be on each page, which have the same number. These tags are tied to the specimens of the same kind, which are collected at the same time and place. The data such as the field number of specimen, date of collection, the locality, the habitat, the botanical name of the plant (if known), the flower color, abundance, variations and the associated plants, the vernacular name and the uses if any are recorded in the field note book at the time of collection.

Field notes

The following is a list of what one can record in the field notebook at the time of collection:

- Collector's name, associated collector's name(s).
- Collector's sequential field number (must be unique).
- Collection date.
- Locality : if possible one can include the longitude and latitude and also a sketch map is useful.
- Habitat and ecology : terrain characteristics, vegetation type, association with other plants.
- Plant habit : the overall size and shape of the plant (tree, bush, epiphyte, etc.) should be described.
- Field identification (family, genus, species if possible).
- Stem and trunk : height and diameter; color, texture, thickness and hardness; the presence of thornes and spines.
- Leaves : deciduous or evergreen; color, texture and overall aspect; orientation; exudate or presence of glands.
- Inflorescence and flowers : note of everything that could be undetected in prepared specimens; color; heterostylous, monoecious or dioecious; different behavior (open / closed) during the day; exudate or glandular; pollinators, etc.
- Fruits and seeds : size, shape, color, texture; smell, etc.
- Underground organs : take some samples or describe them (size and shape, tap root, tubers, bulb, etc.).
- Scent : record any particular scent, especially of cut parts and flowers.
- Sap or latex : note the color, smell, consistency, etc.
- Name : the locally used name(s) may be recorded.
- Uses : the uses may be recorded after getting confirmed.

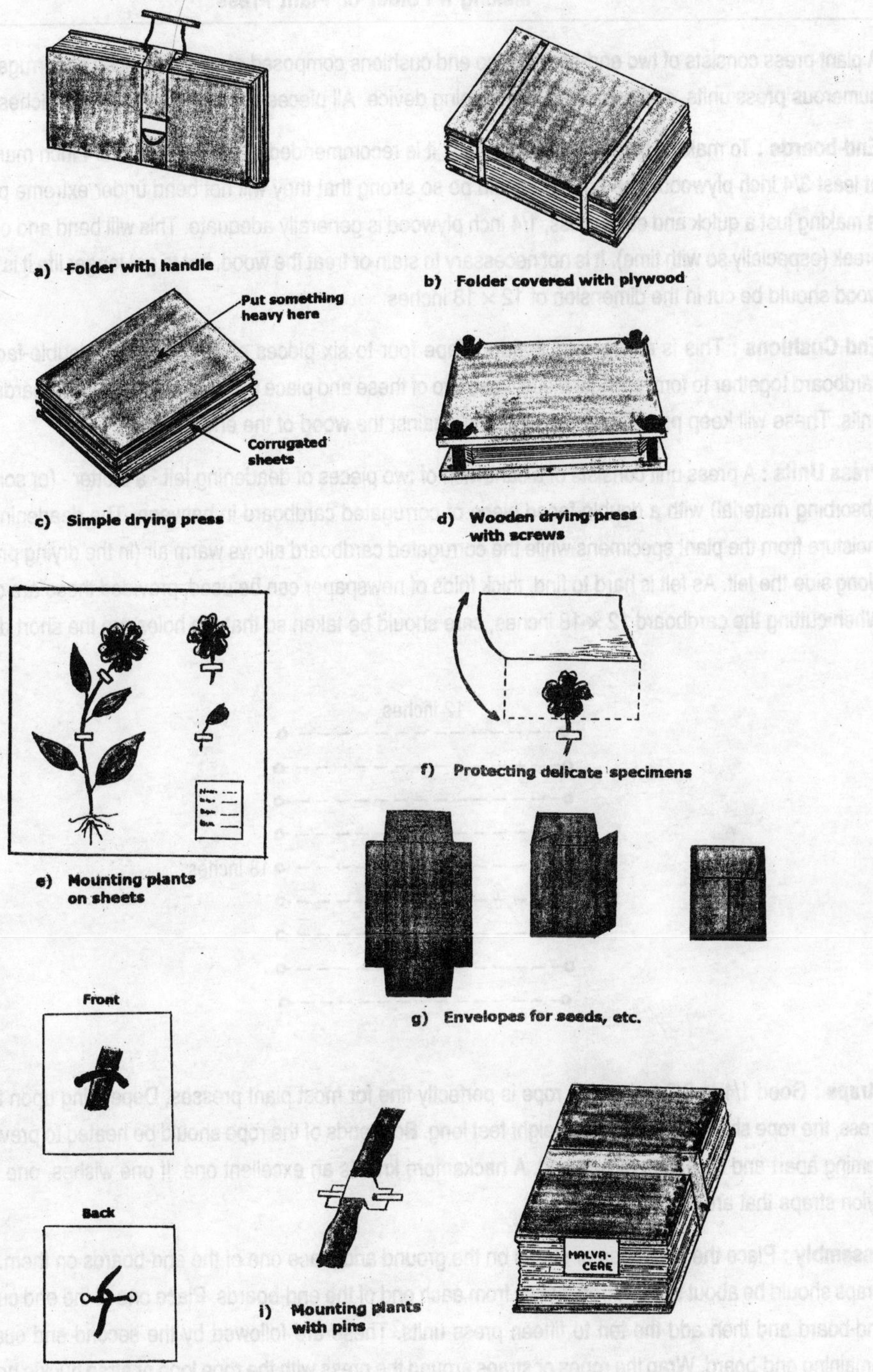

Fig. 4.1 : Tools and equipments for herbarium preparation

Making a Folder or Plant Press

A plant press consists of two end-boards, two end cushions composed of several pieces of corrugated cardboard, numerous press units, and some sort of strapping device. All pieces should be cut 12 × 18 inches.

End-boards : To make a permanent plant press, it is recommended to make them of 3/4 inch marine plywood or at least 3/4 inch plywood. The boards should be so strong that they will not bend under extreme pressure. If one is making just a quick and easy press, 1/4 inch plywood is generally adequate. This will bend and occasionally will break (especially so with time). It is not necessary to stain or treat the wood, but to get longer life it is advisable. The wood should be cut in the dimension of 12 × 18 inches.

End Cushions : This is an optional feature. Tape four to six pieces of 12 × 18 inch, double-faced corrugated cardboard together to form a single piece. Take two of these and place them between the end-boards and the press units. These will keep press units from rubbing against the wood of the end-boards.

Press Units : A press unit consists of a sandwich of two pieces of deadening felt - a blotter - (or some other water absorbing material) with a double-faced piece of corrugated cardboard in between. The deadening felt removes moisture from the plant specimens while the corrugated cardboard allows warm air (in the drying process) to pass along side the felt. As felt is hard to find, thick folds of newspaper can be used, provided these are changed often. When cutting the cardboard 12 × 18 inches, care should be taken so that the holes run the short distance, thus:

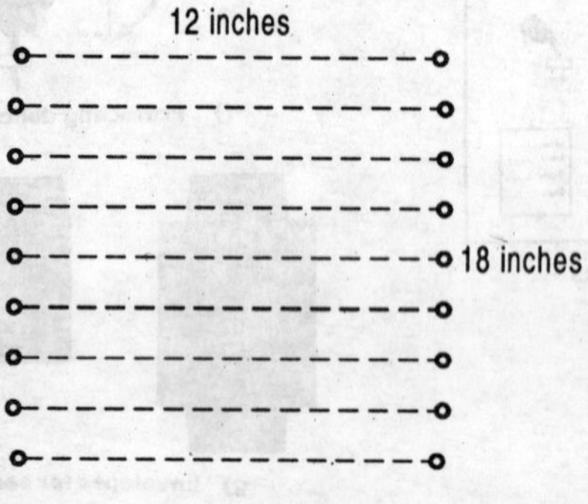

Straps : Good 1/4 to 3/8 inch nylon rope is perfectly fine for most plant presses. Depending upon the size of the press, the rope should be from five to eight feet long. Both ends of the rope should be heated to prevent them from coming apart and tie a loop at one end. A hackamore knot is an excellent one. If one wishes, one can use good nylon straps that are easier to work.

Assembly : Place the two ropes or straps on the ground and place one of the end-boards on them. The ropes or straps should be about two or three inches from each end of the end-boards. Place one of the end cushions on the end-board and then add the ten to fifteen press units. These are followed by the second end cushion and the remaining end-board. Wrap the ropes or straps around the press with the rope loop or strap buckle near the bottom of the stack. Tie up the press. When the press is empty, one should have at least two feet of extra rope.

(*Courtesy :* Prof. James L. Reveal, Northern Brown Herbarium, University of Maryland)

4.5.2 Identification and classification back at home

All collections should be identified and classified as soon as possible. Any seriously considered collection of plants must report the most exact taxonomic denomination for each specimen, along with the date and place of collection. The name one should apply to a collected plant is the scientific name, a Latin noun, which, represents the only denomination that unequivocally identifies the specimen. To classify every plant one picks up, one needs books that will describe all the species (and hopefully the subspecies too) of plants living in the region of one's interest. This kind of book usually includes only black and white pictures, often drawings, and focus on all the parts (fruits, flowers, roots, etc.) that can help to classify a plant. There are various books available for the classification of plants. Some are very complete and expensive, others are handbooks reporting only a particular group of plants (e.g. from a certain geographical area, or from a typical habitat), while others acquaint the reader with every kind of plant with little or any limitations, but they may not actually cover all the specimens one may find during travels. During identification back at home, one has to carefully examine very small and complex parts, like flowers, internal features and little seeds. One will need some very important instruments such as a magnifying glass, tweezers and a cutter for this purpose. Good field notes enables one to recall height of the plant and other observations made while collecting. It is a good idea to record the number of sepals, petals, stamens and note the condition of the ovary. Whether, the corolla is sympetalous, or if the stamens are epipetalous, and also the nature of any zygomorphic condition may be noted. Sketches of the flower made in the field are often extremely useful. It is also important that one finishes the classification work before the plants start to wilt. In fact to compare the specimens to the books descriptions, one must hold fresh plants. Otherwise the identification work could become very difficult and maybe wrong in the end too. Using an appropriate manual, the family, genus and species of the plant should be identified, adding the author to the binomial. If it is a subspecies or variety, then the infraspecific name and authorship should be recorded as well.

4.5.3 Setting the plants up for herbarium

One of the most difficult steps while pressing a plant is gaining the right arrangement of the specimens in the herbarium, as when the specimens are placed into the press they will acquire a shape and appearance that will last after the drying process, and so be similar or identical to the specimen's shape in the herbarium. Because of size and thickness limitations, it will probably be necessary to remove certain branches, leaves and bunches of flowers, or to carefully section them. This delicate job should be carried out with the help of a knife and tweezers, trying to arrange the flowers in the best possible manner, letting them show the most natural shape and all their features. During this process, care should be taken to avoid the elimination of relevant elements, especially those, which were important for the taxonomic classification. Any loose part may be kept inside paper envelopes, which will be placed on the same mounting sheet as the specimen. Each species should also display both sides of its own leaves. If only one leaf is available, one may cut off a part, so that the other side can be observed. If there are more than one flowers available, one can arrange some to show the back of the flower also. Depending on the size of the herbarium folders, one will be able to dry more or less large specimens. Some plants will have to be cut or bent, due to its excessive size. A stem should be bent at an acute angle. While arranging the specimens on the paper layers, various plants can be placed on the same piece of paper, but care should be taken not to superimpose two or more specimens upon each other. It is better to choose specimens of the same thickness so that the pressure on that layer will be equally distributed. If rather thick parts have to be placed in the press, an uniform thickness may be achieved by placing some pieces of papers near the more thick elements, so that the layer with the plants will have approximately the same height.

4.5.4 Special treatments

Some particular group of plants must be specially treated to achieve a good drying and durability.

(a) Cactus and succulent plants must loose their high percentage of water before being put to dry. To do that they must be placed below some blotting paper sheets, on them a hot iron may be passed rapidly. That must be done a few times, always changing the blotting paper. Before the iron treatment, the softening of the cactus can be aided by the immersion in boiling water for half a minute, taking care to avoid immersing the flowers. Instead of boiling water one can employ diluted acetic acid or strong alcohol (20 minutes) or formalin (1.5 parts formalin, 1 part water). Since, most cactus, are really bulky and fleshy, it is appropriate to just keep and dry sections of the whole plant.

(b) Some plants have tubers or bulbs as roots, and they must be treated before drying. As with cactus plants, a few minutes in boiling water softens the parts. Care should be taken to immerse in water only the roots that are to be treated.

(c) Many plants have some (or all) parts, which are so bulky that it will be possible to dry only a longitudinal section (e.g. of the stalk or the inflorescence). Hence it may be necessary to eliminate most of the internal parts and pulp, leaving the external features untouched.

(d) Conifers and heathers generally loose the leaves during desiccation. To prevent this the use of warm paper is suggested, adding some passes with a hot iron. It could be helpful to put the specimens in boiling water first, then drying them as fast a possible. Before the drying process, the small branches and the leaves can be smeared with diluted liquid universal glue like Vinavil (glue used for plastic, wood, cardboard, leather, etc.).

(e) Many flowers change color in some way during the drying process. This may not be any real trouble for the herbarium. But some flowers show a dramatic change, particularly some violet, blue and red petals. To prevent that problem it is vital to achieve a fast desiccation, with many changes of the paper sheets. Other suggested methods are to put the drying plants in contact with salicylic acid sheets (the sheets previously wetted with a 1% salicylic acid solution). Another practice is to immerse the flowers in petrol (American equivalent : gasoline). Then they must be air dried and finally treated with the paper sheets as the other flowers.

(f) Some plants have rather fleshy flowers (some orchids, iris, lilies, etc.) that must be filled with cotton wool before being put to dry. Every time the paper is changed, the cotton wool should be replaced too. The parts can be dried separately and then the flowers may be reassembled with the help of gum Arabic.

(g) If the flowers are likely to stick to the folder and break when handled, one can detach them and dry them apart using tissue or non-absorbent toilet paper, opening their folders only when the flowers are well dry.

4.5.5 Drying plant specimens

After identification of the plant, one must hurry to go on to the drying process. It is best to work in a dry and ventilated room, avoiding damp, which could easily harm the specimens. Flowers and other plant parts must be dried up into a peculiar kind of press. It is not difficult to build a home-made press for that purpose, keeping in mind that what must be accomplished is to keep the specimens squeezed between layers of paper (newspapers or blotting paper), until they are totally devoid of the original content of water, that is dried. As for the folder previously mentioned, the upper and lower parts of the press might be made of heavy cardboard or thick plywood or equivalent material. The specimens should be placed between the layers of paper, which should be pressed, by two pieces of wood. These two pieces must be held tight by putting some heavy body on top, i.e. bricks or large books (Fig. 4.1c).

It is important that the plants are put under sufficient pressure, otherwise more time will be required to achieve a good desiccation, besides they could be damaged by dampness and moulds. Every specimen in the press must be linked with its own written data. The best one can do is to use a tag for each specimen, which can be attached with a string. As circulating air is very important to get the specimens dried in a short time, it is better to add some corrugated cardboard sheets between the paper layers, so that some air can penetrate and speed up the drying process. Corrugated aluminum sheets are even more effective than cardboard. Bulky parts can directly be placed in contact with corrugated material to speed up drying. If no such material is available, the filled press size should be kept small. Instead of newspapers, some recommend the use of blotting paper together with tissue paper.

A more advanced kind of press might be built for an optimal drying of the plants. This press can be made with two wooden boards with screws and nuts placed at each corner. Turning the nuts, the two boards will come closer pushing together the paper with the plants (Fig. 4.1d). The press should be exposed to a gentle heat source, avoiding excessive heat that will "cook" the specimens. If fire is the heat source, keep the press at safe distance to prevent fire starting on the press. The process of drying may be hastened with the aid of artificial heat like an electric heater. Sometimes it is also possible to exploit the heat from the sun. In this case the presses must be rather small.

Changing the paper is surely a very important step whose importance must not be underestimated. In the first three or four days, a paper change should take place every day. Then one can leave more time between two changes. If the change of paper is neglected, the plants will take more time to loose their water content, besides they could be damaged if the paper stays wet for a few days. While changing the paper one must try to keep the specimens intact, besides taking care not to mix up the tags or labels with the name and the other information regarding the drying plants. Once a specimen has become dry and stiff, it can be placed into the herbarium.

Types of drying systems

There are basically two types of drying systems. The first is a convection system, which is easier and cheaper to build, and can use a wider variety of heat sources. The second is a forced-air system, which is more complicated and expensive to build but tends to be safer, and when properly designed is usually more efficient.

Convection System

In the convection system the plant presses are placed over a heat source. The warmed, dry air rises, passes through the channels in the corrugated cardboard, and carries the moisture away. A convection dryer can be made safer by putting screen, chicken wire, or mesh between the press and the heat source. The press should never be put too close to the heat source. As the material in the press dries, it becomes more flammable and may ignite. The heat source may be light bulbs, electric heating strips, kerosene stoves or lamps, propane, or any other source of heat. The system is most efficient, if both the heat source and the sides of the presses, are enclosed by the walls of the drying rack. Heat should not escape except by going through the plant press.

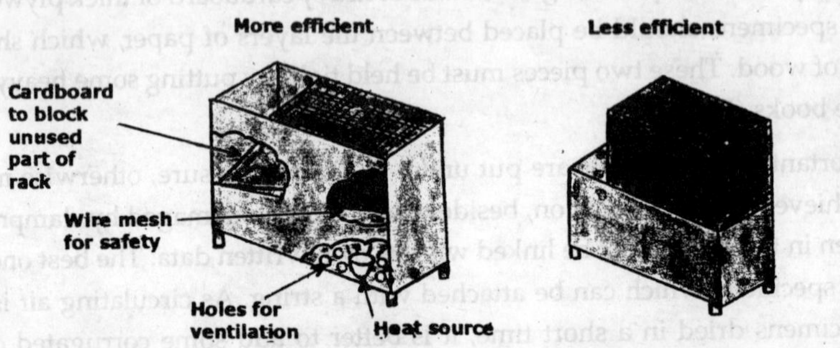

If the heat is not evenly distributed in the dryer, an aluminum panel constructed above the heat source will allow better distribution. This aluminum baffle must either be well supported throughout its length or have the edges turned up to prevent it from sagging. A double baffle system distributes the heat still more evenly or allows a reduced distance between heat source and press.

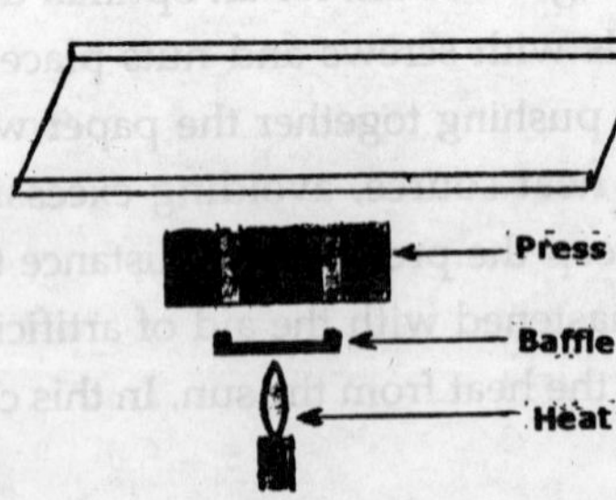

Forced-air System

In the forced-air system the heat source is separate from the specimens and a fan is used to blow the air across the heat source and then through the presses. It is safer because the heat source does not have to be directly under the presses.

(*Courtesy* : Prof. James L. Reveal, Northern Brown Herbarium, University of Maryland)

4.5.6 Mounting of Specimens on herbarium sheets

Once the specimens have been dried, they will have to be mounted on a strong mounting paper sheet to display the specimens and its data in the most clear possible way, and the specimens will be carefully preserved being attached to mounting paper.

4.5.6.1 Arrangement of specimen on mounting paper

It is very important that the plants be arranged properly so that one can immediately see all the main characters of any particular species. The best manner to place the plants on the mounting sheets is to align them with the right side of the page (or diagonally if space is required) and to have the heaviest parts and specimens at the bottom (Fig. 4.1e). The longest specimens can be folded if that can prevent the cutting of the stem into pieces. It is better to leave an empty space at the borders of the mounting sheets. One can either arrange the specimens (along with the labels and envelopes for small parts) in a regular way from page to page, or stagger the specimens at different positions on each sheet, so that each group of sheets will have a more equally distributed pressure.

4.5.6.2 Mounting paper

The best permanent support is good white or cream cartridge-type paper, in particular 100% rag or chemical wood pulp. The size for mounting paper can range from 42x26 cm. to 45x30 cm. One may even need larger sheets for some bulky specimens. The size of the mounting paper also may depend on the cabinet or box, which will contain the bundle of plants. The most delicate mounted specimens may be covered with a protective sheet of translucent paper, which must be rather strong and easy to fold. This paper can also be used to cover only parts of plants, such as fragile petals (Fig. 4.1f).

4.5.6.3 Small envelopes

Small paper envelopes can contain particular portions (e.g. seeds) or very small plants. When these envelopes are folded and glued to the sheet, it should be possible to open them flat and close them without clips. Besides they must securely hold their contents (Fig. 4.1g).

4.5.6.4 Mounting

There are different ways to mount the specimens to the herbarium sheets:

(a) **Strapping :** In this method, the specimens are strapped with linen or cotton thread, that are knotted on the reverse side of the sheet, where it is better to add some gummed paper to avoid contact with underlying specimens (Fig. 4.1h). Also gummed linen tape, like the one used by bookbinders, can be used. It must be placed where the specimens have greater strength, avoiding covering delicate details such as flowers. This method, allows to remove, and examine the specimen every time needed, but will allow a certain degree of movement, which can cause some trouble.

(b) **Gluing :** If glue is chosen to mount the specimens, water-based woodworking adhesive or library pastes or latex adhesives must be used, which must be applied quickly, taking care of not using an excessive amount. One can employ a brush or a nozzle applicator, otherwise the specimen (especially if it is a small one) can be put in contact with a plate full of glue then attached to the sheet. Glued material should then be left under pressure overnight, covering each sheet with waxed paper and with drying paper. For aquatic plants only latex glue must be used for gluing the specimens, as these plants can take up water from the glue.

(c) **Pins :** Specimens can also be mounted using small paper bands, which fix the plant to the sheet with the help of pins. The strip is placed on the stem (or other crucial parts of the specimen) and the pin joins together the mounting sheet, the stem and the strip, passing below the specimen (Fig. 4.1i). As for strapping, this manner leaves a remarkable freedom of movement for further examination.

4.5.6.5 Labels

Each specimen must have a label on its own sheet. It is advisable to use permanent and water resistant ink (black or blue), otherwise a pencil can be used (medium lead). The label should include the following:

- Taxonomic denomination (at least family, genus and species).
- Scientific name: In *italics* or underlined. If there is an infraspecific name it may be included.
- Location: Indicate, in caps, the country or place where the plant was found, followed by a colon (:). Then give the location where the plant was found so that, (a) the spot might be found by someone else, in distant future, and (b), a person with a generalized road map could find the location again. Avoid such references as "hill near my house" or "front of grandfather's barn"
- Habitat : Try to indicate where i.e. the locality where the plant was found. This can often be done, by noting the type of soil or rock, outcrop, exposure, or general condition where the plant was found. One can amplify this by indicating the associated plants. In some cases, elevation is useful. Likewise, a general statement about the abundance of the plant may be helpful, especially to those in the future studying plant distributions.
- Descriptive information : Information about the plant that either cannot be seen or might be lost in the future should be recorded on the label. Some taxonomists add local common names if they learn them and local uses. Observations are often useful and can help future scientists better understand the plant.

- Collector(s) name(s) and number : All person involved with the collecting of specimens should be recorded within reason. Mostly the names of two to five additional collectors are recorded on labels. However, the primary collector's name appears first and the collector's collection number is reported.
- Date : The date the collection was made should be recorded on the label. One should avoid using 08/09/94 as in the United States this usually means 9 Aug., 1994 while in Europe this would mean 8 Sep., 1994. The taxonomic community has largely adopted the mode of giving the day of the month first, followed by month (first three letters if abbreviated and then without a period) and finally the full year (1994, not 94).

4.5.6.6 Herbarium organization

The mounted specimens must be placed in bundles, which could be kept together with the help of strings or straps. Besides it is better to have cardboard or millboard supports at top and bottom of the bundles, to gain further steadiness (Fig. 3j). Each bundle must have its own label so that one will be immediately able to recognize the contents of each particular group of specimens. Depending on the number of plants in each bundle and on the criteria chosen when dividing the collected plants into groups, one may write down the geographic origin or the taxonomic level (family, genus, etc.) or whatever identifies every particular bundle. The mounted specimens are stored systematically in special wooden or steel cabinets with pigeon holes also.

The herbarium must follow a certain arrangement, which will give some logical and scientific order to the collection. There are various systematic arrangements that can be followed. In most of the Indian herbaria the specimens are arranged according to Bentham and Hooker's system of classification. One can also choose to divide the specimens by their families (or even genus) alphabetically, or else one can group them geographically. These collections of specimens should then be kept inside boxes or cabinets, which should remain tightly closed and dust-proof. The best places one can choose for the herbarium should be safe from the danger of fire, floods and other accidents. The temperature should be around 20^0 C and the humidity should not exceed 60%, in order to prevent fungal infections. Adequate ventilation is required, especially if the herbarium itself has a high concentration of insecticide or fungicide. A dry room is the best to avoid proliferation of insects and consequent damage. Naphthalene and paradichlorobenzene (PDB) are chemical repellents often used for preventing infestation. They are however very effective only at high concentrations, which increases the risks for human health. Other products, which have been used in large herbaria are, mercuric chloride, cresol and LPCP, but they are complicated to use, and the results can be dangerous to human health too.

(*Courtesy :* A Herbarium by Andrea Tarozzi, 1996, English version revised by Edward Green; and Prof. James L. Reveal, Northern Brown Herbarium, University of Maryland)

Some important suggestions for arrangement of plants before drying

- The plants should be arranged in such a way so as to clearly show both surfaces of leaves and reproductive structures. Some flowers may be pressed open, and some closed, and others split to show the internal structures. Bracts in the capitula of Asteraceae are very important, and should be pressed, so that some can be clearly seen. Pubescence, stomata, and other characters are frequently more important on lower leaf surfaces than upper. If only one large leaf or fern frond sample is available, it should be folded so that part of both surfaces can be seen.

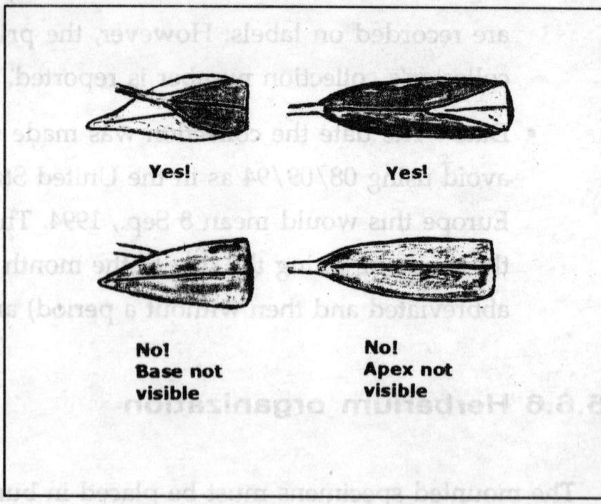

- One should not cover flowers, fruits, or stem with leaves. Either the leaves should be pressed away from the other plant parts, or folded underneath them. When folding leaves, the larger part should be underneath so that one can still measure length, width, etc.

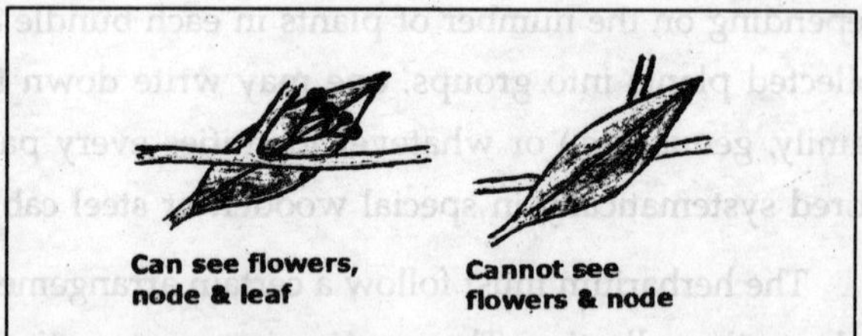

- The excess leaves should be cut off, but always part of the petiole should be kept to show the position. The petiole base should never be cut off and also the stem attachment of a compound leaf. If possible one should keep some of the petiole bases of the other leaves and the apex of the stem. A large compound leaf should not be mistaken for a branch with simple leaves. The twig should not be split, because the opposite or alternate arrangement of the leaf will no longer be evident. A specimen of two sheets or more may be necessary with very large leaves.

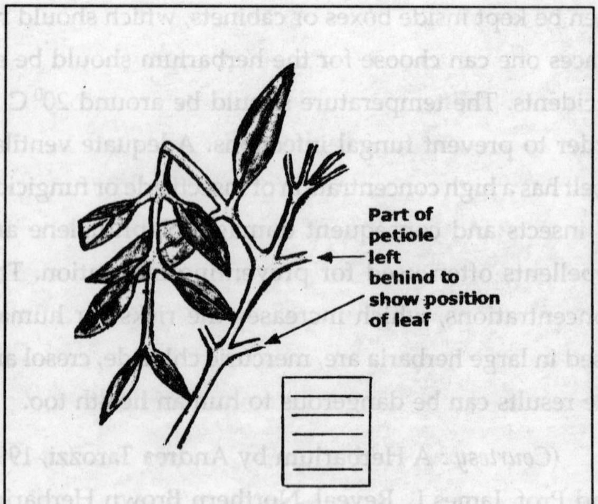

Some important suggestions for arrangement of plants before drying

- It is usually easier to fold the stem than to try to fold all of the leaves. If the leaf tips stick out of the newspaper, then they may be folded over or they will be broken off and lost. If large leaves are folded, one can sometimes put two large leaves on a sheet rather than one. Length, width, shape, upper surface, and lower surface are still observable.

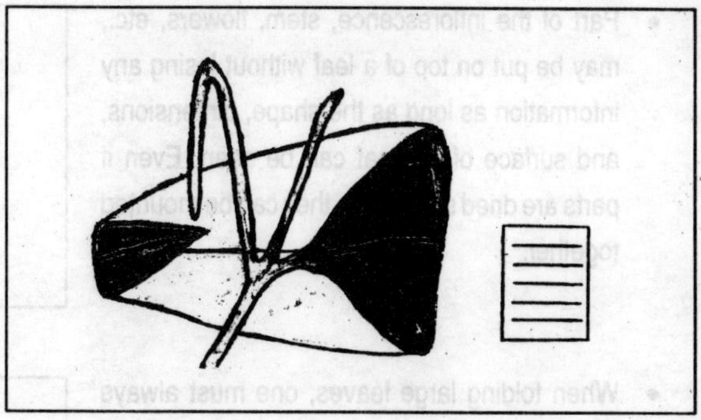

- With the long leaves of palms and large ferns, an apical section, a mid-section, and a basal portion with pinnae should be taken. The length of the leaf, length of the petiole, number of pinnae, and the arrangement of the pinnae (i.e., regular, staggered, or irregularly spaced) should be stated in the field notes if the specimen does not show it. For palms, the position of the inflorescence should be described in relation to the leaves, also whether species are solitary or colonial, and samples of stem, spines and bark should be included. Each specimen, at a minimum, should consist of an apex, a base, a mid-section, selected parts of the inflorescence, stem, and petiole base. Photographs are very useful.

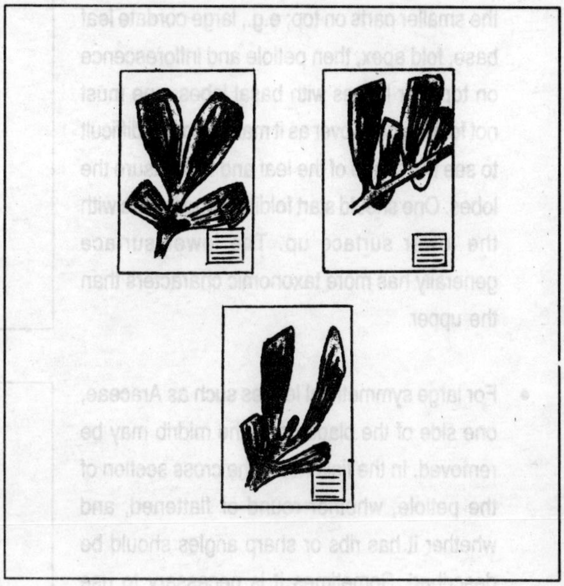

- A representative specimen of large plant may require additional sheets for completeness. In general the sheets should not be overcrowded. Multiple-sheet specimens contain far more information than fewer overcrowded sheets. Large fruits should be cut into one inch thick slices, both longitudinally and transversely. It is preferable to have the fruits attached to the branches when the specimen is mounted. A large leaf should be folded, so that base and apex can be seen.

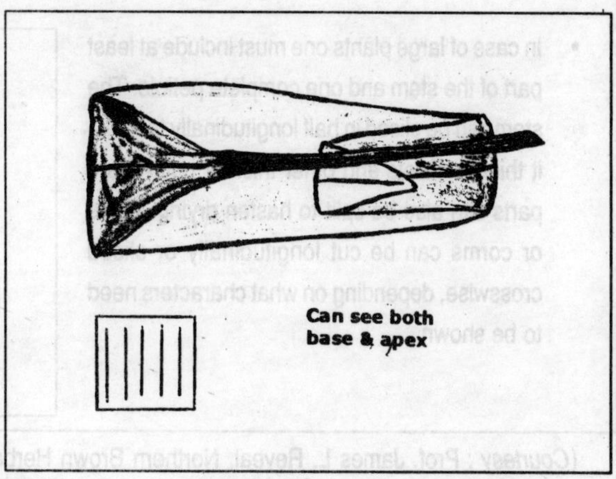

Some important suggestions for arrangement of plants before drying

- Part of the inflorescence, stem, flowers, etc., may be put on top of a leaf without losing any information as long as the shape, dimensions, and surface of the leaf can be seen. Even if parts are dried separately, they can be mounted together.

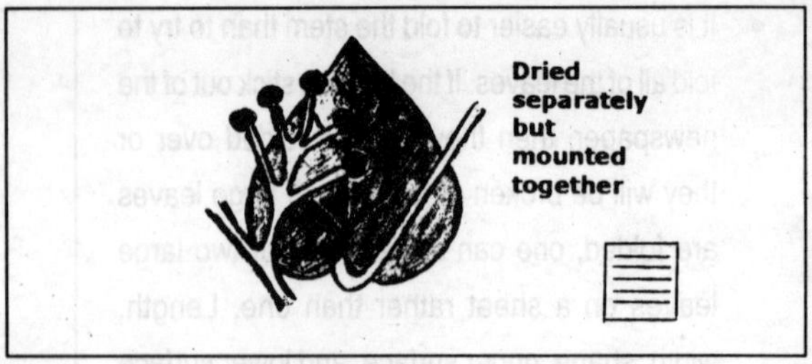

- When folding large leaves, one must always start with the largest portion of the leaf, and then the smaller parts on top; e.g., large cordate leaf base, fold apex, then petiole and inflorescence on top. For leaves with basal lobes, one must not fold the base over as it makes it more difficult to see the shape of the leaf and to measure the lobes. One should start folding large leaves with the lower surface up. The lower surface generally has more taxonomic characters than the upper.

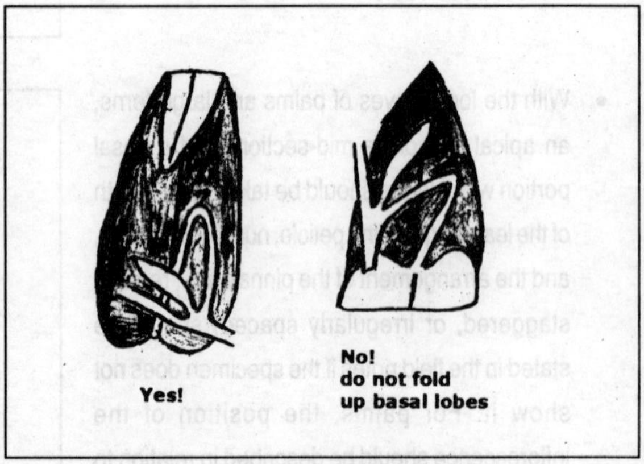

- For large symmetrical leaves such as Araceae, one side of the blade up to the midrib may be removed. In the field notes the cross section of the petiole, whether round or flattened, and whether it has ribs or sharp angles should be described. Sometimes it is necessary to use multiple sheets.

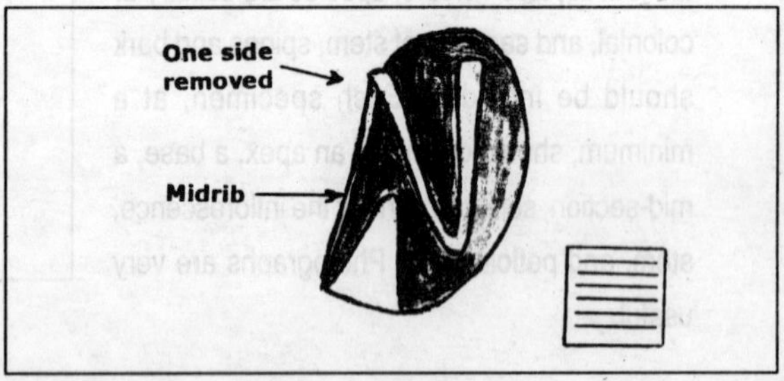

- In case of large plants one must include at least part of the stem and one complete petiole. The stem can be sliced in half longitudinally to make it thinner. Roots and other thick or bulky plant parts can also be split to hasten drying. Bulbs or corms can be cut longitudinally or sliced crosswise, depending on what characters need to be shown.

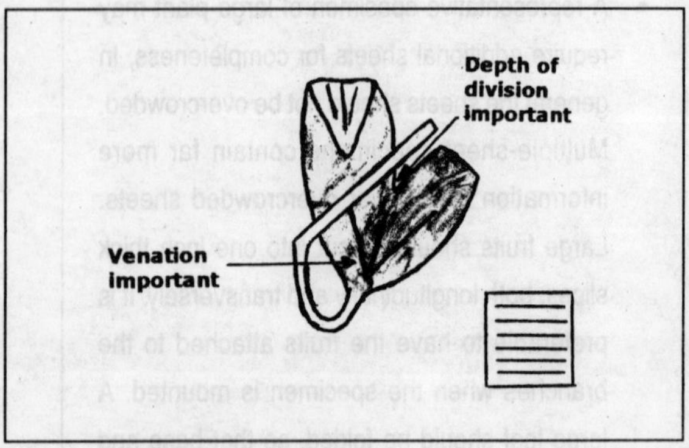

(*Courtesy* : Prof. James L. Reveal, Northern Brown Herbarium, University of Maryland)

4.6 Herbarium of other groups of plants

It is also possible to build a herbarium with ferns and other non-vascular plants such as lichens and fungi. Here is a brief description of the main topics regarding this particular kind of herbarium.

4.6.1 Pteridophytes

The ferns and allied plants are grouped together in the pteridophytes. Typically, pteridophytes have photosynthetic organs called fronds, which resemble the leaves of vascular plants. Most of the ferns have their spores in sporangia usually found in the margins or under the fronds. This is a very important feature, as these plants must be collected when they are fertile, otherwise no identification will be possible. In many allied plants the spores are arranged or aggregated in different manners. If one is interested to collect these plants, one should carry a x10 hand-lens to look for the presence of sporangia. To gain a successful identification it is also advisable to collect part of the rhizomes, examining their covering elements and the appearance they give to the plant. Smaller ferns sometimes have long rhizomes, which can form mats even with different species. So they must be carefully distinguished. With larger ferns it is important to collect and dry all the most significant features, even if some will be cut into pieces. It is important to make a note or a picture of all the aspects, which cannot be seen in the collected parts.

4.6.2 Bryophytes

These small non-vascular plants include mosses (Musci), liverworts (Hepaticae) and hornworts (Anthocerotae). Their life cycle includes a long-lived green phase (gametophyte, which reproduces sexually) and a short-lived sporophyte, which reproduces by spores. It is better to collect them into paper envelopes and to avoid polythene [plastic], which can badly interact with humidity. To find many interesting species, one must look in particular habitats like rocks, trunks, soil and other plants, remembering to carry a magnifying glass. During drying, which should be begun quickly, one must be careful not to overcook the specimen if heat is used.

4.6.3 Fungi and Lichens

For collecting fungi it is better to use a basket along with paper to keep the specimens separated. Only fresh specimens in good conditions should be collected. It is possible to pick many specimens of the same species if they show different development stages. One must always look carefully before picking a fungus to detect the presence of parts hidden in the soil (pick them up too) and take a note of the habitat. If the fungus is parasitic, one must look where it grew. Besides, some relevant characteristics could be changed after collection of the specimen. Hence it is very important to observe whether color changes occur and note whether there is presence of latex, the consistency, the odor and so on. An interesting feature to add to the collection is the spore print, which is done leaving the specimen overnight with its hymeneal surface downwards onto a white sheet of paper. A quick drying must soon follow, keeping the specimens at a temperature of approximately 40° C, avoiding lower temperatures and dampness. Larger fungi should be cut in two or more parts to allow drying. Before placing the fungi in the herbarium it is better to leave them in a freezer for two days, to kill any insects or eggs. The specimens, if not particularly fragile, can be kept in paper envelopes (18 × 12 cm or less) with their data and then glued to standard herbarium sheets. The delicate species, as many fungi are, should be placed in a small cardboard box (approx. 7.5 × 4.5 × 1 cm) within the paper packets. Lichens should not be pressed and, once dry, they must be placed in packets on sheets with their data.

4.6.4 Ancillary Collections

The preservation of plants (or parts of them) in liquid medium allow the maintenance of the natural three dimensional aspect of the specimen. Hence various techniques have been developed to keep plants in spirit. Some groups like orchids or succulents are always better kept in spirit rather than on mounting sheet. Different preservative liquids can be formed using alcohol, formalin or other components, which can be rather dangerous to handle. The Kew preservative is a mixture of 53% industrial methylated spirit (i.e. ethanol + 2-4% methanol + 9% water), 37% water, 5% formalin solution in water and 5% glycerol. The glass jars for the ancillary collection can range from 70 to 3000 cc and they must have wide necks.

(*Courtesy* : A Herbarium by Andrea Tarozzi, 1996, English version revised by Edward Green).

SUGGESTED READING

- Forman, L. & D. Bridson, 1989, *The Herbarium Handbook*, Royal Botanic Gardens, Kew.
- Franks, J.W., 1965, *A Guide To Herbarium Practice*, Handbook for Museum Curators, Museum Association, London.
- Holmgren, P.K., W. Keuken, E.K. Shofield, 1990, *Index Herbariorum, Part I. The Herbaria of the World*, 8th Ed., Regnum Veg., New York,120.
- Jain, S.K. and R.R. Rao, 1977, *Hand Book of Field and Herbarium Methods*, Today and Tommorrows, New Delhi.
- Santapau, H., 1955, *Botanical collector's Manual*, Botanical Survey of India, Calcutta.
- Womersley, J.S., 1981, *Plant Collecting and Herbarium Development*, Fao Plant Production and Protection Paper, 33, Xi, Rome.

5 BOTANICAL GARDENS

AT A GLANCE

Botanical gardens are those where living plants of different varieties, collected from different parts of the world, are grown in a scientific and systematic manner. These gardens are thus, botanical institutions, which not only includes trees, shrubs, herbs, climbers and other plants of wild varieties, but also ornamental and cultivated ones as well as medicinal and economically important plants of various geographical regions. They also contain green houses, glass houses, a herbarium, a library, research laboratories, lakes and other miscellaneous resources. Hence, it is a place for ex-*situ* conservation, where plants are displayed for the purposes of research and education. A modern botanical garden is thus, a center for documentation, research, education, reference, conservation and data storage and of immense importance to botanists, horticulturists and foresters. Well established, botanical gardens are members of the International Association of Botanic Gardens (IABC), established in 1954.

A botanical garden differs from a recreation park in that, public parks and gardens are for recreation and primary centers of aesthetic beauty and public amenity, whereas a botanical garden primarily serves scientific functions, although well planned botanical gardens may also serve as centers of aesthetic attraction. Both the terms 'botanic' and 'botanical' can be used, as both are grammatically correct.

5.1 History of Botanical Gardens

The history of gardens dates back to our ancient Indian History, where such gardens have been mentioned in our epics likes Mahabharata, Ramayana etc. Our kings and emperors grew gardens around their palaces for aesthetic beauty and there gardens included plants of different kinds of flowers, fruits and vegetables. In other parts of the world, the Hanging Gardens of Babylon are considered to be the most ancient wonders. The Romans maintained gardens as sources of medicine and even the Chinese, Persians and Aztecs had shown interest in gardens, which grew plants with beautiful flowers and those, which yielded perfumes.

The Hanging Gardens of Babylon

The Hanging Gardens of Babylon, considered to be one of the Seven Wonders of the Ancient World, were the most revered and awesome structures in all of history. They were built by the ruler Nebuchadnezzar, the son of Nabopolassor, between the years of 604-562 B.C. He built it to please his homesick wife Amyitis, who was from Media and he had married her to make an alliance between Media and Babylonia. She longed for the meadows and mountains of her homeland. Although the Hanging Gardens no longer exist, the idea of such a magnificent feat of engineering still fascinates people today.

The garden had exotic flourishing plants, which were cultivated above ground level. These plants were imported from foreign lands and may have included cedar, cypress, myrtle, juniper, almond, date palm, ebony, olive, oak, terebirth, nuts, ash, firs, nightshade, willow, pomegranate, plum, pear, quince, fig and grapevine. The gardens formed a quadrilateral shape. The plants were suspended over the heads of observers on terraces and the brilliantly colored trees and flowers dangled and draped over the terraced walls, creating a lush and magical environment. Arches were underneath these terraces. Since Babylon rarely received rain, the gardens had to be irrigated and probably the technique of hydroponics was used for irrigation, which involves, irrigation with water containing nutrients, with no use of soil, as excavations have found an elaborate tunnel and pulley system, which brought ground water to the top terrace. Water was dispersed by means of a chain of pumps. Thus, the Hanging Gardens of Babylon were an impressive example of architecture.

It was during the 16th century, that the first botanical garden of the world was established by an Italian Professor of Botany, Luca Ghini (A.D. 1490-1556) at **Pisa**, Italy in 1543. He was the director of this garden and was later succeeded by Andrea Caesalpino (1554-1558). It was during this century that the herbalists grew medicinal plants collected from different parts of the world in their gardens. Other important botanical gardens established during this century include the following :

(a) **Padua Botanical Garden** which was established by Professor Francis Bonafede at Padua, Italy in 1545 and is considered to be the oldest Botanical Garden and is named "Orto Botanico". It covers an area of about 20,000 square meters and has a large herbarium and a library attached with it. Some of the major plants of this garden include *Cedrus deodora, Ginkgo biloba, Mesembryanthemum, Platanus orientalis, Robinia preudoacacia,* etc. This garden has the portraits of over 600 botanists that have been displayed over here.

(b) **University Botanical Garden,** which was established by Luca Ghini at Florence, Italy in 1545. It was founded on 1st December, 1545 when Grand Duke Cosimo dei Medici purchased the land from the Dominican sisters. The traditional name of this garden is "Giardino dei Simplici" because of the fact that it was used to cultivate and raise medicinal plants. The garden was later improved and embellished with the 18th century collections and presently has its own green houses and hot houses for tropical plants, medicinal plants and some superb examples of sequoid trees, cedars and oaks from the early nineteenth century. There are even sections dedicated to medicinal plants, cactuses and carnivorous plants which are very interesting from the didactic point of view.

(c) **Vatican Garden**, established by Michele Mercati at Rome in 1556.

(d) **Bologna Botanical Garden** established at Bologna, Italy in 1568 with Ulisse Aldrovandi (1567-1605) as its first director. Having being transferred various times, the Gardens were finally located in the old Collegio Ferrario with over 5,000 specimens of exotic and local plants. This garden now has three different green houses, two of which houses tropical plants, especially ferns, Orchideaceae, Bromeliaceae and trees of nutritional interest, while the third houses a collection of succulents. It also has various medicinal and aromatic plants and other ornamentals.

(e) **Leiden Botanical Garden**, which was established in 1587 at Leiden, Netherlands. The famous herbalist Carolus Clusius was its first director and contributed a lot to the growth of this garden. The first green house was established in this garden in 1599 for the protection of plants, which were introduced from Cape of Good Hope. This garden mostly includes the plants that were collected by the East India Company.

(f) **Botanical Garden, Heidelberg**, Germany established in 1593.

(g) **Montpellier University Botanical Garden** at Montpellier, France, which was founded in 1593.

During the 17th **Century**, medicine had shown tremendous development and a large number of Botanical Gardens were further established in different parts of the world including various universities, mainly with the purpose of displaying for medical use. Some of the important Botanical Gardens of this century are as follows :

(a) **Botanical Garden of Strasbourg,** France, established in 1619.

(b) **Oxford University Botanic Garden**, Oxford, England, which was initiated in 1621. This garden possesses the first green house ever established in England, which was founded in 1734.

(c) **National Museum of Natural History, Paris,** France established in 1635.

(d) **Jardin des Plantes,** a botanical garden established in 1640 at Paris for the purpose of growing all kinds of medicinal herbs.

(e) **Botanical Garden of Groningen,** Netherlands established in 1642.

(f) **Botanischer Garten in Berlin,** Germany, which was founded in 1646, and A.W. Eichler, Adolf Engler and L. Diels were its directors and is one of the greatest botanical gardens.

(g) **Botanical Garden of Uppasala,** Sweden founded in 1655 by O.J. Rudbeck at Sweden. This garden has the privilege of the works of Linnaeus, who worked at this garden from 1741 to 1777.

(h) **Royal Botanic Garden, Edinburg,** Scotland, which was founded in 1670 with J. Sutherland as its first director, is famous for *Rhododendron* and Azaleas. The Rock garden present over here is one of the first in the world, which is remarkable for its Himalayan alpine plants. This garden also has a glass house, a herbarium and a library.

(i) **The Chelsea Physic Garden, London,** England, was established in 1673, by the Society of Apothecaries.

(j) **Amsterdam Botanical Garden,** Netherlands, founded in 1682.

(k) **Botanical Gardens of Tokyo Imperial University,** Tokyo, which was established in 1684.

During the next three centuries more famous botanical gardens were established. Some of the important ones are as follows :

Name of the Garden	Place	Year of establishment
Botanic Garden of University of Moscow	Moscow, Russia	1707
Druggist's Garden of St. Petersburg renamed as V.L. Komarov Botanical Institute of the Academy of Sciences	Leningrad, Russia	1713
Botanical Garden of Philadelphia	Philadelphia	1713
Botanical Garden of Vienna	Vienna, Austria	1754
Jardin Botanico de Madrid	Spain	1755
Royal Botanic Garden, Kew	Kew, England	1760
Botanic Garden of Cambridge University	Cambridge, England	1762
Botanical Garden of Coimbra	Portugal	1773
Indian Botanical Garden	Kolkata, India	1787
The Jardin Botanico de Rio de Janeiro	Rio de Janeiro, Brazil	1808
Universitets Botaniske Have	Oslo, Norway	1814
Botanic Gardens of New South Wales	Sydney, Australia	1816
Botanic Garden at Buitenzorg	Java	1817
Conservatoire et Jardin Botaniques de Geneva	Geneva	1817
Botanical Garden of Cape Town	Cape Town	1848

Name of the Garden	Place	Year of establishment
Government Botanical Garden, Kingston	Jamaica	1857
Botanical Garden of Catania (Orto Botanico dell Universita)	Catania, Sicily	1858
Singapore Botanic Gardens	Singapore	1859
Missouri Botanical Garden	St. Louis, USA	1859
Hanbury Botanical Garden	La. Montola, Ventimiglia, (Liguria), Italy	1867
Arnold Arboretum	Harvard University, USA	1872
Botanic Garden of University of Rome	Rome	1883
New York Botanical Garden	New York	1891
Jardin Botanico Municipal of Buenos Aires	Argentina	1892
Botanical Garden, National Museum of Natural History	Beijing	1930
Botanical Garden, Forest Research Institute, Dehradun	Dehradun, India	1934
The Jardin Botanique de Montreal	Canada	1936
National Botanical Garden, Lucknow	Lucknow, India	1946

At present there are about 1400 botanical gardens and arboreta in the world receiving over 100 million visitors per year. A large number are located in Europe and more than 30 are found in Italy, including the University botanical gardens. Thus, Italy boasts a historic first with respect to botanical gardens, most of which were founded in the second half of the 18th century and in the 19th century.

5.2 Role of Botanical Gardens

The main functions of botanical gardens can be summarized as follows:

(a) Botanical gardens are in fact exceptional acclimatization areas, where exotic plants arriving from all regions of the world, live together in the open air, even though out of their natural environment.

(b) The unique and diversified collections of living plants provide basis for research and modern taxonomic studies on morphology, ecology, genetics, systematics and evolution. They supply living plant resources for academic studies on cytology, embryology, anatomy, phytochemistry, etc.

(c) They provide information on food plants, ornamental plants, medicinal plants, etc.

(d) Botanical gardens provide germplasm for hybridization to produce better varieties of flowers, fruits and vegetable plants commercially from these plants.

(e) The various facilities of glass-houses, green houses, etc. at botanical gardens provide protection to many endangered plants and means of propagation of rare plants.

(f) Different botanical gardens of the world have served as acclimatization centers and provided adequate facilities for protection of economic plants. Many are responsible for commercial utilization and expansion of many valuable plants such as tea, rubber, coffee, cotton, teak, cinchona, hemp, vanilla, etc., which are natives of one place and have been introduced in other countries.

(g) Botanical gardens also provide seeds and sapling of important plants on exchange basis to other gardens, to encourage the introduction of some exotic, useful and economic plants.

(h) Botanical gardens practice and provide training for conservation of nature and are useful for training people in growing horticultural plants, landscape and gardening.

(i) They are centers for recreation and aesthetic beauty.

5.3 Special Types of Botanical Gardens

(a) **Arboretum** (Arboreta) — They are botanical gardens or parts of botanical gardens where only trees i.e. woody species are grown, e.g. The Carleton Arboretum, Northfield, USA.

(b) **Pinetum** (Pineta) — It is an arboretum where only conifers are grown, e.g. Indian Botanical Garden, Kolkata where such as arboretum is maintained.

(c) **Orchidarium** (Orchidaria) — They are gardens where only orchids are grown, e.g. National Orchidaria at Botanical Survey of India, Shillong.

(d) **Bambusetum** (Bambuseta) → It is a garden where the main collection is of bamboos, e.g. Indian Botanical Garden, Kolkata where a Bambusetum is maintained.

5.4 World's largest Botanical Garden (Royal Botanical Garden, Kew)

The World's largest Botanical Garden is the Royal Botanical Garden, Kew. Originally it belonged to the Kew house owned by Richand Bennet, but after the death of Princess Augusta in 1772, it was united with the gardens of the Palace of Richmond, and the extensive area of the two gardens was later taken over by the Government in 1838, and became named as Royal Botanical Garden, which was originally established in 1760, but officially opened in 1841. Its first Curator was William Aiton. Sir William Jackson Hooker later took over as Director of the Garden in the same year. He was succeeded by Joseph Dalton Hooker, and later by Thistleton Dyer. At present, this garden includes a herbarium consisting of over 5 million specimens, which was established by W.J. Hooker, who also introduced a green house and a palm house. A *Rhododendron* garden was introduced by J.D. Hooker and an Alpine house, Rose Garden, Bamboo Garden and a Lily pond by T. Dyer. Its arboretum has over 7,000 species and glasshouses over 13,000 species. It also has a Jodrell Laboratory, which provides research facilities to botanists from various fields. Some of the other remarkable features of this garden includes a Chelsea Physic Garden and the world's best rock gardens and the main inclusion in its plant layout are the oak, cedar, elm, willows, lilacs, pine, Azaleas, Cypress, Thuja, Red Wood, ferns, etc. Some of the famous publications of this garden include *Flora of British India, Index Kewensis* and *Kew Bulletin*.

5.5 Some Major Botanical Gardens of India

(a) Indian Botanical Garden, Kolkata

This famous botanical garden which is the largest and oldest of its kind in India, and considered to be the oldest botanical garden in South East Asia, is located at Shibpur near Kolkata, on the west bank of the river Hooghly. It was founded by Lt. Col. Robert Kyd in 1787, with an aim to establish a stock of plants which may be disseminated and prove beneficial to the inhabitants, rather than with a purpose of collecting rare plants as things of mere curiosity or furnishing articles for the gratification of luxury. Thus, this garden has long been known for its distinct role in the introduction, naturalization and development, as well as expanding cultivation of tea, jute, cinchona, teak and mahogany in India. Lt. Kyd was later succeeded by William Roxburg in 1787. He is considered as the Father of Indian Botany, and held the post of the second director till 1813. It was he who founded the famous herbarium of this garden. Presently, the garden is under the control of Botanical Survey of India and Dr. K. Biswas, was the first Indian to be appointed as Superintendent of this garden in 1937.

Covering an area of 110 hectares of land, this garden has nearly 12,000 living plants from various countries of the world. The chief attraction of this garden is the great Banyan tree, which is considered to be more than 200 years old and has over 1600 aerial roots actually rooted to the ground. This gives it a look of a miniature forest with a circumference of about 404 meter, and maximum height of about 30 meter.

The other major attractions of this garden are its well-maintained Bambusatum, the Orchid house, Pinetum, Cacti-collection, Fernary, the Palm house including the creeper *Porana paniculata* and the branching palm *Hyphaene thebaca.* One of the latest inclusions of this palm house is the double coconut (*Lodoicea maldivica*), palm like tree [Family – Arecaceae (Palmae)], also known as *coco de mer*, of the Seychelles, which produces a two-lobed edible nut, one of the largest known fruits from Maldives. This species grows only on a very small island named Praslin, located in the Seychelles Archipielago, in the Indian Ocean and has the world's largest and heaviest seed, measuring 12 inches in length and nearly 3 feet in circumference and with a weight of about 40 pounds, which looks like two coconuts fused together. The giant water lily of the Amazon, *Victoria amazonica*, which can hold a baby of 5 kilograms, is grown in the ponds of this garden, which attracts visitors and scientists all round the year. With its library, herbarium, laboratories and offices located within the garden, it has become a center for valuable scientific communication and research.

(b) National Botanical Garden, Lucknow

This garden situated on the banks of river Gomti, was established in 1789, by the emperor Nawab Sadat Ali Khan. It was named as Sikander Bagh by Nawab Wajid Ali Shah, in rememberance of his beloved wife Begum Sikander Mahal. It was later converted into a Botanical Garden in 1946 with Professor K.N. Kaul as its first director and is now known as the National Botanical Research Institute, Lucknow, which is one of the numerous national laboratories under the Council of Scientific and Industrial Research. It also has a sub-centre at Banthra, about 20 kms from Lucknow, covering an area of 120 hectares, where economic plants are grown on large scale.

The main garden covers an area of 30 hectares and is famous for its Palm house, Fern house, Rosarium, Cactus house, Orchid house, orchards of mango, guava and *Citrus* and medicinal plants section. Attached with this garden are a herbarium, library and laboratories carrying on extensive research in various fields of Botany.

(c) Lalbagh or The Mysore State Botanical Garden, Bangalore

This is the best botanical garden in South India, which was named 'Lalbagh' by Hyder Ali in 1760 because of its beautiful rose garden and red flowers. Its first director was Major Waugh (1799-1819), who introduced a number of exotic plants from various countries into this garden. Later, this garden was converted into a proper botanical garden in 1856 and Rao Bahadur H.C. Jayaraja was its first Indian director.

This garden is now famous for its beautiful layout and as a big center of horticultural activities. It has a grape orchard, economic garden, a herbal garden, a tropical nursery and well-equipped laboratories for seed testing and soil testing.

(d) Botanical garden of Forest Research Institute, Dehradun

This garden was established at Dehradun in 1934 by C.E. Parkinson. It covers an area of 20 acres and is the major center of research in problems related to plant introduction. It now has about 700 different species of plants from different countries of the world, a green house, a cactus house and a Plant Introductory Nursery. It also has a large herbarium attached with it.

(e) Lloyd Botanic Garden, Darjeeling

This garden situated at an altitude of about 2100 meters in the heart of Darjeeling town, West Bengal was established as a branch of the then Royal Botanic Garden, Kolkata, in 1878 by William Lloyd, after whom it was named. Today, it stretches over an area of 40 acres. Initially it served as a nursery for the cultivation of many exotic ornamental plants. It now has a rich collection of many Indian and foreign plants from the temperate region. The garden has been divided into seven divisions: an orchid house having more than 46 orchids, a large conservatory and a small conservatory, a cactus house having about 40 species of Cacti, a herbarium, a living fossil and a rock garden which is a botanist's paradise conserving some rare species of plants, most of them with ethno-medicinal importance. The most frequented spots on this track are the areas housing the two living fossils – *Ginkgo biloba* and *Metasequoia glyptostroboides*. The conservatories have some of the most endangered plant species from different parts of the world. The most interesting plant over here is the 118 year-old woody climber *Wisteria chinensis*. The herbarium contains about 3230 species of different plants belonging to 173 families.

5.6 Directory of Arboreta and Botanical Gardens in Other Parts of the World

(a) Arboreta

- American College Arboretum, Bryn Mawr, Pennsylvania, USA.
- Arboretum and City Park in Novorossiysk, Russia.

- The Arboretum at Flagstaff, Arizona, USA. The mission of The Arboretum at Flagstaff to foster and maintain a physical and intellectual environment dedicated to the conservation, understanding, presentation, and interpretation of plants and plant communities appropriate to the Colorado Plateau.
- The Arboretum of Los Angeles County, USA. The mission of the Arboretum of Los Angeles County is to foster an awareness and appreciation of the role of plants in nature and in human affairs. This is accomplished through labelled plant collections, reference materials, and educational programs. The mission also includes preserving, displaying and interpreting historically important buildings and artifacts on the property.
- Arboretum National des Barres, Domaine des Barres, France.
- Arboretum Stadt Freiburg, Germany.
- Arboreto appenninico del Parco Nazionale d'Abruzzo, Italy.
- The Arboretum at the University of California at Davis, USA.
- The Arnold Arboretum of Harvard University, Jamaica Plain, Massachusetts, USA. The Arnold Arboretum is a collection of hardy trees, shrubs, and vines located on 265 acres in Boston, Massachusetts, USA. The grounds were planned and designed by the Arboretum's first director, Charles Sprague Sargent, in collaboration with the landscape architect, Frederick Law Olmsted. Begun in 1872 and named for its benefactor James Arnold, the Arboretum continues to collect woody plants from around the world and offer horticultural education in accordance with its mission.
- The Bartlett Arboretum, University of Connecticutt, Stamford, USA.
- Bernheim Arboretum and Research Forest, Bullitt County, Kentucky, USA.
- Boyce Thompson Southwestern Arboretum, near Superior, just outside the metropolitan Phoenix area, Arizona, USA.
- Canadian Botanical Gardens and Arboreta. Maintained by the Canadian Botanical Conservation Network at Royal Botanical Gardens, Hamilton, Ontario, Canada.
- California Arboretum Foundation. CAF is a non-profit corporation benefitting the Arboretum of Los Angeles County, a 127-acre horticultural and botanical center for Southern California.
- The Carleton Arboretum, Northfield, USA.
- Concordia Arboretum - Concordia, College, Seward, Nebraska, USA.
- Connecticut College Arboretum, USA. The Connecticut College Arboretum is focused on native plants, wetland conservation, and energy conservation. The Arboretum offers a variety of different plant collections including; the Native Woody Plant Collection, the Caroline Black Garden and the Greenhouse Collection.
- The Cora Hartshorn Arboretum and Bird Sanctuary, Short Hills, New Jersey, USA
- Core Arboretum, West Virginia University, USA.
- Cox Arboretum, USA.
- Currency Creek Arboretum, Adelaide, Australia. A specialist eucalypt (Angophora, Eucalyptus and Corymbia) arboretum with its main purpose being research into the morphology, ecology and physiology of eucalypts.
- D.A. Murphy Panhandle Arboretum, University of Nebraska, USA.

- The Dallas Arboretum, Dallas, Texas, USA.
- The Dawes Arboretum (Newark, Ohio, USA). Founded in 1929 by Beman and Bertie Dawes, The Dawes Arboretum includes 1,149 acres of plant collections and natural areas.
- Davis Arboretum, University of California, USA. The Davis Arboretum, established in 1936, occupies approximately 125 acres on the banks of Putah Creek, near the center of the Great Central Valley of California.
- Dyck Arboretum of the Plains, Hesston, Kansas, USA.
- The Eddy Arboretum (Placerville, California, USA). The Eddy Arboretum is part of the Institute of Forest Genetics (IFG), operated by the USDA Forest Service, Pacific Southwest Forest and Range Experiment Station.
- Frelinghuysen Arboretum, Morristown, NJ, USA.
- Fullerton Arboretum, California State University, USA.
- Forstbotanischer Garten und Arboretum der Universität Göttingen, Germany.
- Gilman Park Arboretum, Pierce, Nebraska, USA.
- Gödöllö Arboretum, Hungary.
- Hayes Regional Arboretum, Richmond, Indiana, USA.
- The Henry Schmieder Arboretum of Delaware Valley College, USA, Doylestown, PA.
- Highstead Arboretun, Redding, Connecticutt, USA.
- Hoersholm Arboretum, Institute of Botany, Dendrology and Forest Genetics, Denmark
- Holden Arboretum, Kirtland, Ohio, USA. Located in Kirtland, Ohio, twenty five miles east of Cleveland, the Holden Arboretum encompasses over 3,100 acres and is the largest arboretum in the United States.
- Holmdel Arboretum, Freehold, New Jersey, USA.
- Houston Arboretum and Nature Center, Texas, USA.
- Hoyt Arboretum, Portland, Oregon, USA.
- Humber College Arboretum, Toronto, Canada.
- Huron County Nature Center, Wilderness Arboretum, Michigan, USA.
- Idaho State Arboretum, Pocatello, Idaho, USA.
- James Madison University Arboretum, Harrisonburg, Virginia, USA.
- JC Raulston Arboretum, North Carolina State University, Raleigh, NC, USA.
- John C. Gifford Arboretum, University of Miami - Florida, USA.
- Lasdon Park and Arboretum, Somers, Westchester County, NY, USA.
- Laurentian University's Arboretum, Sudbury, Ontario, Canada.
- Linnaeus Arboretum; Gustavus Adolphus College, Saint Peter, Minnesota, USA.
- Louisiana State Arboretum, Ville Platte, USA.
- Lyon Arboretum, The Harold L. Lyon Arboretum, nestled deep in Manoa Valley is a unit of the University of Hawaii (USA), devoted to research, instruction and public service.
- Maxwell Arboretum, University of Nebraska-Lincoln, USA.

- Mercer Arboretum and Botanic Gardens, Texas, USA.
- Morden Arboretum, Agriculture and Agri-Food Canada, Manitoba.
- Morris Arboretum of the University of Pennsylvania. Morris Arboretum is located in the Chestnut Hill section of Philadelphia, USA.
- The Morton Arboretum, Lisle, Illinois, USA.
- Mount Pisgah Arboretum, Eugene, Oregon, USA.
- Nebraska Statewide Arboretum, USA.
- Nichols Arboretum (University of Michigan, USA) lies on the outskirts of The Ann Arbor Campus and covers 123 acres.
- Odell Park and Arboretum, Fredericton, New Brunswick, Canada.
- Ohio State University Chadwick Arboretum, Columbus, Ohio, USA.
- Paine Art Center & Arboretum, Oshkosh, Wisconsin, USA.
- Palmerton Arboretum, Roger River, Oregon, USA.
- Pierce Arboretum, Nebraska, USA.
- Planting Fields Arboretum in Oyster Bay, New York (USA). 409 acres with many different collections, one acre under glass, 70 room mansion, Synoptic garden, etc.
- Queen Elizabeth Arboretum, Vancouver, British Columbia, Canada.
- R.A. Stranahan Arboretum, University Of Toledo, Ohio, USA.
- Reflection Riding Arboretum and Botanical Garden, Chattanooga, Tennessee, USA.
- S. Mary Grace Burns Arboretum of Georgian Court College - Lakewood, New Jersey, USA.
- Schedel Foundation Arboretum and Gardens, Elmore, Ohio, USA.
- Scott Arboretum, Swarthmore College, USA.
- Shaw Arboretum of the Missouri Botanical Garden, USA. Shaw Arboretum of the Missouri Botanical Garden includes four square miles of Ozark Border landscape and managed plant collections. It is dedicated to environmental education, to the maintenance and restoration of the region's native flora and fauna, and to public enjoyment of the natural world.
- Sherwood Fox Arboretum, University of Western Ontario, London, Canada
- Slayton Arboretum, Hillsdale College, MI, USA
- The Stephen F. Austin State University Arboretum, Nacodoches, Texas, USA.
- Svinvik Arboretum, Norway.
- Taylor Memorial Arboretum, Wallingford, Pennsylvania, USA.
- Tyler Arboretum Homepage, Media, Pennsylvania, USA.
- The United States National Arboretum - Washington D.C., USA. The U.S. National Arboretum is a U.S. Department of Agriculture research and education facility and a living museum. It is dedicated to serving the public and improving environment by developing and promoting improved floral and landscape plants and new technologies

through scientific research, educational programs, display gardens, and germplasm conservation. The U.S. National Arboretum is located in northeast Washington, D.C. The U.S. National Arboretum's mission is to conduct research, provide education, and conserve and display trees, shrubs, flowers, and other plants to enhance the environment.

- University of California, Irvine - Arboretum and Gene Bank, USA.
- University of California Santa Cruz Arboretum, USA.
- University of Central Florida Arboretum, USA.
- University of Guelph Arboretum (Ontario, Canada). The Arboretum at the University of Guelph facilitates understanding and respect for the world by designing and providing environmental education activities, developing and maintaining plant collections, evaluating flora and supporting related research.
- University of Illinois at Urbana-Champaign Arboretum, USA.
- University of Minnesota Landscape Arboretum, Chanhassen, USA. The Arboretum maintains 14 display and specialty gardens, 60 generic collections, 5,000 species of plants, and 8 native environments on 905 acres.
- University of Nebraska - Lincoln: Botanical Garden and Arboretum, USA.
- University of Wisconsin Arboretum, Madison, USA.
- Washington Park Arboretum, Seattle, WA, USA.
- Wayne State College Arboretum, Wayne, Nebraska, USA.
- Westonbirt Arboretum, Tetbury, UK.
- Yakima Area Arboretum, Yakima, Washington, USA.

(b) Botanical Gardens

General

- Botanic Gardens Conservation International (BGCI), maintained in the Royal Botanic Gardens Kew, UK.

Australia

- Australian National Botanic Gardens (ANBG). ANBG is part of the Biodiversity Directorate of the Australian Nature Conservation Agency (ANCA) and is located on the slopes of Black Mountain in Canberra, ACT. It is one of a number of agencies working on the botanical biodiversity of Australia.
- Botanic Gardens and and State Herbarium of Adelaide, Australia
- Brisbane Botanic Gardens - Mt Coot-tha, Mt Coot-tha Rd, Toowong, Brisbane, Australia. The Brisbane Botanic Gardens Mt Coot-tha, officially opened in 1976, is situated in the Queensland state capital, 8km from the city centre. The 52 hectare complex contains an extensive collection of native and exotic plants as well as an administration centre, a botanical/horticultural library, an education centre, a botanical laboratory, a 300 seat auditorium, the Sir Thomas Brisbane Planetarium and the Queensland Herbarium.

- Directory of Australian Botanic Gardens and Arboreta. Maintained by M. Fagg & J.Wilson, Australian National Botanic Gardens.
- The Grevillea Park, Illawarra, south coast of NSW, Australia.
- Jervis Bay (Booderee) Botanic Gardens, Australia.
- Kings Park & Botanic Garden, Perth, Australia.
- National Rhododendron Gardens Olinda, Victoria, Australia.
- Norfolk Island Botanic Gardens, Australia.
- Olive Pink Botanic Garden, Alice Springs, Northern Territory, Australia.
- Royal Botanic Gardens Melbourne, Australia.
- Royal Botanic Gardens Sydney, Australia.
- Royal Tasmanian Botanical Gardens Appeal, Australia. The Royal Tasmanian Botanical Gardens in Australia are under threat from the Armillaria luteobubafina fungus, which is a type of root rot endemic to Australia. The gardens, which were established in 1818 have already lost 33 trees since the problem was diagnosed in March 1994. The entire gardens may be devastated unless a solution is found soon.

Austria

- The Botanical Garden of the University of Vienna, Austria.
- Botanischer Garten der Stadt Linz, Austria.
- University of Salzburg, Botanical Garden, Austria.

Barbados

- Andromeda Botanic Gardens, St Joseph, Barbados.
- The Flower Forest of Barbados has over 50 acres of tropical plants from all over the world, most labelled with their Latin name, common name and place of origin.

Belgium

- Botanical Garden of the University of Gent, Belgium.
- L'Observatoire du Monde des Plantes, Liège, Belgium.
- National Botanic Garden of Belgium.

Bermuda

- The Bermuda Botanical Gardens, Hamilton, Bermuda

Brazil

- Jardim Botânico do Rio de Janeiro, Brazil.

British Virgin Islands

- Joseph Reynold O'Neal Botanic Gardens, Tortola, British Virgin Islands.

Canada

- The Butchart Gardens, Victoria, British Columbia, Canada.
- Canadian Botanical Gardens and Arboreta. Maintained by the Canadian Botanical Conservation Network at Royal Botanical Gardens, Hamilton, Ontario, Canada.
- Cedar Valley Botanical Gardens, Brighton, Ontario, Canada.
- Dr. Sun Yat-Sen Classical Chinese Garden, Vancouver, Canada, the first full scale classical Chinese garden built outside of China.
- Les Grands Jardins de Normandin, Québec, Canada.
- Hatley Park: the Public Gardens of Royal Roads University, Victoria, British Columbia, Canada. 1908 Edwardian Estate Architecture, Native Old Growth Forest, Bordering on a tidal lagoon which is a wildlife sanctuary, views across Straits of Juan de Fuca to Olympic Mts of USA, Japanese Gardens, Italian Garden, mature specimen landscape trees, alpine and rock gardens, bog garden, rose garden, croquet lawn with perennial borders, native plant woodland garden, and many smaller sized plantings.
- Le Jardin botanique du Nouveau-Brunswick, Canada.
- The Memorial University of Newfoundland Botanical Garden, Canada.
- Minter Gardens, Columbia, British Canada.
- The Montréal Botanical Garden and Insectarium, Québec, Canada / Le Jardin botanique de Montréal, Quebec, Canada.
- Niagara Parks Botanical Gardens and School of Horticulture, Niagara Falls, Canada
- Olympic Park, Botanical Garden, Biodôme, Montreal, Québec, Canada.
- Parc Marie-Victorin, Marie-Victorin, Kingsey Falls, Québec, Canada
- Royal Botanical Gardens, Hamilton/Burlington, Ontario, Canada. McMaster University, The Royal Botanic Gardens, Ontario, Canada.
- The University of Alberta, Devonian Botanic Garden, Canada. The Devonian Botanic Garden, established in 1959 by the University of Alberta, includes an authentic Japanese garden, a tropical butterfly house, attractive floral gardens, collections of native and alpine plants and ecological reserves, situated within an attractive rolling landscape of pine trees and large ponds. Connected gardens extend over 80 acres with an additional 110 acres of natural area for visitors to enjoy.

- University of British Columbia Botanical Garden, Vancouver, B.C.,Canada. Perched on the cliffs of Point Grey overlooking the Strait of Georgia, the UBC Botanical Garden is a superb place to visit at any time of year. Seventy acres, planted with over 10,000 different trees, shrubs and flowers, are presented in the Alpine Garden, Asian Garden, Winter Garden, B.C.Native Garden, Physick Garden, Food Garden and Perennial Borders. A true botanical garden, it is designed primarily for education and research, with emphasis on its diverse collections and seasonal displays.
- Université de Moncton - Campus d'Edmundston, Jardin Botanique, Canada.
- University of Waterloo Botanical Garden, Ontario, Canada.
- VanDusen Botanical Garden, Vancouver, British Columbia, Canada.

Cayman Islands

- The Queen Elizabeth II Botanic Park, Grand Cayman, Cayman Islands.

China

- Shanghai Botanic Garden, Shanghai, China. Public garden in Shanghai, the newly built Penjing Exhibit is the first in China.
- Yuan Ming Yuan - Garden of Centred Wisdom, China, was burned in 1860.

Costa Rica

- Arenal Botanical Gardens, Costa Rica.
- Lankester Botanical Garden, Costa Rica, Universidad de Costa Rica. Since 1973, Lankester Botanical Garden has dedicated itself to the conservation of the epiphytic flora of Costa Rica. Today, the Garden includes one of the largest collections of orchids in Latin America.
- The Robert & Catherine Wilson Botanical Garden & Las Cruces Biological Station, Costa Rica.

Croatia

- Botanical Garden of Zagreb, Croatia.
- Botanical Gardens in Croatia.

Czech Republic

- Palacky University, Department of Botany and Botanical Garden, Olomouc, Czech Republic
- Zoologicka a Botanicka zahrada mesta Plzen, Zoological and Botanical Garden of Plzen, Czech Republic.

Denmark

- Botanic Garden at the University of Copenhagen, Denmark.
- The Botanical Garden of KVL, Royal Veterinary and Agricultural University in Copenhagen, Denmark.

Dominican Republic

- National Botanical Garden, Dominican Republic.

Ecuador

- Parque Pedagógico Etnobotánico Omaere, Puyo, Ecuador.

Estonia

- The Botanical Garden of Tartu University, Estonia.
- Tallinn Botanic Garden, Estonia. The task of the Botanic Garden is to diversify the Estonian garden flora with ornamental plants, to protect the rare and endangered plant species and to carry out ecological research.

Finland

- Botanical Garden, University of Helsinki, Finland.
- Botanical Gardens of the University of Oulu, Finland.
- Botanical Garden, University of Turku, Finland.

France

- The Botanical Garden of Nice (Nizza), France.
- The Botanical Garden of Orchaise, France.
- The Botanical Garden of the Villa Thuret (Antibes, France). Jardin Botanique de la villa Thuret, INRA Institut National de la Recherche Agronomique, Antibes, France. Botanical garden of 4 ha.
- Claude Monet's Garden (Giverny, France).
- Conservatoire et Jardins Botaniques de Nancy, France. Regrouping two botanic gardens: Jardin Botanique du Montet (carnivorous and tropical plants), Jardin d'Altitude du Haut Chitelet (alpine plants).
- Le Jardin Botanique de Bordeaux, France.
- Jardin Botanique de Gondremer, Vosges, France.

- Jardin botanique du Montet, Nancy, France.
- Les "jardins d'exception" de la Ville de Menton, France.
- Les Jardins de Provence - Gardens of Provence, France.
- Le Parc Zoologique et Botanique de Mulhouse, France.
- Parks and Gardens in Normandy, France.
- Tête d'Or Park, Lyon, France. This 290 acres park was laid out en 1862 by the Town of Lyon in the English style.

Germany

- Botanical Garden and Botanical Museum Berlin-Dahlem Information Server, Germany.
- Botanischer Garten der Rheinischen Friedrich-Wilhelms-Universität Bonn, Germany.
- Botanischer Garten der TH Darmstadt, Germany.
- Botanischer Garten der Technische Universität Dresden, Germany.
- Botanischer Garten der Universität Düsseldorf, Germany.
- Forstlehrgarten Universität Freiburg, Germany.
- Botanischer Garten der Justus-Liebig-Universität Gießen, Germany.
- Der Botanische Garten in Göttingen, Universität Göttingen, Germany.
- Botanischer Garten, Ruprecht-Karls-Universität Heidelberg, Heidelberg, Germany.
- Botanischer Garten der Universität Heidelberg (Botanical Garden of the University of Heidelberg), Germany.
- Botanisches Garten der Universität Hohenheim (Botanical Garden of the University of Hohenheim), Germany.
- Botanisches Institut und Botanischer Garten der Christian-Albrechts-Universitat zu Kiel, (Botanical Institute of the University of Kiel), Germany.
- Botanischer Garten Universität Leipzig (Botanical Garden of the University of Leipzig), Germany.
- Botanischer Garten München-Nymphenburg (Botanical Garden of the University of München-Nymphenburg), Germany.
- Botanischer Garten der Universität Erlangen-Nürnberg, Germany.
- Botanischer Garten der Universität Oldenburg, Germany.
- Botanischer Garten der Universität Osnabrück, Germany.
- Botanischer Garten der Universität Regensburg, Germany.
- Botanical Garden, University of Tübingen, Germany.
- Botanischer Garten der Universität Ulm, Germany.

- Forstbotanischer Garten Tharandt (Technische Universität Dresden, Germany). Botanical garden concentrating on the cultivation of trees.
- Institut für Allgemeine Botanik und Botanischer Garten, Universität Hamburg, Germany.
- Institute of Geobotany and Botanical Garden, University of Halle, Germany.
- Institute of Systematic Botany and Botanic Garden, Johannes Gutenberg University, Mainz, Germany.
- International Phenological Gardens, Berlin, Germany. The idea of the IPG is to obtain comparable phenological data across Europe, from plants not influenced by different genetic conditions.
- Julius-von-Sachs-Institut für Biowissenschaften mit Botanischem Garten Würzburg, Botanical Garden of the University of Würzburg, Germany.
- Palmengarten der Stadt Frankfurt (Main), Germany.
- Verband Botanischer Gärten e.V., Germany.
- Westfälische Wilhelms-Universität, Münster: Botanischer Garten, Germany. The Botanical Garden of the University of Münster specializes in the systematic of Geraniaceae and Asclepiadaceae (Prof. Dr. F. Albers).
- Ökologisch-Botanischer Garten der Universität Bayreuth, Germany.

Honduras

- Lancetilla: Un Jardin Botanico, Honduras.

Hungary

- The Botanical Garden and botanical regions of Szabolcs-Szatmár-Bereg County, Hungary.

Ireland

- The Dillon Garden, Dublin, Ireland.
- National Botanic Gardens Glasnevin, Dublin, Ireland.
- Talbot Botanic Garden, Malahide, Ireland.
- Trinity College Botanic Garden, Dublin, Ireland.

Israel

- Jerusalem Botanical Garden, Israel.
- Tel Aviv University Botanic Gardens, Israel.

Italy

- Accademia dei Fisiocritici, Orto Botanico, Siena, Italy.
- Alpine Botanical Garden (Gran Sasso), University of L'Aquila, Italy. The alpine botanical garden was founded in the 1954 to accomodate the main floristic entities of the karstic Gran Sasso mountain. It is located 2110 m a.s.l., at the base of the "Corno Grande", which is the highest tip of the Apennines, 2914 m a.s.l.
- Botanical Garden of Abetone, Italy.
- Botanical Garden of the University of Bari, Italy.
- Botanical Garden of Bologna, Italy. Università degli Studi di Bologna, L'Orto Botanico (Italy)
- L'area archeologica dell'orto botanico Cagliari.
- Botanical Garden of Catania, Italy (from Pisa).
- Botanical Garden of Firenze, (Florence) Italy.
- Botanical Garden of Lucca, Italy.
- Botanical Garden of the University of Modena - L'Orto Botanico dell'Universita' di Modena, Italy.
- Botanical Garden of Napoli, Italy. Orto Botanico della Facoltà di Scienze, Università degli Studi di Napoli "Federico II, Italy.
- Botanical Garden of the University of Parma, Italy.
- Botanical Garden of the University of Perugia, Italy.
- Botanical Garden of Pisa, Italy.
- Botanical Garden of Viterbo, Italy. L'Orto Botanico dell'Universitá della Tuscia.
- Giardino Botanico Alpino delle Viotte del Monte Bondone, Trento, northern Italy.
- Giardini botanici Hanbury, Italy, Botanic Garden of Genoa, Italy. Istituto ed Orto Botanico 'Hanbury'- Università degli Studi - Genova, Italia (from Pisa).
- Giardino Botanico "Lorenzo Rota" di Bergamo, Italy.
- Giardino Botanico, Istituto Tecnico Agrario Statale "Celso Ulpiani", Ascoli Piceno, Italy.
- Giardino Botanico della Majella Italy.
- Giardino Botanico "Ruggero Tomaselli", Campo dei Fiori (Varese), Italy.
- Il Giardino Botanico della Scuola Media Statale "E.Toti" di Musile di Piave (Venezia), Italy.
- L'Orto Botanico dell'Università di Camerino, Italy.
- L'Orto Botanico dell'Università di Torino, Italy.
- Orto Botanico, Arcavacata di Rende (Cosenza), Italy.
- Orto Botanico Locatelli (Quartiere Carpenedo-Bissuola - Mestre), Italy.
- Orto Botanico e Collezioni Botaniche Herbarium Patavinum - Padova, Italy.

- Orto Botanico 'Pietro Pellegrini', Italy.
- Orto Botanico, Università di Pavia, Italy.
- Orto dei Semplici Elbano (Italy). On Monte Serra near Rio nell'Elba there is now a new Botanic Garden with wildflowers, medicinal and endemic plants of the Tuscan Archipelago, situated in a most beautiful location in the Eastern part of Elba, the Hermitage S.Caterina.
- University of Lecce, Villa Tresca Botanical Garden, Italy.
- Università degli Studi di Milano: Botanical Gardens, Italy.
- Università degli Studi di Roma "La Sapienza" - Orto Botanico, Rome, Italy.
- Università Degli Studi Della Tuscia, L'Orto Botanico, Italy.

Japan

- Fukuoka City Zoological and Botanical Gardens, Japan.
- Hakone: Full of Flowers and Dreams; Botanical Gardens and Parks, Japan
- Hiroshima Botanical Garden, Japan.
- Kyoto Prefectual Botanic Garden, Japan.
- Medicinal Plant Garden, Faculty of Pharmaceutical Sciences, Kumamoto University, Japan.
- Ochiai Park/Botanical Garden, Kasugai, Japan.
- Tama Forest Science Garden, Japan.
- Tsukuba Botanical Garden, Japan.

Latvia

- Latvian Academy of Sciences, National Botanical Garden (Salaspils, Latvia).
- University of Latvia Botanical Gardens.

Lithuania

- Botanical garden of Vilnius University, Lithuania.
- Kaunas Botanical garden, University of Lithuania.

Madagascar

- Parc Botanique et Zoologique de Tsimbazaza, Antananarivo, Madagascar.

Mexico

- El jardín botánico Dr. Alfredo Barrera Marín, Mexico.
- Jardin Borda, Cuernavaca, Mexico.

- Jardín Botánico Francisco Xavier Clavijero, Mexico.
- Jardín Botánico del Instituto de Biología (UNAM), Mexico.
- Toluca de Lerdo, Cosmovitral Jardin Botanico, Mexico.
- Universidad Autonoma Chapingo, Unidad Regional Universitaria de Zonas Aridas, Jardin Botanico, Mexico.
- Universidad Nacional Autonoma de Mexico, Jardin Botanico.

Monaco

- Jardin Exotique de Monaco.

The Netherlands

- The Botanical Garden of the University of Nijmegen, the Netherlands. The Botanical Garden of the University of Nijmegen pays special attention to the family of Solanaceae and maintains the most extensive ex situ plant collection of non-tuberous Solanaceae species in the world.
- Hortus Botanicus Botanic Gardens, Amsterdam, The Netherlands.
- Rijksherbarium Hortus Botanicus, Leiden, The Netherlands.
- Utrecht University Botanic Gardens, The Netherlands.

New Zealand

- Christchurch Botanic Gardens, New Zealand.

Norway

- Botanical Garden, University of Oslo, Norway.
- Ringve Botanical Garden, Trondheim, Norway.
- Tromsø Botanic Gardens, Norway. Tromsø Botanic Gardens was officially opened on 2 July 1994. It comprises an Arctic- Alpine Garden at the University area of Tromsø and an Arboretum under development at Holt Research Station. The position of Tromsø is 69°41′ N latitude and 18°57′ E longitude. The Botanic Gardens are the most northerly in Norway, as well as in the world.

Poland

- Botanical Garden, University of Wroclaw, Poland.

Portugal

- Jardim Botânico da Madeira, Portugal.

Reunion

- Natural Botanic Garden, Reunion Island.

Russia

- Botanic Garden of the Irkutsk State University, Russia.
- Petrozavodsk State University Botanic Garden (PSUBG), Karelia, Russia.

Singapore

- Singapore Botanic Gardens.

Slovakia

- Botanical Garden of P. J. Safárik University, Kosice, Slovakia.

Slovenia

- Alpine Botanical Garden Juliana in Trenta Valley, Slovenia.
- Botanical Garden of the University of Ljubljana, Slovenia.

South Africa

- Botanical Garden, University of Pretoria, South Africa.
- Botanical Garden, University of Stellenbosch, South Africa.
- Free State National Botanical Garden, Bloemfontein, South Africa.
- Harold Porter National Botanical Garden, Betty's Bay, Western Cape Province, South Africa.
- Karoo National Botanical Garden, Worcester, Western Cape Province, South Africa.
- Kirstenbosch National Botanical Garden, Cape Town, South Africa.
- Lowveld National Botanical Garden, Nelspruit, Mpumalanga Province, South Africa.
- Natal National Botanical Garden, Pietermaritzburg, KwaZulu-Natal, South Africa.
- Pretoria National Botanical Garden, Pretoria, South Africa.
- The Witwatersrand National Botanical Garden, South Africa.

Spain

- Jardí Botànic Mar i Murtra, Spain.
- Jardín Botanico Canario Viera y Clavijo, Las Palmas, Gran Canaria, Spain.

- Jardín Botánico de Córdoba, Spain.
- Jardín Botánico de Coria, Spain.
- Jardín Botánico Pinya de Rosa, Spain.
- Jardí Botànic de Sóller, Baleares, Spain.
- Real Jardin Botanica Juan Carlos I, Universidad de Alcalá, Alcalá en Henares, Madrid, Spain.
- Real Jardin Botanico, Madrid, Spain.
- Valencia Botanical Garden, University of Valencia, Spain.

St Vincent and the Grenadines

- St Vincent and the Grenadines Botanic Gardens, Kingstown.

Sweden

- Bergius Botanical Garden (Bergianska Botaniska Trädgården), Stockholm, Sweden.
- The Göteborg Botanical Gardens (= Gothenburg), Sweden.
- The Linnaeus Garden, Uppsala, Sweden. The Linnaeus Garden served as Uppsala University botanical garden for 150 years. Today, it is a historical garden restored according to Carl Linnaeus own plan from 1745.
- Linnaeus Hammarby, Uppsala, Sweden. Hammarby is the summerhouse of Carl Linnaeus. It lies on the countryside outside Uppsala, surrounded by a garden and park.
- Uppsala University Botanical Garden, Sweden. The Uppsala University Botanical Garden is the oldest botanical garden in Sweden being c. 350 years old. It consists of 13 ha of gardens, an orangery and a greenhouse. Well-known previous directors of the garden include Carl Linnaeus, Carl Peter Thunberg and Elias Fries.

Switzerland

- Botanischer Garten in Brüglingen, Basel, Switzerland.
- Botanischer Garten der Universität Basel, Switzerland.
- Botanical Garden, University of Fribourg, Switzerland.
- Conservatoire & Botanical Garden of the City of Geneva, Switzerland.
- Jardin botanique de Lausanne, Switzerland.
- Le jardin botanique de Porrentruy, Switzerland.
- University of Zürich, Institute of Systematic Botany and Botanic Garden, Switzerland.

Togo

- Le Jardin Botanique de l'Université du Bénin, Lomé, Togo.

UK

- Borde Hill Garden, Sussex, UK. Botanical gardens set in 200 acres of parkland. Best private collection of Champion Trees in Britain.
- Brickmans Botanical Country Garden, UK.
- Cambridge University Botanic Garden, UK.
- Cruickshank Botanic Gardens, Aberdeen, Scotland, UK.
- Durham University Botanic Garden, City of Durham, England.
- Harlow Carr Botanical Gardens, Harrogate, England. Harlow Carr Botanical Gardens set in Harrogate, North Yorkshire, United Kingdom. Home of the Northern Horticultural Society. Education and plant trials centre and home of National Collections, rhubarb, dryopteris and polypodium, calluna and hypericum. Also new Marie Curie Cancer Care daffodil garden. Small Museum of Gardening History, well appointed Library and Study Centre.
- Harris Garden, University of Reading, UK.
- Kew, The Royal Botanic Gardens, UK.
- Logan Botanic Garden, Port Logan, UK.
- National Botanic Garden of Wales, Middleton, UK.
- Ness Botanic Garden, University of Liverpool, UK.
- Newcastle University, Moor Bank Garden, UK.
- Paignton Zoological and Botanical Gardens, Devon, UK.
- Pine Lodge Gardens, St. Austell, Cornwall - National UK.
- The Royal Botanic Garden Edinburgh, UK. The Royal Botanic Garden Edinburgh was founded in the 17th century on an area the size of a tennis court. It now extends to 31 hectares (at Inverleith in Edinburgh), and incorporates Specialist Gardens at three very different locations in Scotland (Younger, Logan and Dawyck; 50ha, 12ha and 25ha). The Garden is first and foremost a scientific institution, dedicated to the pursuit of research of the highest quality on the systematics and biology of plants.
- Sir Harold Hillier Gardens and Arboretum, Ampfield, UK.
- St. Andrews Botanic Gardens. St. Andrews Botanic Garden in Scotland, UK, was founded in 1889 to provide botanical specimens for teaching and research in the University of St. Andrews.

- Tresco Abbey Gardens, Tresco, UK.
- University of Wales Aberystwyth Botany Gardens - The Tropical Glasshouse, Wales, UK.
- Ventnor Botanic Garden, Isle of Wight, UK.
- Younger Botanic Garden, Benmore, UK.

USA

- Alice Keck Garden Photo Tour, California, USA.
- Alfred B. Maclay State Gardens, Tallahassee, Florida, USA.
- Amy Greenwell Ethnobotanical Garden, Captain Cook, Hawaii, USA.
- Arizona Cactus Botanical Garden, Bisbee, AZ, USA.
- Atlanta Botanical Garden, Georgia, USA.
- Bartram Gardens, Philadelphia, Pennsylvania, USA.
- Bellevue Botanical Garden, Washington, USA.
- Bellingrath Gardens and Home, Alabama, USA.
- Berkshire Botanical Garden, Stockbridge, Massachusetts, USA.
- The Berry Botanic Garden, Portland, Oregon, USA. A non profit, private botanic garden located in Portland, Oregon, founded in 1977. Six and one half acre site includes native woodland, a collection of species rhododendrons, a half acre rock garden and alpine collection, unusual perennials and woodland plants, and specimen shrubs and trees.
- Betty Ford Alpine Garden, Vail, Colorado, USA.
- Birmingham Botanical Gardens, Alabama, USA. Established in 1962 by the City of Birmingham, the Gardens sits on 671/2 acres of land which were once Indian campgrounds.
- Boerner Botanical Gardens, Milwaukee, USA.
- The Botanic Garden of Smith College and Lyman Plant House, Northampton, MA, USA.
- Botanica, The Wichita Gardens, Wichita, Kansas, USA.
- Boxerwoord Gardens, Lexington, Virginia, USA.
- Brooklyn Botanic Garden, USA.
- California Living Museum, Bakersfield, USA. CALM is a zoo, botanical garden, and natural history museum. CALM is located along the Kern River, east of Bakersfield, California between Hart Park and Lake Ming, with over 250 unreleasable animals native to California and 2000 native plants.
- Cecil B. Day Butterfly Center - Calloway Gardens, Pine Mountain, Georgia, USA.
- Cedar Valley Arboretum & Botanic Garden, Waterloo, Iowa, USA.
- Chaparral Gardens. Chaparral Gardens, in the Sierra Foothills, California, USA, specializes in modern hybrid daylilies.

- The Cheyenne Botanic Gardens, WY, USA.
- Chicago Botanic Garden, Illinois, USA.
- Cincinnati Area Parks, USA
- Cincinnati Zoo & Botanical Garden, USA.
- Cleveland Botanical Garden, Cleveland, Ohio, USA.
- Clinton Community Garden, New York, USA.
- Columbus Zoological Gardens, Powell, Ohio, USA.
- Como Park Conservatory - Como Park St Paul, Minnesota.
- The Conservatory in Golden Gate Park, San Francisco, California, USA.
- Cornell Plantations Path Guide, Ithaca, New York, USA.
- Daniel Stowe Botanical Garden, Belmont, North Carolina, USA. Daniel Stowe Botanical Garden, where "nature and horticulture blend into one," is a developing facility in the heart of the North Carolina Piedmont Region of North America.
- Denver Botanic Gardens, USA.
- Descanso Gardens, California, USA. Descanso Gardens is a 160-acre public garden located twenty minutes from downtown Los Angeles in La Cañada Flintridge.
- Desert Botanical Garden, Phoenix, Arizona, USA. The Desert Botanical Garden, founded in 1937, is a living museum dedicated to desert lands research and education. It is located at the southeastern edge of the Salt River Valley in Phoenix, Arizona, near the cities of Scottsdale and Tempe, on 145 acres in the midst of the red rock buttes and giant cactus plants of Papago Park.
- Des Moines Botanical Center.
- Dixon Gallery and Gardens, Memphis, Tennessee, USA.
- Dothan Area Botanical Gardens, Alabama, USA.
- Dow Gardens, Midland, Michigan, USA.
- Dubuque Arboretum & Botanical Gardens at Marshall Park, Iowa, USA.
- The Elizabeth F. Gamble Garden Center, Palo Alto, California, USA.
- Fairchild Tropical Garden, Coral Gables, Florida, USA.
- Fellows Riverside Gardens, Youngstown, Ohio, USA.
- Florida Community College at Jacksonville Botanical Garden, USA.
- Franklin Park Conservatory and Botanical Garden, Columbus, Ohio, USA.
- Frederik Meijer Gardens, Michigan, USA.
- George Eastman's Gardens, Rochester, New York, USA.
- The Georgeson Botanical Garden, Alaska, USA. The Georgeson Botanical Garden is a research program dedicated to the propagation, cultivation and conservation of native and introduced plant species in the subarctic North. It is a repository for all ornamental

and edible plant materials hardy in subarctic Alaska and a definitive source of horticultural information.

- Harry P. Leu Gardens, Orlando, Florida, USA.
- The Hawaii Botanical and Public Gardens, USA.
- Hawaii Tropical Botanical Garden, Papakou, USA.
- Huntington Library, Art Collections, & Botanical Gardens, San Marino, California, USA. An educational, display, and research institution with 150 acres of gardens and extensive holdings of living plants, including cacti and other succulents, camellias, roses, exotic trees and shrubs.
- Huntsville-Madison Botanical Garden, Alabama, USA.
- The Idaho Botanical Garden, Boise, ID, USA.
- Japanese Garden Database. Maintained by Robert Cheetham, University of Pennsylvania, USA.
- Japanese Garden in Portland Oregon, USA.
- Krohn Conservatory, Cincinnati, USA.
- London Town House & Gardens, Edgewater, Maryland, USA.
- Longwood Gardens, PA, USA.
- Luther Burbank Home and Gardens, Santa Rosa, California, USA.
- Magnolia Plantation and Its Gardens, Charleston, South Carolina, USA.
- Marie Selby Botanical Gardens - Sarasota, Florida, USA.
- Matthaei Botanical Gardens, University of Michigan, Ann Arbor, Michigan, USA.
- Meerkerk Rhododendron Gardens, USA. Located on beautiful Whidbey Island in Washington State's Puget Sound region, Meerkerk Gardens features mature rhododendron hybrids, rare and unusual rhododendron species, azaleas, spring bulbs and naturalized companion plantings in a peaceful woodland setting.
- The Memphis Botanic Gardens, Memphis, Tennessee, USA.
- Missouri Botanical Garden, St. Louis, USA
- Missouri State Parks and Historic Sites, USA.
- Mitchell Park Horticultural Conservatory The Domes, Milwaukee, USA.
- Mockernut Hill Botanical Garden, Marion County, Florida, USA.
- Moody Gardens, Galveston Island, Texas, USA.
- Mounts Botanical Garden, Florida, U.S.A. Located in West Palm Beach, Florida, Mounts Botanical Garden is thirteen acres of tropical and subtropical gardens. Collections of special note include tropical fruit, native plants, and bromeliads. Research gardens evaluate perennials, annuals, and bulbs for South Florida.
- Mynelle Gardens, Jackson, Mississippi, USA.
- Myriad Botanical Gardens & Crystal Bridge Tropical Conservatory, Oklahoma City, USA.

- The National Garden, Washington D.C., USA and The U. S. Botanic Garden, Washington, DC, USA. Established by Congress in 1820, the United States Botanic Garden is the oldest botanic garden in North America. The garden's first greenhouse was constructed in 1842; since 1849 the Garden has been located at the eastern end of the Mall.
- National Tropical Botanical Garden (NTBG), USA. Non-profit botanical gardens in Hawaii and Florida, with active programs in conservation, research and education.
- The New York Botanical Garden , The Bronx, New York, USA.
- Norfolk Botanic Garden, Virginia, USA.
- North American Botanical Conservatories.
- The North Carolina Botanical Garden, USA. Made up of three tracts in Chapel Hill: the Coker Arboretum, the Mason Farm Biological Reserve, and the North Carolina Botanical Garden proper.
- Oklahoma State University (OSU) Botanical Garden and Arboretum (OBGA), USA.
- Olbrich Botanical Gardens, Madison, Wisconsin, USA.
- Parrot Jungle And Gardens, Miami, Florida, USA.
- Phipps Conservatory in Pittsburgh, PA. USA.
- Powell Gardens, Kansas City, USA.
- Presby Memorial Iris Gardens, New Jersey, USA.
- Quail Botanical Gardens, San Diego, California, USA.
- Rancho Santa Ana Botanic Garden, Claremont, California, USA.
- Red Butte Garden and Arboretum, University of Utah, Salt Lake City, Utah, USA.
- Reiman Gardens, Iowa State University, Ames, Iowa, USA.
- Rhododendron Species Botanical Garden, USA.
- The Rio Grande Botanic Gardens and Conservatory, Albuquerque, USA.
- Riverbanks Zoo and Botanical Garden, Columbia, South Carolina, USA.
- Rockwood Museum, Wilmington, Delaware, USA. Rockwood, a Rural Gothic manor home surrounded by Gardenesque grounds, was built in 1851 by Joseph Shipley.
- Rotary Gardens, Inc., Janesville, Wisconsin, USA.
- The Ruth Bancroft Garden, Walnut Creek, California, USA. A three acre garden in Walnut Creek, California (CA) emphasizing xerophytes, particularly cacti (cactus) and succulents (succulent).
- San Antonio Botanical Gardens, Texas, USA. A 33 acre botanical garden in San Antonio, Texas. The Lucile Halsell Conservatory complex features plants from alpine regions to tropical rainforests. Collections are housed in underground environments including alpine, epiphyte, desert, insectivorous, hot tropical, palm, cycad, aroid, fern, fruit, and aquatic collections. The formal gardens include rose, formal, herb, old fashioned, Japanese, etc. A large area of the gardens is arranged into a regional plant exhibit for diverse regions across Texas, including examples of early architecture.

- San Jose Gardens, California, USA.
- Santa Barbara Botanic Garden, California, USA.
- Sherman Gardens and Library Photo Tour, California, USA.
- Shofuso: The Pine Breeze Villa, Philadelphia, Pennsylvania, USA.
- Sonnenberg Gardens, Canandaigua, NY, USA. A beautiful Victorian mansion and gardens, located at the foot of Canandaigua Lake, in the heart of the Finger Lakes Region in Upstate New York.
- The South Carolina Botanical Garden, USA. The South Carolina Botanical Garden is an interdisciplinary resource center for Clemson University students studying horticulture, forestry, natural sciences, landscape design, architecture and the arts.
- Botanical Garden, University of South Florida, Tampa, Florida, USA.
- The State Botanical Garden of Georgia, USA. The State Botanical Garden of Georgia is a 313-acre horticultural preserve, featuring display gardens, a conservatory, and five miles of nature trails. The Garden is a unit of the University of Georgia. The Garden currently maintains eleven specialty gardens: the Rose, Rhododendron, Dahlia, Shade, Azalea, Annual/ Perennial, Herb, Daffodil, Ground Cover, Trial, and Native Flora gardens are located throughout the central portion of the Garden premises.
- The Staten Island Botanical Gardens at Snug Harbor, New York, USA.
- Strybing Arboretum & Botanical Gardens, San Francisco, California, USA.
- Surreybrooke, Middletown, Maryland, USA
- Tohono Chul Park, Tucson, Arizona, USA
- Tower Hill Botanic Garden, Boylston, Massachusetts, USA. New England's newest Botanic Garden, featuring gardens,educational facilities and a diverse collection of woody plants, native to New England.
- Tucson Botanical Gardens, Arizona, USA.
- Twentieth Century Gardens, Hot Springs National Park, Arkansas, USA.
- University of California Botanical Garden, Berkeley, CA, USA.
- University of Connecticut, Department of Ecology & Evolutionary Biology - Greenhouses, Storrs, CT, USA.
- University of Delaware Botanic Gardens. The Gardens are located on the grounds of the College of Agriculture at the University of Delaware in Newark, DE, USA.
- University of Georgia, Botany Department - Greenhouse Tour, Athens, Georgia, USA.
- University of Nebraska - Lincoln: Botanical Garden and Arboretum, USA.
- University of South Florida Botanical Garden, Tampa, Florida, USA. The garden is affiliated with the Biology Department, the herbarium and the Institute for Systematic Botany. Includes educational displays of Florida riparian and sandhill native plants, palms, carnivorous plants, herbs, begonias and bromeliads.
- University of Washington, Medicinal Herb Garden, Seattle, Washington, USA.
- University of Wisconsin - Madison Botanical Garden, USA.

- Utah State University Botanical Gardens, Farmington, Utah, USA. Utah State University Extension and the Utah Agriculture Experiment Station created and maintains a beautiful botanical garden in Farmington, Utah, USA.
- Virginia Tech Horticulture Garden, Blacksburg, Virginia, USA.
- Volunteer Park Conservatory - Volunteer Park, Seattle, Washington, USA.
- Warsaw Biblical Gardens, Mishawaka, Indiana, USA.
- WaveHill, Bronx, New York, USA. One of New York's 32 City-owned cultural institutions, Wave Hill is a 28-acre public garden dedicated to fostering the dynamic inter-relationship between natural processes and human culture.
- Wilder Park Conservatory - Elmhurst Park, Elmhurst, Illinois, USA.
- William Paca Garden, Annapolis, Maryland, USA. The mission of the William Paca Garden is to stimulate public support for preservation by maintaining an historically accurate, pre- Revolutionary, urban garden reconstruction and by assembling, documenting, and propagating a living collection of period plants as a central part of the environmentally sensitive garden exhibit. The adjacent Middendorf Conservation Center promotes the preservation of Maryland's native plants through a library, greenhouses, and educational programs.
- A World of Orchids, Kissimmee, Florida, USA.
- W.J. Beal Botanical Garden is one of the principal centers of plant interest within the arboretum-like campus of Michigan State University. More than 5,000 different kinds of plants can be found on the five-acre site.
- The Wrigley Memorial and Botanical Gardens, Catalina Island, California, USA.
- Zilker Botanical Garden, Austin, Texas, USA.

[*Courtesy* : The Internet Directory of Botany available on the Internet, compiled by Anthony R. Brach (Harvard University Herbarium, Cambridge / Missouri Botanical Garden, St. Louis, USA), Raino Lampinen (Botanical Museum, Finnish Museum of Natural History, University of Helsinki, Finland), Shunguo Liu (SHL Systemhouse, Edmonton, Canada) and Keith McCree (Oakridge, Oregon)].

SUGGESTED READING

- Avery, G.S. Jr., 1957, Botanic gardens – What role today? *Amer. J. Bot.*, 44(3):268-271.
- Hill, A.W., 1915, The history and functions of botanic garden. *Annl. Miss. Bot. Gard.*, 2:185-223.
- Hyams, E. and W. Macquitty, 1969, *Great Botanic Gardens of the World*, Nelson, New York.
- Thompson, P.A., 1972, The role of botanic garden. *Taxon*, 21(1):115-119.

PLATE 1

The Great Banyan Tree at the Indian Botanical Garden, Shibpur, Kolkata

The Gabang Palm, *Corypha indica* (Infront of Botany Dept., The Univ. of Burdwan, Golapbug)

The branching palm, *Hypaene sp.* (Indian Botanical Garden, Kolkata)

PLATE 2

Victoria amazonica — The giant water lily of the Amazon (Indian Botanical Garden, Shibpur, Kolkata)

The huge leaves of *Victoria amazonica* (Indian Botanical Garden, Shibpur, Kolkata)

PLATE 3

The tree fern *Cyathea spinulosa* at Lloyd Botanic Garden, Darjeeling

Some important cactuses growing inside the glass house at Lalbagh, Bangalore

PLATE 4

The map of the Lloyd Botanic Garden, Darjeeling

The double coconut (*Lodoicea maldivica*) at Indian Botanical Garden, Shibpur, Kolkata

PLATE 5

Gingo biloba — a living fossil at Lloyd Botanic Garden, Darjeeling

Metasequoia glytostropboides — another living fossil at Lloyd Botanic Garden, Darjeeling

PLATE 6

Cymbidium sp. — in full bloom at the Orchid house at Lloyd Botanic Garden, Darjeeling

Paphiopedium hirtissium — another orchid bloom at the Orchid house at Lloyd Botanic Garden, Darjeeling

The herbarium at the Lloyd Botanic Garden, Darjeeling

6 HISTORY OF PLANT CLASSIFICATION

AT A GLANCE

Plant classification is the arrangement of plants into groups and categories for a clear understanding, proper study and effective organization. According to Radford (1986) **"classification is the arrangement of groups of plants with particular circumscriptions by rank and position according to artificial criteria, phenetic similarities, or phylogenetic relationships"**.

The early civilization of India, Egypt and China who had a fair knowledge of plants used both as food as well as for medicinal purposes, also practiced a sort of plant taxonomy. *Vrikshayurveda* compiled by Parasara before the beginning of the Christian era is one of the earliest Indian works, which deals with plants in a scientific manner and follows a classification largely based on comparative morphology of plants, and it was considered to be more advanced than the ones developed in Europe before the 18th century. Several families known as *ganas* are clearly distinguished in *Vrikshayurveda*, which are easily recognizable even today. As for example, it mentions the flowers of Samiganyam as bypogynous with a gamosepalous calyx, five petals of different sizes and the fruit a legume with the seeds on the side. This evidently indicates to the family Leguminosae.

A classification system has predictive value and provides an index to information storage and retrieval on that plant. Since systematics today is very much a reflection of the past, a historical review revealing the historical period and the technologies available during the historical periods needs to be summarized.

6.1 Preliterate Humans

Ancient men who made their living by gathering food from the landscape, knew many plants by necessity and are considered to be practical plant taxonomists. They described and classified a plant as useful or harmful, so that others could recognize it, and identified other plants belonging to a particular category and assigned a name to the category, so that, it could be easily referred to. Since this is pre-written record, we have little direct knowledge of this period. However, there are much archaeological evidence and even information from ancient paintings, drawings or utensils and virtually all civilizations have taken advantage of the medicinal properties of plants, the closely guarded secrets of which were passed on from medicine man to medicine man. Even today, we can learn a lot about plants from the so-called "primitive" cultures. For example, present day indigenous people such as the *brujos* in the American tropics can recognize thousands of plants, which they use for food or medicine, having their own names, and we can learn a lot from them.

6.2 Ancient Greeks

One of the first documented efforts to systematize a local flora was that of Theophrastus. Theophrastus and Dioscorides are considered to be the main contributors among the ancient Greeks.

(i) **Theophrastus** (370-285 B.C.) who was a student of Plato and Aristotle, and later became chief of the Lyceum (gardens and library) in Athens, is often called the "Father of Botany". He named about 480 species in his famous works *Enquiry into Plants* and *The Causes of Plants* (both translated in English). He classified plants by form into trees, shrubs, undershrubs and herbs. He also recognized annuals, biennials, perennials and floral morphology such as superior and inferior ovaries, gross anatomy i.e., whether the calyx and corolla are modified leaves, free and sympetalous corollas, etc. He noted the distinction between different kinds of tissues and between flowering plants (Phanerogams) and non flowering plants (Cryptogams). A few generic names currently in use such as *Asparagus, Crataegus, Daucus* (carrot) and *Narcissus* (daffodil) originated during this time.

(ii) **Pedanios Dioscorides** (62-128 A.D.) was a Greek physician in the Roman Army. His most famous work was *Materia Medica*, in which he described the medicinal qualities of 600 kinds of plants. Many members of currently recognized families were classified together such as mints and umbels i.e., he grouped plants together with at least a superficial degree of natural relationships.

6.3 Middle Ages

There was little original science or observations between the fall of the Roman Empire and the Renaissance. Intellectual stagnation of the middle ages resulted in minimal original work in plant systematics. The Ancients were thought to have possessed all knowledge, and their works were copied repeatedly. One of the taxonomists during this period named **Albertus Magnus** (1193-1280) in his work *De Vegetabilis*, recognized monocots and dicots and divided the dicots into vascular and nonvascular plants.

6.4 Herbalists (15th through 17th century)

People during this period, sought originality, and studied plants and not just accepted and copied what the ancients said. This period of renaissance was marked by extensive exploration and learning and with the invention and implementation of the printing press (1440), many large volumes, about plants and their uses, known as herbals, were produced throughout Europe for use, particularly by the physicians. Many natural and well-defined genera and families were established during this period and for the first time in history, the key characteristics of plants were depicted via woodcut or metal plate engravings. Some of the well-known herbalists and their famous works during this period include:

(i) ***Otto Brunfels*** (1464-1534) – *Herbarium Vivae Eicones*

(ii) ***Jerome Bock*** (1469-1554) – *Neu Kreuterbuch*

(iii) ***Leonhart Fuchs*** (1501-1566) – *De historia stirpium*

(iv) ***John Gerard*** (1542-612) – *The Herball*, or, *Generall Historie of Plantes*

(v) ***Rembert Dodoens*** (1517-1585) – *Cruydeboeck*

(vi) ***Casper Bauhin*** (1560-1631) – *Pinax theatri botanici*

6.5 The Pre-Linnaean Period

(i) **Andrea Caesalpino** (1519-1603) : He is considered to be the first plant taxonomist as he was the first botanist to try to base a taxonomic scheme upon reason and logic rather than purely on utilitarian concepts. His most famous work is *De plantis Libri*, in which he named 1520 plants. His classification system used an Aristotelian approach, which is based upon habit (trees, shrubs, herbs), but it also used a whole series of floral, fruit, and seed characters. It recognized what today we know as the mustard, aster, carrot, and herbaceous and woody legume families.

(ii) **Joseph Pitton de Tournefort** (1656-1708) : His most famous work is *Institutiones Rei Herbariae*. He is the first in a series of important French botanists and his artificial system of classification for 10,146 plants includes 698 genera. He is the first to give a real concept of genera. He divided flowering plants into 2 large categories, trees and herbs, which were further subdivided into groups on the basis of characters such as petalous or apetalous, regular or irregular and polypetalous or gamopetalous.

(iii) **John Ray** (1628-1705) : He was a British botanist and his most famous work is *Synopsis Methodica Stirpium Britannicarum*, which is the first British flora. He also published *Methodus Plantarum Nova* in 1682 and *Historia Plantarum* (1696-1704). His natural system of classification in which he grouped together plants that looked alike, was complicated, as it used many different characters and split natural groups and lumped together unnatural ones. He believed in "essential" characters. The last edition of *Methodicum* treated 18,000 species, and the main divisions were woody and herbaceous plants and he recognized the taxa of monocots and dicots, the classes were based on fruit type and the subdivisions on the basis of leaf and flower characters.

6.6 The Linnaean Period

By the middle of the 18th century, further developments in science, advances in navigation, early microscopes, widespread travel, use of herbarium specimens, widely available printed books, ground work of herbalist, made the science of botany ready for a person to synthesize information.

(i) **Carl von Linné or Carolus Linnaeus** (Latinized form of his name) [1707-1778] : He was a Swedish and his famous work included *Systema Naturae*. His classification system is called artificial, as it relied upon a few features of the flower (mainly the number of stamens and pistils) as the primary basis for division. His system was correct in some cases, but not in others, as he put clearly unrelated plants together. He developed the sexual system of classification and called it *Nuptiae Plantarum* (marriages of plants). Written in metaphorical terms in *Systema Naturae* is "husband and wife have same bed," or "five husbands in the same marriage," etc. referring to the arrangement of stamens and carpels in the flower. He named 12,000 species (7,700 plants, 4,300 animals) and 1,105 genera. His other important works include, *Genera Plantarum* and *Species Plantarum*. The latter, which was published in May, 1753, is the starting point for botanical nomenclature. Species names are given in the margin and descriptions were like the old polynomials. Linnaeus is most known today for the following reasons:

- He popularized a system of nomenclature now called the Binomial System, which replaced the polynomials used in his original work, by a "shorthand" way of referring to the plant by two names – a generic name and a specific epithet.
- His works included brief diagnoses (short descriptions) of the plants.
- He standardized the synonomy (duplicate names) of many plants.
- He developed a number of terms used in plant morphology.

6.7 Post-Linnaean Period

The French never really adopted the system of Linnaeus, and based on their experience in the botanical gardens of Paris (*Jardin de Botanique de Paris*), they soon developed their own, first natural system of classification. Some leading botanists of that time were:

(i) **Michel Adanson** (1727-1806) : He is considered as the 'Grandfather of Numerical Taxonomy'. His famous work is *Familles des Plantes*. He rejected artificial classification and developed ways to logically deal with large number of characters and his methodology foreshadowed numerical taxonomy (phenetics). He felt that using as many characters as possible, would give the most natural and useful classification. His system is often called Adansonian Taxonomy.

(ii) **Antoine Laurent de Jussieu** (1748-1836) : His work *Genera Plantarum* led to a Natural Classification System, based on common sense. He divided the flowering plants into 15 classes, which were further subdivided into 100 natural orders corresponding to most major families, each of which was clearly differentiated, named and provided with a description. Many of the families recognized then are still recognized today. Botanists of the time readily accepted this system over that of Linnaeus, as the plants which looked alike were grouped together.

(iii) **Jean Baptist de Lamarck** (1744-1829) : His work includes *Flore Françoise*. He is best known for an early theory of evolution by inheritance of acquired traits. He used identification keys similar to dichotomous keys of today.

(iv) **Augustin Pyramús de Candolle** (1778-1841) : His work *Prodromus Systematis and Naturalis Regni Vegetabilis* was an attempt to do a world flora, which was never finished. It included 58,000 species and 161 families. It depicted the catkin-bearing dicots (e.g. Betulaceae, Fagaceae, etc.) as being related to gymnosperms and his system of classification began the dicots with Ranunculaceae.

6.8 Evolutionary Classifications

(i) **Charles Darwin** (1809-1882) : His famous work includes *On the Origin of Species by Means of Natural Selection* (1859). Darwin was strongly influenced by observations during the voyage of the H.M.S. Beagle to the Galapagos Islands. His observation on the differences in animals (e.g. fishes) and plants on different islands prompted his mechanism for evolutionary change, i.e. natural selection. Darwin thought then that the earth was much older than 6000 years and that evolutionary change occurs through gradual, accumulated differences. At the same time another scientist, Alfred R. Wallace, was working in the East Indies and developed a very similar theory i.e. theory of evolution. Both theories were presented at the meetings of the London Linnaean Society in 1858, and this concept of evolutionary change was embraced by most scientists, including Haekel whose *Tree of Life* (1866) depicted all taxa (Protists, Plants, and Animals).

(ii) **George Bentham** (1800-1884) and **Sir Joseph D. Hooker** (1817-1911) : *Genera Plantarum* was published after Darwin's *Origin of Species*, but it was not possible to change the system to reflect new evolutionary concepts despite the fact that both Bentham and

Hooker were great champions of this theory. Bentham and Hooker's system is post-Darwinian in chronology but pre-Darwinian in concept. They named 200 families and 7,569 genera, with fabulously detailed, often original descriptions. Many herbaria in the world are still arranged according to this system. They recognized monocots and dicots and began the latter with the polypetalous plants.

6.9 Phylogenetic Classifications

(i) **Asa Gray** (1810-1888) : He is not considered a major figure in systems of classification, but he was the first professor of botany at U.S. University (Harvard, 1842). His most famous work includes *A Manual of Botany of the Northern United States* (1848). Gray's *Manual of Botany* was published later by M. L. Fernald (8th edition, 1950).

(ii) **Adolph Engler** (1844-1930) and **K. A. E. Prantl** (1849-1893) : The first phylogenetic systems came from Botanical Garden in Berlin. Engler and Prantl believed that the classification system should reflect the evolutionary history and their greatest work *Die Naturlichen Pflanzenfamilien*, adopted a system that proceeded from simple structures to complex. Unisexual, cone-like flowers (catkins) were considered to be primitive (alder, birch, etc.). This was the first major phylogenetic classification. This system is used today in many herbaria and floras and very useful for a modern taxonomist, but the concepts of what is primitive and advanced has changed today.

(iii) **Charles Edwin Bessey** (1845-1915): He was a student of Asa Gray and he developed a set of "dicta" (rules) stating which characters are primitive and which are advanced in flowering plants. Not all are considered correct today but many are. The *Magnolia* type flower was considered primitive, not the unisexual catkin-bearing plants. His published work is *The Phylogenetic Taxonomy of Flowering Plants*.

(iv) **John Hutchinson** (1884-1972) : He worked at Kew Gardens and published *The Families of Flowering Plants* and *Genera of Flowering Plants*. He considered woody and herbaceous to be of fundamentally different evolutionary trends. Not much of this classification is used today.

(v) **Robert F. Thorne** (1920 to present) : He works at Rancho Santa Anna Botanical Garden in Claremont, CA. His work includes *A Phylogenetic Classification of the Angiospermae*.

(vi) **Armen Takhtajan** (1910 to present) : He was initially at Lenningrad, but now spends most time at New York Botanical Garden. His work *Flowering Plants: Origin and Dispersal* was an influential early work. A revision of this classification appeared in 1997 as *Diversity and Classification of Flowering Plants*.

(vii) **Rolf Dahlgren** (1932-1985) : He worked at the University of Copenhagen. His work includes *A Revised System of Classification of the Angiosperms*.

(viii) **Arthur Cronquist** (1919-1991) : He worked most of his life at the New York Botanical Garden. Many of his ideas are Besseyan with some influence from Takhtajan. His most famous work includes *An Integrated System of Classification of Flowering Plants* (1981), which has been revised in 1988 and is much adopted today.

6.10 Modern Phylogeneticists

Vast majority of plant taxonomists doing phylogenetic work today are using molecular techniques, i.e. studying protein and isoenzyme profiles, sequencing DNA (chloroplast DNA, mitochondrial DNA), etc. The term Molecular Phylogeny is the study of evolutionary relationships among organisms using DNA and RNA.

Summary of approaches to Classification
Utilitarian (Uses) — Dioscorides, Herbalists
Form (habit of plant) — Theophrastis, Caesalpino
Artificial (few obvious characters) — Tournefort, Linnaeus
Natural (overall resemblances) — Bauhin, Ray, de Jussieu, de Candolle, Bentham & Hooker
Phylogenetic (reflects evolution) — Engler & Prantl, Bessey, Hutchinson, Cronquist, Takhtajan, Thorne, Dahlgren

SUGGESTED READING

- Bentham, G. and J.D. Hooker, 1862-1883, *Genera Plantarum*, 3 Vols., London.
- Bessey, C.E., 1915, Phylogenetic taxonomy of flowering plants. *Ann. Mo. Bot. Gard.*, 2:109-164.
- Candolle, A.P. de., 1813, *Theorie elementaire de la botanique*, Paris.
- Candolle, A.P. de., 1824-1873, *Prodromus systematics naturalis regni Vegetabilis*, 17 Vols., Paris.
- Cronquist, A., 1981, *An Integrated System of Classification of Flowering Plants*, Columbia University Press, New York.
- Dahlgren, R., 1980, A revised system of classification of angiosperms, *Bot. J. Linn. Soc.*, 80:91-124.
- Dahlgren, R., S. Rusendal-Jenses and B.J. Nielsen,1981, A revised classification of the angiosperms with comments on the co-relation between chemical and other characters. In: *Phytochemistry and Angiosperm Phylogeny* (D.A. Young and D.S. Seigler, eds.), Praeger, New York.
- Dahlgren, R., S. Rusendal-Jenses and B.J. Nielsen,1983, General aspects of angiosperm evolution and microsystematics. *Nord. J. Bot.*, 3:119-149.
- Eichler, A.W., 1875-1878, *Bluthendiagramme construirt und erlautert*, 2 Vols., Leipzig.
- Engler, A. and K. Prantl, 1887-1915, *Die Naturlichen Pflanzenfamilien*, 23 Vols., Leipzig.
- Hutchinson, J., 1964-1967, *The Genera of Flowering Plants*, 2 Vols., Claredon, Oxford.
- Hutchinson, J., 1969, *Evolution and Phylogeny of Flowering Plants*, Academic Press, Oxford.
- Hutchinson, J., 1973, *The Families of Flowering Plants*, 3rd ed., Claredon, Oxford.
- Jussieu, A.L. de., 1789, *Genera Plantarum Secundum Ordines Naturals Disposita*, Paris.
- Linnaeus, C., 1735, *Systema Naturae*, Lugduni Batavorum.
- Linnaeus, C., 1737, *Genera Plantarum*, Lugduni Batavorum.
- Linnaeus, C., 1753, *Species Plantarum*, 2 Vols., Stockholm.
- Takhtajan, A., 1980, Outline of the classification of flowering plants (Magnoliophyta), *Bot. Rev.*, 46:225-359.
- Thorne, R.F., 1976, A phylogenetic classification of the angiospermae. *Evol. Biol.*, 9:35-106.
- Thorne, R.F., 1983, Proposed new realignments in the angiosperms. *Nord. J. Bot.*, 3:85-117.

7 SYSTEMS OF PLANT CLASSIFICATION

AT A GLANCE

7.1 Types of Classification

The past taxonomic literature described three types of systems of classification. These include the artificial system, natural system and phylogenetic system. However Radford (1986) has described a fourth type of classification viz. mechanical classification.

(a) Artificial Classification

The earliest systems of classification which remained dominant from 300 B.C. up to about 1830 were artificial systems, which were based on one or a few easily observable characters of plants, such as habit (trees, shrubs, herbs, etc.) or floral characters (particularly the number of stamens and carpels). Such types of classification using some arbitrary or at least easily observable characters, often irrespective of their affinity, is called artificial. The sexual system of Linnaeus is a good example of artificial classification, which uses only one attribute i.e. the number of stamens for grouping plants into 24 Classes as a result of which, various unrelated taxa, which are not at all related but, similar in one respect only, have been placed under the same Class.

(b) Natural Classification

These systems of classifications are based upon overall resemblances, mostly in gross morphology, thus, utilizing as many taxonomic characters as possible, to group taxa. Charles Darwin's proposed theory of evolution (1859) postulates that, the present day plants have descended from those existing in the ancient past, through a series of modifications in response to changing environmental conditions, which means that all present day plants are related to each other in one way or another. Thus, the closely related plants should naturally be grouped together. This is called natural classification. Thus, larger the number of characters shared by different taxa, the more closely related they are to each other. This is the basis of modern classification.

(c) Phylogenetic Classification

The classification systems proposed after Darwin's theory are mostly phylogenetic i.e. they use as many taxonomic characters as possible in addition to the phylogenetic (evolutionary) interpretations. These are expressed in the form of phylogenetic trees or shrubs showing presumed evolution of the groups. The natural systems are two-dimensional i.e. based on the data available at any time and is known as **Horizontal Classification**, whereas the addition of the third dimension i.e. past history or ancestral history results in phylogenetic classification also known as **Vertical Classification or Evolutionary Classification.**

According to **Radford (1986)** however there are four systems of classifications. They are :

(a) Artificial Classifications — These systems use the habit and importance to man as taxonomic characters e.g. systems of Theophrastus, Dioscorides, etc.

(b) Mechanical Classifications — These systems use one or a few selected taxonomic characters to group taxa e.g. systems of Caesalpino, Bauhin, Ray, Tounefort and Linnaeus.

(c) Natural Classifications

(d) Phylogenetic Classifications

The latter two are the same as described earlier.

7.2 Some Important Systems of Plant Classification

7.2.1 Classification system of Carolus Linnaeus (1707-1778)

Carl Linnaeus (1707-1778)

7.2.1.1 Carolus Linnaeus (1707-1778)

Carl Linnaeus, also known as Carl von Linné or Carolus Linnaeus, is also called the *Father of Taxonomy*. He developed a system for naming, ranking, and classifying organisms which is still in wide use today, however with many changes.

He was born on May 23, 1707, in Rashult in Southern Sweden and he showed a deep love and fascination for plants right from a very early age. He joined the University of Lund to study medicine in 1727, and a year later, he transferred to the University of Uppsala, the most prestigious university in Sweden, where he spent most of his time collecting and studying plants, his true love, as at that time, training in botany was part of the medical curriculum and every doctor had to prepare and prescribe drugs derived from medicinal plants. Linnaeus mounted a botanical and ethnographical expedition to Lapland in 1731 and to central Sweden in 1734. A year later in 1735, he went to the Netherlands and promptly finished his medical degree at the University of Harderwijk. Later, he enrolled in the University of Leiden for further studies and that same year, he published the first edition of his classification of living things, the *Systema Naturae*. In 1738, he returned back to Sweden where he practiced medicine (specializing in the treatment of syphilis) and lectured in Stockholm, before being awarded a professorship at Uppsala in 1741. At Uppsala, he restored the University's botanical garden (arranging the plants according to his system of classification) and made three more expeditions to various parts of Sweden. Inspired by him, many of his students made exploration voyages to different parts of the world such as Australia, Europe, north-eastern American colonies, Japan, South America, south-east Asia, Africa and the Middle East.

In the meantime, Linnaeus continued to revise his *Systema Naturae*, which grew to a multi-volume work, as his concepts were modified and as more and more plant and animal specimens were sent to him from different parts of the world by his students and other European botanists. In 1758, he bought the manor estate of Hammarby, outside Uppsala, where he built a small museum for his extensive personal collections and in 1761, he was granted nobility, and became Carl von Linné. Linnaeus died in 1778 and his son, also named Carl, succeeded him as a professor at Uppsala. But he never proved to be noteworthy as a botanist. The younger Carl died five years later with no heirs, and following his death, his mother and sisters sold the elder Linnaeus's library, manuscripts, and natural history collections to the English natural historian Sir James Edward Smith, who founded the Linnean Society of London in 1788, to take care of them.

7.2.1.2 Carolus Linnaeus's system of classification

His system of classification is an artificial sexual system, which is based solely on the number and arrangement of the reproductive organs. A plant's Class was determined by its stamens (male organs), and its Order by its pistils (female organs). This however, resulted in many groupings that seemed unnatural. For example, the Order Monadelphia under the Class Monoecia includes plants with separate male and female flowers on the same plant and with multiple male organs joined onto one common base. This Order also includes conifers such as pines, firs, and cypresses, but also a few true flowering plants, such as the castor bean. Plants without obvious sex organs were classified in the Class Cryptogamia, or "plants with a hidden marriage," which lumped together the algae, lichens, fungi, mosses and other bryophytes, and ferns. Linnaeus thus, attached great significance to plant sexual reproduction and divided plants into 24 Classes, mainly on the basis of number, union and length of stamens.

Linnaeus was also the first person to use the binomial system of nomenclature. For him, species of organisms were real entities, which could be grouped into higher categories called genera (singular, genus). Part of Linnaeus's innovation was the grouping of genera into higher taxa that were also based on shared similarities. In his original system, Genera were grouped into Orders, Orders into Classes, and Classes into Kingdoms. Earlier he believed that the Species was not only real, but unchangeable, but after observing how different species of plants might hybridize, to give rise to new species, he abandoned the concept that species were fixed and invariable, and suggested that some, perhaps most species in a genus might have arisen after the creation of the world, through hybridization.

7.2.1.3 An outline of Linnaeus system (Sexual system) of classification

Class	Subclass (Ordo)	Distinguishing Characters	Examples
1. **Monandria**	1. Monogynia 2. Digynia	Single stamen	*Lemna, Scirpus, Salicomia, Canna*
2. **Diandria**	1. Monogynia 2. Digynia 3. Trigynia	Two stamens	*Salvia, Veronica, Lepidium, Olea*
3. **Triandria**	1. Monogynia 2. Digynia 3. Trigynia	Three stamens	Many grasses, *Iris, Sisyrinchium*

Class	Subclass (Ordo)	Distinguishing Characters	Examples
4. **Tetrandria**	1. Monogynia 2. Digynia 3. Tetragynia	Four stamens	*Ulmus, Mentha, Galium, Protea*
5. **Pentandria**	1. Monogynia 2. Digynia 3. Trigynia 4. Tetragynia 5. Pentagynia	Five stamens	*Primula, Ipomoea, Campanula*
6. **Hexandria**	1. Monogynia 2. Digynia 3. Trigynia 4. Tetragynia 5. Polygynia	Six stamens	*Alisma, Lilium, Narcissus, Rumex, Barberis*
7. **Heptandria**	1. Monogynia	Seven stamens	*Aesculus, Trientalis*
8. **Octandria**	1. Monogynia 2. Digynia 3. Trigynia 4. Tetragynia	Eight stamens	*Fagopyrum, Vaccinium, Dirca*
9. **Enneandria**	1. Monogynia 2. Trigynia 3. Hexagynia	Nine stamens	*Ranunculus, Rheum, Laurus, Butomus*
10. **Decandria**	1. Monogynia 2. Digynia 3. Trigynia 4. Pentagynia 5. Decagynia	Ten stamens	*Oxalis, Acer, Rhododendron*
11. **Dodecandria**	1. Monogynia 2. Digynia 3. Trigynia 4. Pentagynia 5. Dodecagynia	11 to 19 stamens	*Asarum, Euphorbia, Reseda, Calla*
12. **Icosandria**	1. Monogynia 2. Digynia 3. Trigynia 4. Pentagynia 5. Polygynia	20 or more stamens attached to the calyx	*Rubus, Rosa, Siraea*
13. **Polyandria**	1. Monogynia 2. Digynia 3. Trigynia 4. Tetragynia 5. Pentagynia 6. Hexagynia 7. Polygynia	20 or more Stamens attached to the receptacle	*Nymphaea, Tilia, Papaver*

Class	Subclass (Ordo)	Distinguishing Characters	Examples
14. **Didynamia**	1. Gymnospermia 2. Angiospermia	Didynamous stamens	*Monarda, Linaria, Linnaea*
15. **Tetradynamia**	1. Siliculosa 2. Siliquosa	Stamens tetradynamous	Cruciferae
16. **Monadelphia**	1. Pentandria 2. Decandria 3. Polyandria	Stamens in one bundle	Members of Geraniaceae and Malvaceae
17. **Diadelphia**	1. Hexandria 2. Octandria 3. Decandria	Stamens in two bundles	All legumes
18. **Polyadelphia**	1. Pentandria 2. Icosandria 3. Polyandria	Stamens in many bundles	*Theobronia, Hypericum*
19. **Syngenesia**	1. Polygamia Aequalis 2. Polygamia Superflua 3. Polygamia Frustranea 4. Polygamia Necessaria 5. Monogamia	Stamens syngenesious (i.e. anthers united)	Asteraceae family, *Lobelia, Viola*
20. **Gynandria**	1. Diandria 2. Triandria 3. Tetrandria 4. Pentandria 5. Hexandria 6. Decandria 7. Polyandria	Stamens adnate to the pistil	*Aristolochia*, Orchids
21. **Monoecia**	1. Monoandria 2. Diandria 3. Triandria 4. Tetrandria 5. Pentandria 6. Hexandria	Plants monoecious	*Morus, Typha, Quercus, Carex*

Class	Subclass (Ordo)	Distinguishing Characters	Examples
	7. Heptandria 8. Polyandria 9. Monoadelphia 10. Syngenesia 11. Gynandria		
22. **Dioecia**	1. Monoandria 2. Diandria 3. Triandria 4. Tetrandria 5. Hexandria 6. Octandria 7. Ennaeandria 8. Decandria 9. Polyandria 10. Monoadelphia 11. Syngenesia 12. Gynandria	Plants dioecious	*Salix, Urtica, Juniperus*
23. **Polygamia**	1. Monoecia 2. Dioecia 3. Polyoecia	Plants polygamous	Many Asteraceae, *Acer, Empetrum, Nyssa*
24. **Cryptogamia**	1. Filices 2. Musci 3. Algae 4. Fungi	Flowerless plants	Algae, Fungi, Bryophyta Ferns, Lichens

7.2.1.4 Merits of Linnaeus's system of classification

The only merit of Linnaeus system lies in the fact that it aids in the quick and easy identification by means of one or a few characters.

7.2.1.5 Demerits of Linnaeus's system of classification

(a) Since this system is based on differences rather than on similarities of the sex organs, very often closely related plants have been placed in widely separated groups, while those quite different from each other have been placed in the same group.

(b) Affinities and relationships among the taxa have not been considered.

(c) There is no distinction between the Dicotyledones and Monocotyledones and have been mixed up and put together.

(d) The placement of Gymnosperms in the Class 14 (Didynamia) along with the angiosperm family Labiatae is most unsatisfactory.

7.2.2 Classification system of John Ray (1628-1705)

John Ray (1628-1705)

7.2.2.1 John Ray (1628-1705)

John Ray was born in the village of Black Notley, Essex, England on November 29, 1627. His father was a blacksmith, and his mother was known as a healer and herbalist. His love for nature, and especially for plants was perhaps gained from his mother. He joined the Cambridge University in 1644, and rapidly became expert in languages, mathematics, and natural science. He later became a Fellow of Cambridge University in 1649, a Lecturer in 1651, and a junior Dean in 1658. In 1660 he was ordained a priest in the Anglican Church. However, due to political reasons, he left Cambridge and between 1660 and 1671 he made many trips throughout England to collect plants, animals, and rocks. He also did experimental work in embryology and plant physiology. He was the first person to prove that the wood of a living tree conducts water and this work won him a place in the newly formed Royal Society of London, one of the world's first scientific societies, in 1667. He spent the last decades of his life writing books on languages, theology, and natural history. He died on January 17, 1705. Some of his famous works include *Catalogue of Cambridge Plants* (1660) and *Synopsis Methodica Avium et Piscium* (1713). Ray also published systematic works on plants, birds, mammals, fishes, and insects.

7.2.2.2 John Ray's system of plant classification

Ray's system was a "natural" system in which the classification of organisms reflected the Divine Order of creation. He classified plants on the basis of overall morphology: Thus, plants, which looked alike were grouped together and he recognized the fact that individual characters in themselves may or may not be worthwhile taxonomic features. The classification in his book *Methodus Plantarum Nova* (1682) was based on flowers, seeds, fruits, and roots. His plant classification system was the first to divide flowering plants into Monocots and Dicots. Since this method considered more than one feature, it expressed the similarities between species more fully. Ray was a devout Christian, who expounded his belief in "natural theology," the doctrine that the wisdom and power of God could be understood by studying His creation, the natural world.

Broadly Ray divided the plants as under :

- I. Herbae (Herbs)
  - → Imperfectae (Flowerless plants)
  - → Perfectae (Flowering plants)
    - → Dicotyledones
    - → Monocotyledones
- II. Arborae (Shrubs and Trees)
  - → Monocotyledones
  - → Dicotyledones

Ray's division of plants

7.2.3 Classification system of Charles Edwin Bessey (1845-1915)

Charles Edwin Bessey (1845-1915)

7.2.3.1 Charles Edwin Bessey (1845-1915)

C.E. Bessey who trained under Asay Gray and then became a Professor of Botany at the University of Nebraska, proposed a phylogenetic system of classification in 1894, which reflected the evolutionary ideas of Darwin which had greatly impressed him. His system was however published in the final form in 1915 in a paper entitled " The phylogenetic taxonomy of flowering plants" (Ann. Mo. Bot. Gard. 2:108-164). In his earlier works he had followed Bentham and Hooker's system of classification, but finally came to the synthesis of his own system, which was based on certain guiding principles of primitive and advanced characters called *dicta,* in which the most primitive modern plants were the Ranunculales. He believed that both the monocots and dicots arose from within the Ranunculales, along three major lines. This system is often called **Ranalian concept of evolution.** On being shown in the diagrammatic form, his system took the form of a cactus plant, which came to be known as **"Bessey's Cactus"** (Fig. 7.1) or as **"Opuntia besseyi".**

Bessey believed in the monophyletic origin of angiosperms and his ideas were based on the belief that flowering plants arose from the extinct, Mesozoic cycad-like

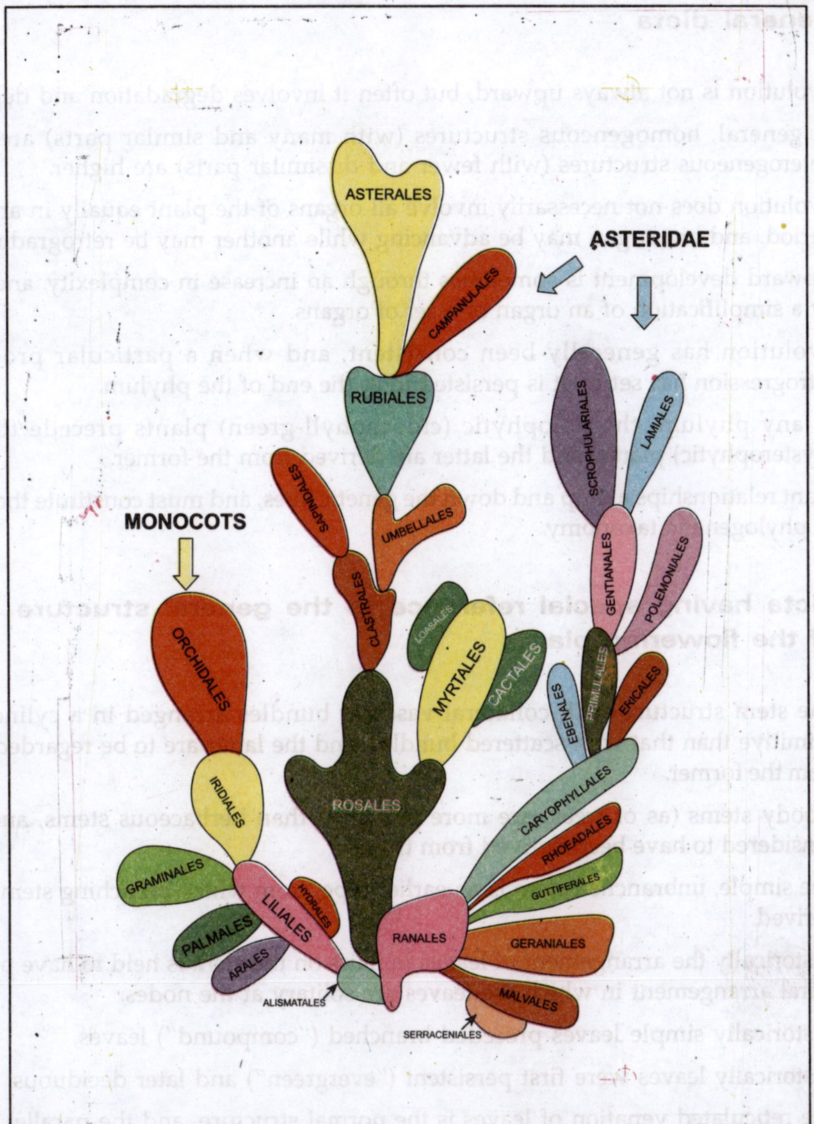

Fig. 7.1 : Bessey's Cactus — Ranalian concept of evolution

Bennetitales (= Cycadeoids). In these plants the reproductive structures consisted of cone-like strobili, with spirally arranged bracts, microsporophylls, and megasporophylls. Bessey was of the view that these organs became transformed into, respectively, the perianth, stamens, and pistils of flowering plants. This hypothesis, and Bessey's system generally, was largely accepted and taught in North America as it provided an useful illustration of how evolution had presumably taken place.

7.2.3.2 General dicta

1. Evolution is not always upward, but often it involves degradation and degeneration.
2. In general, homogeneous structures (with many and similar parts) are lower, and heterogeneous structures (with fewer and dissimilar parts) are higher.
3. Evolution does not necessarily involve all organs of the plant equally in any particular period, and one organ may be advancing while another may be retrograding.
4. Upward development is sometimes through an increase in complexity, and sometimes by a simplification of an organ or a set of organs.
5. Evolution has generally been consistent, and when a particular progression or retrogression has set in, it is persisted in to the end of the phylum.
6. In any phylum, the holophytic (chlorophyll-green) plants precede the colorless (hysterophytic) plants, and the latter are derived from the former.
7. Plant relationships are up and down the genetic lines, and must constitute the framework of phylogenetic taxonomy.

7.2.3.3 Dicta having special reference to the general structure of the flowering plants

8. The stem structure with collateral vascular bundles arranged in a cylinder is more primitive than that with scattered bundles, and the latter are to be regarded as derived from the former.
9. Woody stems (as of trees) are more primitive than herbaceous stems, and herbs are considered to have been derived from trees.
10. The simple, unbranched stem is an earlier type, from which branching stems have been derived.
11. Historically the arrangement of leaves in pairs on the stem is held to have preceded the spiral arrangement in which the leaves are solitary at the nodes.
12. Historically simple leaves preceded branched ("compound") leaves.
13. Historically leaves were first persistent ("evergreen") and later deciduous.
14. The reticulated venation of leaves is the normal structure, and the parallel venation of some leaves is derived from it.

7.2.3.4 Dicta having reference to the flowers of flowering plants

15. The polymerous flower structure precedes, and the oligomerous structure follows from it, and this is accompanied by a progressive sterilization of sporophylls.
16. Petaly is the normal perianth structure, and apetaly is the result of perianth reduction (aphanisis).
17. The apochlamydeous perianth is earlier and the gamochlamydeous perianth is derived from it by symphysis of the members of perianth whorls.
18. Actinomorphy is an earlier structure than zygomorphy, and the latter results from a change from similar to dissimilar growth of the members of the perianth whorls.

19. Hypogyny is the more primitive structure, and from it epigyny was derived later.
20. Apocarpy is the primitive structure, and from it syncarpy was derived later.
21. Polycarpy is the earlier condition, and oligocarpy was derived from it later.
22. The endospermous seed is primitive and lower, while the seed without endosperm is derived and higher.
23. Consequently, the seed with a small embryo (in endosperm) is more primitive than the seed with a large embryo (in scanty or no endosperm).
24. In earlier (primitive) flowers, there are many stamens (polystemonous), while in later flowers there are fewer stamens (oligostemonous).
25. The stamens of primitive flowers are separate (apostemonous), while those of derived flowers are often united (synstemonous).
26. The condition of powdery pollen is more primitive than that with coherent or massed pollen.
27. Flowers with both stamens and carpels (monoclinous) precede those in which these occur in separate flowers (diclinous).
28. In diclinous plants the monoecious condition is the earlier, and the dioecious later.

7.2.3.5 Bessey's system of classification

In Bessey's system of classification, the angiosperms have been divided into two Classes – Alternifoliae (Monocotyledoneae) and Oppositifoliae (Dicotyledoneae). The former has 2 Subclasses and 2 Orders and the latter has 2 Subclasses, 5 Superorders and 5 Orders.

Following in an outline of Bessey's system of classification :

Class — Alternifoliae (Monocotyledoneae)
- Subclass — Strobiloideae
  - Orders — Alismatales (Families – 1-9)
    - Liliales (Families – 10-22)
    - Arales (Families – 23-25)
    - Palmales (Family – 26)
    - Graminales (Families – 27-31)
- Subclass — Cotyloideae
  - Orders — Hydrales (Family 32)
    - Iridales (Families – 33-43)
    - Orchidales (Families – 44-45)

Class — Oppositifoliae (Dicotyledoneae)
- Subclass — Strobiloideae
  - Superorder — Apopetalae — Polycarpellatae
    - Orders — Ranales (Families – 46-69)
      - Malvales (Families – 70-81)
      - Sarraceniales (Families – 82-83)
      - Geraniales (Families – 84-105)
      - Guttiferales (Families – 106-125)
      - Rhoeadales (Families – 126-132)
      - Caryophyllales (Families – 133-149)

Superorder — Sympetalae – Polycarpellatae

Orders — Ebenales (Families – 150-154)

Ericales (Families – 155-160)

Primulales (Families – 161-165)

Superorder — Sympetalae – Dicarpellatae

Orders — Gentianales (Families – 166-171)

Polemoniales (Families – 172-177)

Scrophulariales (Families – 178-187)

Lamiales (Families – 188-191)

Subclass — Cotyloideae

Superorder — Apopetalae

Orders — Rosales (Families – 192-214)

Myrtales (Families – 215-229)

Loasales (Families – 230-234)

Cactales (Families – 235)

Celastrales (Families – 236-259)

Sapindales (Families – 260-274)

Umbellales (Families – 275-277)

Superorder — Sympetalae

Orders — Rubiales (Families – 278-282)

Campanulales (Families – 283-286)

Asterales (Families – 287-300)

7.2.3.6 Merits of Bessey's system of classification

(a) Although Bessey initially followed Bentham and Hooker's system of classification, he however did not place the Gymnosperms in between Dicotyledons and Monocotyledons as in Bentham and Hooker's system.

(b) Bessey abolished Monochlamydeae from his system and distributed the families of Monochlamydeae in Oppositifoliae of Dicotyledons.

(c) Ranales, which are accepted as the primitive forms form the starting point of Bessey's system.

(d) The families with superior ovary precede families with inferior ovary in this system.

7.2.3.7 Demerits of Bessey's system of classification

(a) In this system, the monocots have been discussed prior to dicots, which is its greatest demerit.

(b) Hypogyny, perigyny and epigyny have been emphasized too much in this system.

7.2.4 Classification system of George Bentham (1800-1884) and Sir Joseph Hooker (1817-1911)

George Bentham (1800-1884)

Sir Joseph Hooker (1817-1911)

7.2.4.1 George Bentham (1800-1884) and Sir Joseph Hooker (1817-1911)

George Bentham, a self-trained British Botanist and Sir Joseph Dalton Hooker, Director of the Royal Botanic Gardens, Kew (England), proposed a natural system of classification in their book *Genera Plantarum*, which includes three volumes containing the description of all known genera of seed plants in Latin and was published during July 1862 and April 1883. This system of classification, which is clearly derived from the systems of de Jussieu and de Candolle, is the best and most accepted natural system and is still used and followed in several herbaria of the world, including India.

The Bentham and Hooker system of classification divided all the Phanerograms or seed plants into Dicotyledons, Gymnosperms and Monocotyledons, with Ranales placed in the beginning of the classification and grasses at the end. The *Genera Plantarum* includes the generic descriptions of 92,205 species belonging to 7,569 genera of 200 families of flowering plants based on the personal observations of Bentham and Hooker. Large genera have been divided into sections and sub-sections and the description of the genera are complete and accurate.

Following is a table showing the number of Orders, Genera and Families described by Bentham and Hooker :

Group	No. of Orders	No. of Genera	No. of Species
Dicotyledons			
Polypetalae	82	2,610	31,874
Gamopetalae	45	2,619	34,556
Monochlamydae	36	801	11,784
Gymnosperms	3	44	415
Monocotyledons	34	1,495	18,576
Total	**200**	**7,569**	**97,205**

7.2.4.2 A summary of Bentham and Hooker's system of classification (Genera Plantarum, 1862-1883)

Sub-Group : Dicotyledones (Leaves with reticulate venation, pentamerous flowers and seeds with two cotyledons, tap root system)

Division I : Polypetalae (Corolla consisting of separate petals)

Series I : Thalamiflorae (Sepals distinct, stamens many, hypogynous, disc absent)

Order 1. Ranales (Carpels free, floral parts arranged in spiral, cyclic or spirocyclic manner)

Families : 1. Ranunculaceae, 2. Dilleniaceae, 3. Calycanthaceae, 4. Magno- liaceae, 5. Annonaceae, 6. Menispermaceae, 7.Berberidaceae, 8. Nymphaeaceae

Order 2. Parietales (Carpels united to form an unilocular ovary, parietal placentation)

Families : 9. Sarraceniaceae, 10. Papaveraceae, 11. Cruciferae, 12. Cappa-ridaceae, 13. Resedaceae, 14. Cistineae, 15. Violarieae, 16. Canellaceae, 17. Bixineae

Order 3. Polygalinae (Syncarpous, ovary 2-locular)

Families : 18. Pittosporeae, 19. Tremandreae, 20. Polygaleae, 21. Vochysiaceae

Order 4. Caryophyllinae (Free central placentation)

Families : 22. Frankeniaceae, 23. Caryophylleae, 24. Portulacaceae, 25. Tamariscinae

Order 5. Guttiferales (Sepals imbricate, stamens many)

Families : 26. Elatinae, 27. Hypericinae, 28. Guttiferae, 29. Ternstroemiaceae, 30. Dipterocarpae, 31. Chlaenaceae

Order 6. Malvales (Sepals valvate, usually numerous stamens, ovary with 3-8 carpels, axile placentation)

Families : 32. Malvaceae, 33. Sterculiaceae, 34. Tiliaceae

Series II. Disciflorae (Calyx consisting of united or distinct sepals which may be free or adnate to ovary, stamens usually definite in number, hypogynous, disc present)

Order 7. Geraniales (Pentamerous flowers, disc annular or broken into glands, androecium obdiplostemonous, multicarpellary ovary, syncarpous, axile placentation, ovules pendulous, raphe usually ventral)

Families : 35. Lineae, 36. Humiriaceae, 37. Malphighiaceae, 38. Zygo-phyllaceae, 39. Geraniaceae, 40. Rutaceae, 41. Simarubeae, 42. Ochnaceae, 43. Burseraceae, 44. Meliaceae, 45. Chailletiaceae

Order 8. Olacales (Androecium diplostemonous, ovules pendulous, raphe usually dorsal)

Families : 46. Olacineae, 47. Ilicineae, 48. Cyrillaceae

Order 9. Celastrales (Androecium haplostemonous, ovules erect, raphe ventral)

Families : 49. Celastrineae, 50. Stackhousieae, 51. Rhamneae, 52. Ampelidae

Order 10. Sapindales (Androecium diplostemonous, ovules ascending, often 1-2 in each locule, raphe ventral or inverted, usually compound leaves, exstipulate, flowers usually in panicles)

Families : 53. Sapindaceae, 54. Sabiaceae, 55. Anacardiaceae

Ordines anomali: 56. Coriareae, -57. Moringaceae

Series III. Calyciflorae (Sepals united, rarely free, petals in one or two series, perigynous or epigynous, ovary generally inferior)

Order 11. Rosales (Indefinite stamens, often twice or more the number of petals, styles distinct)

Families : 58. Connaraceae, 59. Leguminosae, 60. Rosaceae, 61. Saxifra- gaceae, 62. Crassulaceae, 63. Droseraceae, 64. Hamamelideae, 65. Bruniaceae, 66. Halorageae

Order 12. Myrtales (Usually definite stamens, Perigynous or epigynous, axile placentation)

Families : 67. Rhizophoraceae, 68. Combretaceae, 69. Myrtaceae, 70. Melastomaceae, 71. Lythraceae, 72. Onagraceae

Order 13. Passiflorales (Leaves often much lobed and divided, tendril climbers, stamens 5, parietal placentation)

Families : 73. Samydaceae, 74. Loasaceae, 75. Turneraceae, 76. Passifloraceae, 77. Cucurbitaceae, 78. Bego-niaceae, 79. Datisceae

Order 14. Ficoidales (Perianth undifferentiated, Ovules in basal or axile placentation)

Families : 80. Cactaceae, 81. Ficoideae

Order 15. Umbellales (Umbel inflorescence)

Families : 82. Umbelliferae, 83. Araliaceae, 84. Cornaceae

Division II. Gamopetalae (Corolla partially or completely fused, perianth dichlamydeous, heterochlamydeous)

Series I. Infereae (Inferior ovary, stamens as many as the petals and alternating with them)

Order 1. Rubiales (Stamens epipetalous, distinct anthers, ovary with 2-many locules each with 1-many ovules)

Families : 85. Caprifoliaceae, 86. Rubiaceae

Order 2. Asterales (Stamens epipetalous, distinct anthers or united, ovary with 1 locule with 1 ovule)

Families : 87. Valerianeae, 88. Dipsacaceae, 89. Calycereae, 90. Compositae

Order 3. Campanales (Stamens not adnate to corolla, ovary with 2-6 locules each with many ovules)

Families : 91. Stylidaceae, 92. Goodenovieae, 93. Campa-nulaceae

Series II. Heteromereae (Stamens in two or more series, epipetalous or free from corolla, superior ovary, carpels more than two)

Order 4. Ericales (Stamens twice the number of corolla, or isomerous and alternating with them)

Families : 94. Ericaceae, 95. Vaccinieae, 96. Monotropeae, 97. Epacrideae, 98. Diapensiaceae, 99. Lennoaceae

Order 5. Primulales (Stamens as many as corolla lobes and opposite them, ovary with 1 locule)

Families : 100. Plumbagineae, 101. Primulaceae, 102. Myrsineae

Order 6. Ebenales (Stamens as many as corolla lobes or many and opposite them, ovary with 2-many locules)

Families : 103. Sapotaceae, 104. Ebenaceae, 105. Styraceae

Series III. Bicarpellatae (Stamens as many as corolla lobes and alternating with them or fewer, carpels two, ovary superior)

Order 7. Gentianales (Leaves opposite, corolla actinomorphic)

Families : 106. Oleaceae, 107. Salvadoraceae, 108. Apocynaceae, 109. Asclepiadaceae, 110. Loganiaceae, 111. Gentianaceae

Order 8. Polemoniales (Leaves alternate, corolla actinomorphic)

Families : 112. Polemoniaceae, 113. Hydrophyllaceae, 114. Boragineae, 115. Convolvulaceae, 116. Solanaceae

Order 9. Personales (Corolla zygomorphic, posterior stamen often reduced to staminode, ovules usually more than 4)

Families : 117. Scrophulariaceae, 118. Orobanchaceae, 119. Lentibulariaceae, 120. Columelliaceae, 121. Gesneriaceae, 122. Bignoniaceae, 123. Pedaliaceae, 124. Acanthaceae

Order 10. Lamiales (Corolla zygomorphic, posterior stamen reduced, ovules 4)

Families : 125. Myoporineae, 126. Selagineae, 127. Verbenaceae, 128. Labiatae

Ordines anomaly: 129. Plantagineae

Division III. Monochlamydeae or Incompletae (Presence of one or incomplete whorl of perianth or perianth altogether absent, sepaloid perianth, petals absent)

It has no sub-division into cohorts (orders) but includes eight series.

Series 1. Curvembryeae (Ovule generally one, embryo coiled, seeds mostly with mealy endosperm)

Families : 130. Nyctaginaceae, 131. Illecebraceae, 132. Amaranthaceae, 133. Chenopodiaceae, 134. Phytolaccaceae, 135. Batideae, 136. Polygonaceae

Series 2. Multiovulatae Aquaticae (Ovary syncarpous, ovules many, plants aquatic)

Family : 137. Podostemaceae

Series 3. Multiovulatae Terrestries (Ovary syncarpous, ovules many, plants terrestrial)

Families : 138. Nepenthaceae, 139. Cytinaceae, 140. Aristolochiaceae

Series 4. Microembryeae (Embryo very minute, ovary apocarpous or syncarpous with single ovule)

Families : 141. Piperaceae, 142. Chloranthaceae, 143. Myristichaceae, 144. Monimiaceae

Series 5. Daphnales (Woody or herbaceous plants, sepaloid perianth, stamens perigynous, ovary with one carpel and 2-4 ovuled)

Families : 145. Laurineae, 146. Proteaceae, 147. Thymeleaceae, 148. Penaeaceae, 149. Elaeagnaceae

Series 6. Achlamydosporeae (Usually inferior ovary, one locule, with 1-3 ovules, seeds endospermic, without testa)

Families : 150. Loranthaceae, 151. Santalaceae, 152. Balanophoraceae

Series 7. Unisexuales (Flowers unisexual, perianth usually absent)

Families : 153. Euphorbiaceae, 154. Balanopseae, 155. Urticaceae, 156. Platanaceae, 157. Leitneriaceae, 158. Juglandaceae, 159. Myricaceae, 160. Casuarinaceae, 161. Cupuliferae

Ordines anomali (families of uncertain relationship)

Families : 162. Salicinae, 163. Lacistemaceae, 164. Empetraceae, 165. Ceratopbyllaceae

Sub-Group : Gymnosperms (naked seeded plants)

Families : 166. Gnetaceae, 167. Coniferae, 168. Cycadaceae

Sub-Group : Monocotyledones (leaves with parallel venation, trimerous flowers, seeds with one cotyledon)

Series 1. Microspermae (Inner perianth lobes petaloid, inferior ovary, seeds minute, many, exalbuminous)

Families : 169. Hydrocharitaeae, 170. Burmanniaceae, 171. Orchidaceae

Series 2. Epigynae (Perianth biseriate, inner petaloid, inferior ovary, seeds large, few to many)

Families : 172. Scitamineae, 173. Bromeliaceae, 174. Haemodoraceae, 175. Irideae, 176. Amaryllideae, 177. Taccaceae, 178. Dioscoreaceae

Series 3. Coronarieae (Perianth biseriate, inner whorl petaloid, superior ovary, seeds albuminous)

Families : 179. Roxburghiaceae, 180. Liliaceae, 181. Pontederiaceae, 182. Philydraceae, 183. Xyridaceae, 184. Mayaceae, 185. Commelinaceae, 186. Rapataceae

Series 4. Calycinae (Perianth biseriate, sepaloid, superior ovary, seeds albuminous)

Families : 187. Flagellarieae, 188. Juncaceae, 189. Palmae

Series 5. Nudiflorae (Perianth usually absent or reduced, superior ovary, seeds albuminous)

Families : 190. Pandanea, 191. Cyclanthaceae, 192. Typhaceae, 193. Aroideae, 194. Lemnaceae

Series 6. Apocarpae (Perianth in 1-2 series or absent, superior ovary, carpels free, one or more, apocarpous, seeds exalbuminous)

Families : 195. Triurideae, 196. Alismataceae, 197. Naiadaceae

Series 7. Glumaceae (Minute flowers, perianth reduced, bracts large, scaly, ovary or locules with 1 ovule).

Families : 198. Eriocaulaceae, 199. Centrolepidiaceae, 200. Restiaceae, 201. Cyperaceae, 202. Gramineae

7.2.4.3 Merits of Bentham and Hooker's system of classification

(a) Bentham and Hooker system is the first great natural system of plant classification.

(b) Though this system is not very natural, yet it is very easily workable, and is important from the point of view of its field applications.

(c) This system has been worked out as a result of very careful comparative observations of all known genera of phanerogams and not copied from other available literature.

(d) It was accepted by the entire British Empire, USA and other European countries.

(e) Greater emphasis has been laid on the contrast between free and united petals in this system e.g. the class Dicotyledones is divided into three sub-classes- 1. Polypetalae, 2. Gamopetalae and 3. Monochlamydeae.

(f) A special feature of this system is the addition of Disciflorae, and a totally new arrangement of dividing certain groups on the basis of aquatic and terrestrial characteristics of the plants.

(g) In this system, the Monocots are derived from Dicots and in the class Monocotyledones more importance is given to the relative position of ovary and perianth characteristics.

7.2.4.4 Demerits of Bentham and Hooker's system of classification

(a) This system does not indicate phylogentic relationship among different taxa of plants.

(b) The position of Gymnosperms between Dicots and Monocots is only for the sake of convenience rather than an indication of affinities, which is its foremost demerit.

(c) This system is based mainly on single and mostly artificial (morphological) characters, with the result that many closely related families are widely separated from each other and have been placed under different Orders (cohorts). Similarly unrelated families have been grouped nearer.

(d) The position of the Orchidaceae and Scitamineae at the beginning of the Monocots is not satisfactory as Orchidaceae is an advanced family, and in this system it has been considered primitive and placed at the beginning.

(e) In Monocots, much stress is laid on the relative position of the ovary and perianth characters in determining the affinities, which is unnatural and unphylogenetic.

(f) The origin of angiosperms is not clear.

7.2.4.5 Broad Outline of the Classification proposed by Bentham and Hooker

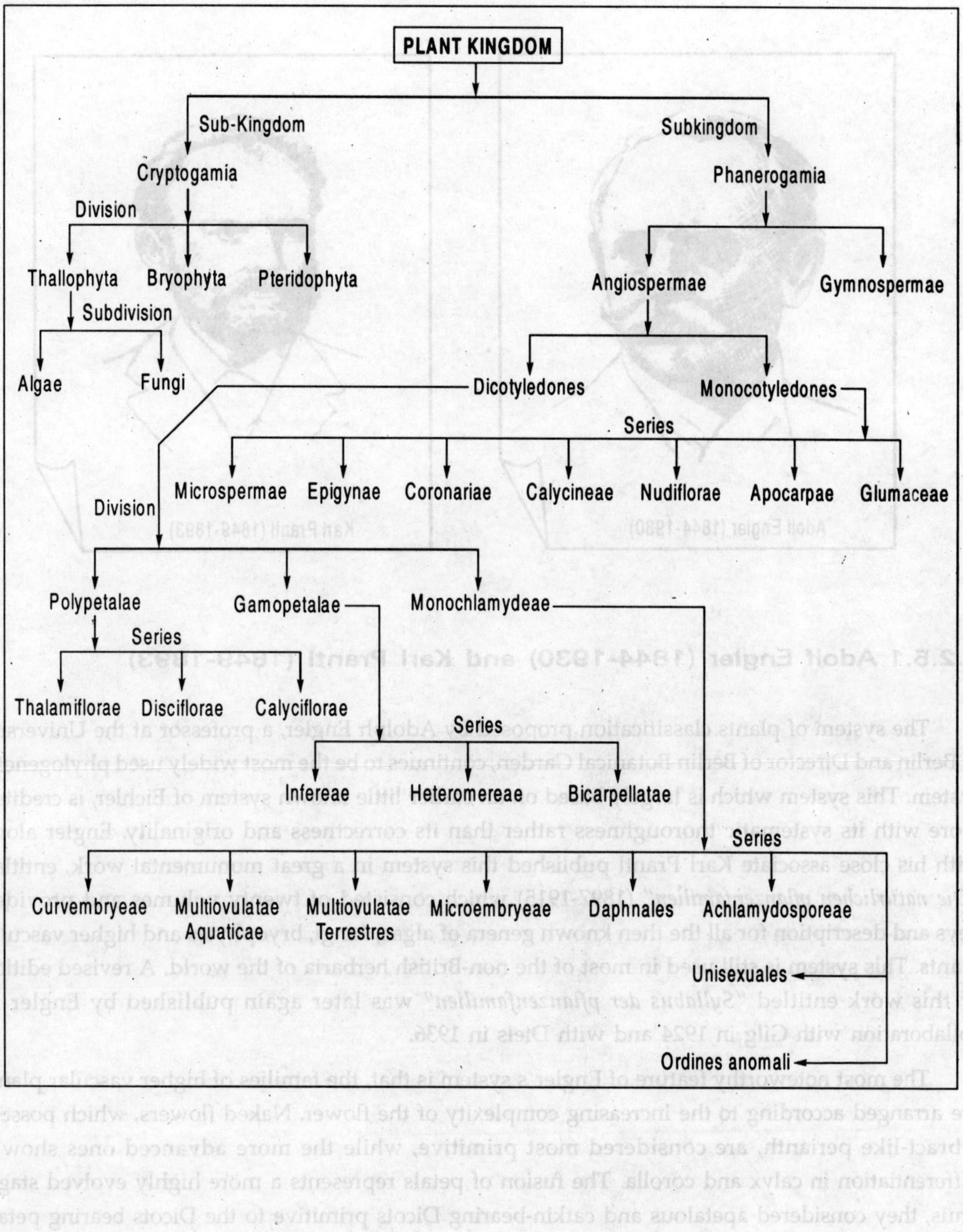

7.2.5 Classification system of Adolf Engler (1844-1930) and Karl Prantl (1849-1893)

Adolf Engler (1844-1930)

Karl Prantl (1849-1893)

7.2.5.1 Adolf Engler (1844-1930) and Karl Prantl (1849-1893)

The system of plants classification proposed by Adolph Engler, a professor at the University of Berlin and Director of Berlin Botanical Garden, continues to be the most widely used phylogenetic system. This system which is largely based on an earlier little known system of Eichler, is credited more with its systematic thoroughness rather than its correctness and originality. Engler along with his close associate Karl Prantl published this system in a great monumental work, entitled *"Die naturlichen pflanzenfamilien"* (1897-1915) which consisted of twenty volumes and provided keys and description for all the then known genera of algae, fungi, bryophytes and higher vascular plants. This system is still used in most of the non-British herbaria of the world. A revised edition of this work entitled *"Syllabus der pflanzenfamilien"* was later again published by Engler in collaboration with Gilg in 1924 and with Diels in 1936.

The most noteworthy feature of Engler's system is that, the families of higher vascular plants are arranged according to the increasing complexity of the flower. Naked flowers, which possess a bract-like perianth, are considered most primitive, while the more advanced ones show a differentiation in calyx and corolla. The fusion of petals represents a more highly evolved stage. Thus, they considered apetalous and catkin-bearing Dicots primitive to the Dicots bearing petals and simple unisexual flowers. In this system the Monocots are placed before Dicots and the orchids, are considered to be more evolved than grasses. They recognized 280 families of flowering plants including those of gymnosperms.

7.2.5.2 Engler and Prantl's system of plant classification (Die Naturlichen Pflanzenfamilien, 1897-1915)

Engler and Prantl divided the plant kingdom into 14 divisions :

1. Schizophyta
2. Myxothallophyta
3. Flagellatae
4. Dinoflagellatae
5. Bacillariophyta
6. Conjugatae
7. Heterocontae
8. Chlorophyceae
9. Charophyta
10. Phaeophyceae
11. Rhodophyceae
12. Eumycetes
13. Archegoniatae or Embryophyta – Asiphonogama
14. Embryophyta – Siphonogama

Subdivision – Angiospermae

Class 1. Monocotyledoneae

Order 1. Pandanales

Families : 1. Typhaceae, 2. Pandanaceae, 3. Sparganiaceae

Order 2. Helobiae

Families : 4. Potamogetonaceae, 5. Najadaceae, 6. Aponogetonaceae, 7. Scheuchzeriaceae, 8. Alismaceae, 9. Butomaceae, 10. Hydrocharitaceae

Order 3. Triuridales

Family : 11. Triuridaceae

Order 4. Glumiflorae

Families : 12. Gramineae, 13. Cyperaceae

Order 5. Principes

Family : 14. Palmae

Order 6. Synanthae

Family : 15. Cyclanthaceae

Order 7. Spathiflorae

Families : 16. Araceae, 17. Lemnaceae

Order 8. Farinosae

Families : 18. Flagllariaceae, 19. Restionaceae, 20. Centrolepidaceae, 21. Mayacaceae, 22. Xyridaceae, 23. Eriocaulaceae, 24. Thurniaceae, 25. Rapateaceae, 26. Bromeliaceae, 27. Commelinaceae, 28. Pontederiaceae, 29. Cyanastraceae, 30. Philydraceae

Order 9. Liliflorae

Families : 31. Juncaceae, 32. Stemonaceae, 33. Liliaceae, 34. Haemodoraceae, 35. Amaryllidaceae, 36. Velloziaceae, 37. Taccaceae, 38. Dioscoreaceae, 39. Iridaceae

Order 10. Scitamineae

Families : 40. Musaceae, 41. Zingiberaceae, 42. Cannaceae, 43. Marantaceae

Order 11. Microspermae

Families : 44. Burmanniaceae, 45. Orchidaceae

Class 2. Dicotyledoneae

Subclass 1. Archichlamydeae

Orders: 1. Verticillatae

Family : 46. Casuarinaceae

Order 2. Piperales

Families : 47. Saururaceae, 48. Piperaceae, 49. Chloranthaceae, 50. Lacistemaceae

Order 3. Salicales

Family : 51. Salicaceae

Order 4. Garryales

Family : 52. Garryaceae

Order 5. Myricales

Family : 53. Myricaceae

Order 6. Balanopsidales

Family : 54. Balanopsidaceae

Order 7. Leitneriales

Family : 55. Leitneriaceae

Order 8. Juglandales

Family : 56. Juglandaceae

Order 9. Batidales

Family : 57. Batidaceae

Order 10. Julianiales

Family : 58. Julianaceae

Order 11. Fagales

Families : 59. Betulaceae, 60. Fagaceae

Order 12. Urticales

Families : 61. Ulmaceae, 62. Moraceae, 63. Urticaceae

Order 13. Proteales

Family : 64. Proteaceae

Order 14. Santanales

Families : 65. Myzodendraceae, 66. Santalaceae, 67. Opiliaceae, 68. Grubbiaceae, 69. Olacaceae, 70. Octoknemataceae, 71. Loranthaceae, 72. Balanophoraceae

Order 15. Aristolochiales

Families : 73. Aristolochiaceae, 74. Rafflesiaceae, 75. Hydnoraceae

Order 16. Polygonales

Family : 76. Polygonaceae

Order 17. Centrospermae

Families : 77. Chenopodiaceae, 78. Amaranthaceae, 79. Nyctaginaceae 80. Cynocrambaceae, 81. Phytolaccaceae, 82. Aizoaceae, 83. Portulacaceae, 84. Basellaceae, 85. Caryophyllaceae

Order 18. Ranales

Families : 86. Nymphaeaceae, 87. Ceratophyllaceae, 88. Trochodendraceae, 89. Cercidiphyllaceae, 90. Ranunculaceae, 91. Lardizabalaceae, 92. Berberidaceae, 93. Menispermaceae, 94. Magnoliaceae, 95. Calycanthaceae, 96. Lactoridaceae, 97. Annonaceae, 98. Eupomatiaceae, 99. Myristicaceae, 100. Gomortegaceae, 101. Monimiaceae, 102. Lauraceae, 103. Hernandiaceae

Order 19. Rhoeadales

Families : 104. Papaveraceae, 105. Capparidaceae, 106. Cruciferae, 107. Tovariaceae, 108. Resedaceae, 109. Moringaceae

Order 20. Sarraceniales

Families : 110. Sarraceniaceae, 111. Nepenthaceae, 112. Droseraceae

Order 21. Rosales

Families : 113. Podostemaceae, 114. Tristichaceae, 115. Hydrostachyaceae, 116. Crassulaceae, 117. Cephalotaceae, 118. Saxifragaceae, 119. Pittosporaceae, 120. Brunelliaceae, 121. Cunoniaceae, 122. Myrothamnaceae, 123. Bruniaceae, 124. Hamamelidaceae, 125. Eucommiaceae, 126. Platanaceae, 127. Crossosomataceae, 128. Rosaceae, 129. Connaraceae, 130. Leguminosae

Order 22. Pandanales

Family : 131. Pandanaceae

Order 23. Geraniales

Families : 132. Geraniaceae, 133. Oxalidaceae, 134. Tropaeolaceae, 135. Linaceae, 136. Humiriaceae, 137. Erythroxylaceae, 138. Zygophyllaceae, 139. Cneoraceae, 140. Rutaceae, 141. Simarubiaceae, 142. Burseraceae, 143. Meliaceae, 144. Malpighiaceae, 145. Trigoniaceae, 146. Vochysiaceae, 147. Tremandraceae, 148. Polygalaceae, 149. Dichapetalaceae, 150. Euphorbiaceae, 151. Calitrichaceae

Order 24. Sapindales

Families : 152. Empetraceae, 153. Coriariaceae, 154. Limnanthaceae, 155. Anacardiaceae, 156. Cyrillaceae, 157. Buxaceae, 158. Pentaphylacaceae, 159. Corynocarpaceae, 160. Aquifoliaceae, 161. Celastraceae, 162. Hippocrateaceae, 163. Salvadoraceae, 164. Stackhousiaceae, 165. Staphyleaceae, 166. Icacinaceae, 167. Aceraceae, 168. Hippocastanaceae, 169. Sapindaceae, 170. Sabiaceae, 171. Melianthaceae, 172. Balsaminaceae

Order 25. Rhamnales

Families : 173. Rhamnaceae, 174. Vitaceae

Order 26. Malvales

Families : 175. Elaeocarpaceae, 176. Chlaenaceae, 177. Gonystilaceae, 178. Tiliaceae, 179. Malvaceae, 180. Bombacaceae, 181. Sterculiaceae, 182. Scytopetalaceae

Order 27. Parietales

Families : 183. Dilleniaceae, 184. Eucryphiaceae, 185. Ochnaceae, 186. Caryocaraceae, 187. Marcgraviaceae, 188. Quiinaceae, 189. Theaceae, 190. Guttiferae, 191. Dipterocarpaceae, 192. Elatinaceae, 193. Frankeniaceae, 194. Tamaricaceae, 195. Fouquieriaceae, 196. Cistaceae, 197. Bixaceae, 198. Cochlospermaceae, 199. Winteraceae, 200. Violaceae, 201. Flacourtiaceae, 202. Stachyuraceae, 203. Turneraceae, 204. Malesherbiaceae, 205. Passifloraceae, 206. Achariaceae, 207. Datiscaceae, 208. Begoniaceae, 209. Ancistrocladaceae

Order 28. Opuntiales

Family : 210. Cactaceae

Order 29. Myrtiflorae

Families : 211. Geissolomataceae, 212. Penaeaceae, 213. Oliniaceae, 214. Thymeleaceae, 215. Elaeagnaceae, 216. Lythraceae, 217. Sonneratiaceae, 218. Punicaceae, 219. Lecythidaceae, 220. Rhizophoraceae, 221. Nyssaceae, 222. Alangiaceae, 223. Combretaceae, 224. Myrtaceae, 225. Melastomaceae, 226. Onagraceae, 227. Haloragidaceae, 228. Hippuridaceae, 229. Cynomoriaceae

Order 30. Umbelliflorae

Families : 230. Araliaceae, 231. Umbeliferae, 232. Cornaceae

Subclass 2. Metachlamydeae (Sympetalae)

Order : 1. Ericales

Families : 233. Clethraceae, 234. Pyrolaceae, 235. Lemnoaceae, 236. Ericaceae, 237. Epacridaceae, 238. Diapensiaceae

Order 2. Primulales

Families : 239. Theophrastaceae, 240. Myrsinaceae, 241. Primulaceae

Order 3. Plumbaginales

Family : 242. Plumbaginaceae

Order 4. Ebenales

Families : 243. Sapotaceae, 244. Ebenaceae, 245. Symplocaceae, 246. Styracaceae

Order 5. Contortae

Families : 247. Oleaceae, 248. Loganiaceae, 249. Gentianaceae, 250. Apocynaceae, 251. Asclepiadaceae

Order 6. Tubiflorae

Families: 252. Convolvulaceae, 253. Polemoniaceae, 254. Hydrophyllaceae, 255. Boraginaceae, 256. Verbenaceae, 257. Labiata, 258. Nolanaceae, 259. Solanaceae, 260. Scrophulariaceae, 261. Bignoniaceae, 262. Pedaliaceae, 263. Martyniaceae, 264. Orobanchaceae, 265. Gesneriaceae, 266. Columelliaceae, 267. Lentibulariaceae, 268. Globulariaceae, 269. Acanthaceae, 270. Myoporaceae, 271. Phrymaceae

Order 7. Plantaginales

Family : 272. Plantaginaceae

Order 8. Rubiales

Families : 273. Rubiaceae, 274. Caprifoliaceae, 275. Adoxaceae, 276. Valerianaceae, 277. Dipsacaceae

Order 9. Cucurbitales

Family : 278. Cucurbitaceae

Order 10. Campanulatae

Families: 279. Campanulaceae, 280. Goodeniaceae, 281. Brunoniaceae, 282. Stylidiaceae, 283. Calyceraceae, 284. Compositae

7.2.5.3 An outline of Engler and Prantl's system of classification

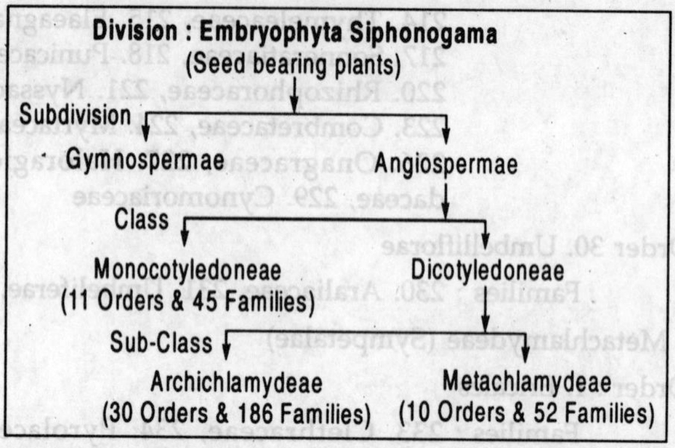

7.2.5.4 Merits of Engler and Prantl's system of classification

(a) Engler's system gives a broad description of the whole plant kingdom and has been excellently arranged into orders and families in a phylogenetic way.

(b) It is very convenient and well known and has been accepted in every part of the world.

(c) The gymnosperms have been treated separately.

(d) Bentham and Hooker's system's Polypetalae and Monochlamydeae have been completely abolished, and have been merged into one single subclass Archichlamydeae in this system.

(e) The treatment of orchids as a more highly evolved group than grasses is satisfactory.

(f) This system provides abundant illustrations and descriptions along with exhaustive keys of families and orders, and the description of each family also contains a summary of embryology, morphology, anatomy and geographical distribution.

7.2.5.5 Demerits of Engler and Prantl's system of classification

(a) This system obscures the phylogeny of angiosperms, which according to Engler, has a polyphyletic origin from an unknown and hypothetical taxon of extinct gymnosperms.

(b) The acceptance of the derivation of dichlamydous flowers from monochlamydous ones is objectionable.

(c) In this system, Monocots have been considered to be more primitive than Dicots and have been placed before Dicots, which does not correspond to the present day knowledge.

(d) Derivation of bisexual flowers from unisexual flowers and parietal placentation from axile placentation are unsatisfactory.

(e) A primitive order Helobiae has been placed between two more advanced orders, the Pandanales and Glumiflorae in this system, which is not satisfactory.

(f) This system places Araceae before Liliaceae, although Araceae are derived from Liliaceae, which is unsatisfactory.

(g) The position of Amentiferae and Centrospermae at the beginning of Dicots before Ranales is unsatisfactory.

7.2.6 Classification system of John Hutchinson (1884-1972)

John Hutchinson (1884-1972)

7.2.6.1 John Hutchinson (1884-1972)

John Hutchinson a British botanist from England, associated with the Royal Botanic Garden, Kew, proposed a phylogenetic system of classification, that in its basic principles somewhat paralleled that of Bessey. He was mainly concerned with the classification of angiosperms. His classification system first appeared in a 2-volume work, *The Families of Flowering Plants* (1926,1934), which underwent several revisions and the final treatment appeared in 1973. In 1969, he published *Evolution and Phylogeny of Flowering Plants.* His other work includes *Genera of Flowering Plants* (1964-1967).

7.2.6.2 John Hutchinson's system of classification

According to Hutchinson's classification, angiosperms were considered monophyletic in origin, which originated from some hypothetical proangiosperms. Initially, the angiosperms were regarded to have evolved along two separate evolutionary lines, Herbaceae (including the predominantly herbaceous families derived from Ranales) and the Lignosae (including fundamentally woody representatives derived from Magnoliales). The Monocotyledons were considered to have derived from the Ranales. Lignosae were further grouped into 54 Orders beginning with Magnoliales and ending with Verbenales while, Herbaceae were further grouped into 28 Orders beginning with Ranales and ending with Lamiales. The Monocotyledons were divided into three distinct groups on the basis of the nature of the perianth - Calyciferae, Corolliferae and Glumiflorae. These were further divided into 29 Orders and 104 Families.

In the first publication (1926), the Dicotyledons included two Subclasses Archichlamydeae (petals free or absent) and Metachlamydeae (petals united). The former were further divided into 59 Orders and 213 Families while the latter were divided into 23 Orders and 51 Families. In the last revision (1973), little alteration was made in the system. Lythrales was transferred from the Herbaceae to the Lignosae, under the Order Myrtales. Little alteration was made in Monocotyledons. Thus, there are 82 Orders and 343 Families of the Dicotyledons and 29 Orders and 69 Families of the Monocotyledons.

(a) Principles of Hutchinson's system of classification

Hutchinson's plant classification was mainly based on the following principles :

(i) Evolution has generally been consistent, and it is both downwards and upwards and does not necessarily involve all organs at the same time.

(ii) The dicots are primitive to the monocots, and the shrubs and trees in any one genus or family are more primitive than herbs. At the same time in any one genus or family, the shrubs and trees are older in comparison with climbers.

(iii) Perennials are older than biennials and annuals.

(iv) The aquatic flowering plants are considered to have been derived from terrestrial ancestors.

(v) Spiral arrangement of vegetative and floral leaves is primitive to cyclic arrangement.

(vi) Simple leaves are more primitive than compound leaves.

(vii) Dioecious plants are more recent than monoecious, and bisexual flowers are primitive than unisexual flowers.

(viii) Aestivation types are evolved from contorted to imbricate to valvate.

(ix) Polymerous flowers (many-parted) precede oligomerous (few-parted) flowers and apetalous flowers have been derived from petalous flowers. Polypetaly in turn is more primitive to gamopetaly.

(x) Actinomorphy and hypogyny is primitive to zygomorphy and perigyny respectively. Epigyny is the most advanced condition.

(xi) Syncarpy is more advanced than apocarpy, while polycarpellary condition is advanced to the condition of a few carpels.

(xii) Flowers with many stamens are primitive to flowers with few stamens. Plants with fused anthers or filaments are advanced to those with separate anthers.

(xiii) Aggregate fruits are advanced to single fruits. As a rule, the capsule precedes the berry or drupe.

(xiv) Endospermic seeds with small embryo are older than non-endospermic seeds with large embryo.

(b) An outline of Hutchinson's system of classification (1959, *Families of Flowering Plants*)

Sub phylum —Dicotyledones

Division 1. Lignosae

Order 1. Magnoliales

Families : 1. Magnoliaceae, 2. Illicidaceae, 3. Winteraceae, 4. Canellaceae, 5. Schisandraceae, 6. Himantandraceae, 7. Lactoridaceae, 8. Trochodendraceae, 9. Cercidiphyllaceae

Order 2. Annonales

Families : 10. Annonaceae, 11. Eupomatiaceae

Order 3. Laurales

Families : 12. Monimiaceae, 13. Austrobaileyaceae, 14. Lauraceae, 15. Trimeniaceae, 16. Hernandiaceae, 17. Gomortegaceae, 18. Myristicaceae

Order 4. Dilleniales

Families : 19. Dilleniaceae, 20. Connaraceae, 21.Crossosomataceae, 22. Brunelliaceae

Order 5. Coriariales

Family : 23. Coriariaceae

Order 6. Rosales

Families : 24. Rosaceae, 25. Dichapetalaceae, 26. Calycanthaceae

Order 7. Leguminales

Families : 27. Caesalpiniaceae, 28. Mimosaceae, 29. Fabaceae (Papilionaceae).

Order 8. Cunoniales

Families : 30. Pterostemonaceae, 31. Cunoniaceae, 32. Philadelphaceae, 33. Hydrangeaceae, 34. Grossulariaceae, 35. Oliniaceae, 36. Greyiaceae, 37. Escalloniaceae, 38. Baueraceae, 39. Crypteroniaceae

Order 9. Styracales

Families : 40. Lissocarpaceae, 41. Styracaceae, 42. Symplocaceae

Order 10. Araliales

Families : 43. Cornaceae, 44. Alangiaceae, 45. Garryaceae, 46. Nyssaceae, 47. Araliaceae, 48. Caprifoliaceae

Order 11. Hamamelidales

Families : 49. Tetracentraceae, 50. Hamamelidaceae, 51. Myrothamnaceae, 52. Platanaceae, 53. Stachyuraceae, 54. Buxaceae, 55. Daphniphyllaceae, 56. Bruniaceae

Order 12. Salicales

Family : 57. Salicaceae

Order 13. Leitneriales

Family : 58. Leitneriaceae

Order 14. Myricales

Family : 59. Myricaceae

Order 15. Balanopales

Family : 60. Balanoporaceae

Order 16. Fagales

Families : 61. Betulaceae, 62. Fagaceae, 63. Corylaceae

Order 17. Juglandales

Families : 64. Rhoipteleaceae, 65. Juglandaceae, 66. Picrodendraceae

Order 18. Casuarinales

Family : 67. Casuarinaceae

Order 19. Urticales

Families : 68. Ulmaceae, 69. Cannabiaceae, 70. Moraceae, 71. Urticaceae, 72. Barbeyaceae, 73. Eucommiaceae

Order 20. Bixales

Families : 74. Bixaceae, 75. Cistaceae, 76. Flacourtiaceae, 77. Cochlospermaceae, 78. Hoplestigmataceae, 79. Achatocarpaceae, 80. Lacistemaceae

Order 21. Thymelaeales

Families : 81. Gonystylaceae, 82. Aaquilariaceae, 83. Geissolomataceae, 84. Penaeaceae, 85. Thymelaeaceae, 86. Nyctaginaceae

Order 22. Proteales

Families : 87. Proteaceae

Order 23. Pittosporales

Families : 88. Pittosporaceae, 89. Byblidaceae, 90. Stegnospermataceae, 91. Vivianiaceae, 92. Tremandraceae

Order 24. Capparales

Families : 93. Capparaceae, 94. Moringaceae, 95. Tovariaceae

Order 25. Tamaricales

Families : 96. Frankeniaceae, 97. Tamaricaceae, 98. Fouquieriaceae

Order 26. Violales

Family : 99. Violaceae

Order 27. Polygalales

Families : 100. Polygalaceae, 101. Krameriaceae, 102. Trigoniaceae, 103. Vochysiaceae.

Order 28. Loasales

Families : 104. Turneraceae, 105. Loasaceae

Order 29. Passiflorales

Families : 106. Malesherbiaceae, 107. Passifloraceae, 108. Achariaceae

Order 30. Cucurbitales

Families : 109. Cucurbitaceae, 110. Begoniaceae, 111. Datiscaceae, 112. Caricaceae

Order 31. Cactales

Family : 113. Cactaceae

Order 32. Tiliales

Families : 114. Dirachmaceae, 115. Scytopetalaceae, 116. Tiliaceae, 117. Sterculiaceae, 118. Peridiscaceae, 119. Bombacaceae

Order 33. Malvales

Family : 120. Malvaceae

Order 34. Malpighiales

Families : 121. Ixonanthaceae, 122. Malpighiaceae, 123. Humiriaceae, 124. Linaceae, 125. Irvingiaceae, 126. Huaceae, 127. Ledocarpaceae, 128. Erythroxylaceae, 129. Ctenolophonaceae, 130. Lepidobotryaceae, 131. Blanitaceae, 132. Zygophyllaceae

Order 35. Euphorbiales

Family : 133. Euphorbiaceae

Order 36. Theales

Families : 134. Bonnetiaceae, 135. Theaceae, 136. Saurauiaceae, 137. Actinidiaceae, 138. Pellicieraceae, 139. Péntaphylaceae, 140. Tetrameristaceae, 141. Marcgraviaceae, 142. Caryocaraceae, 143. Medusagynaceae

Order 37. Ochnales

Families : 144. Straburgeriaceae, 145. Ochnaceae, 146. Sarcolaenaceae, 147. Sphaerosepalaceae, 148. Dipterocarpaceae, 149. Ancistrocladaceae

Order 38. Ericales

Families : 150. Clethraceae, 151. Pyrolaceae, 152. Ericaceae, 153. Epacridaceae, 154. Diapensiaceae, 155. Monotropaceae, 156. Lennoaceae, 157. Vacciniaceae

Order 39. Guttiferales

Families : 158. Hypericaceae, 159. Clusiaceae, 160. Eucryphiaceae, 161. Quiinaceae

Order 40. Myrtales

Families: 162. Myrtaceae, 163. Napoleonaeaceae, 164. Rhizophoraceae, 165. Sonneratiaceae, 166. Punicaceae, 167. Combretaceae, 168. Melastomataceae

Order 41. Celastrales

Families : 169. Pandaceae, 170. Aquifoliaceae, 171. Salvadoraceae, 172. Koeberliniaceae, 173. Cneoraceae, 174. Cardiopteridaceae, 175. Cyrillaceae, 176. Icacinaceae, 177. Empetraceae, 178. Aextoxicaceae, 179. Pentadiplandraceae, 180. Celastraceae, 181. Corynocarpaceae, 182. Stackhousiaceae, 183. Goupiaceae, 184. Hippocateaceae, 185. Erythropalaceae, 186. Syphonodontaceae, 187. Scyphostegiaceae

Order 42. Olacales

Families : 188. Olacaceae, 189. Opiliaceae, 190. Octoknemaceae, 191. Aptandraceae, 192. Dipentodontaceae, 193. Medusandraceae

Order 43. Santalales

Families : 194. Loranthaceae, 195. Grubbiaceae, 196. Santalaceae, 197. Myzodendraceae, 198. Balanophoraceae

Order 44. Rhamnales

Families : 199. Heteropyxidaceae, 200. Elaeagnaceae, 201. Rhamnaceae, 202. Vitaceae

Order 45. Myrsinales

Families : 203. Myrsinaceae, 204. Theophrastaceae, 205. Aegicerataceae

Order 46. Ebenales

Families : 206. Ebaenaceae, 207. Sapotaceae, 208. Sarcospermataceae

Order 47. Rutales

Families : 209. Rutaceae, 210. Simaroubaceae, 211. Burseraceae, 212. Averrhoaceae

Order 48. Meliales

Family : 213. Meliaceae

Order 49. Sapindales

Families : 214. Melianthaceae, 215. Sapindaceae, 216. Podoonaceae, 217. Sabiaceae, 218. Anacardiaceae, 219. Aceraceae, 220. Hippocastanaceae, 221. Staphyleaceae, 222. Akaniaceae, 223. Julianiaceae, 224. Didiereaceae

Order 50. Loganiales

Families : 225. Potaliaceae, 226. Loganiaceae, 227. Buddlejaceae, 228. Antoniaceae, 229. Spigliaceae, 230. Strychnaceae, 231. Oleaceae

Order 51. Apocynales

Families : 232. Plocospermataceae, 233. Apocynaceae, 234. Periplocaceae, 235. Asclepiadaceae

Order 52. Rubiales

Families : 236. Dialypetalanthaceae, 237. Rubiaceae

Order 53. Bignoniales

Families : 238. Cobaeaceae, 239. Bignoniaceae, 240. Pedaliaceae, 241. Martyniaceae

Order 54. Verbenales

Families : 242. Ehretiaceae, 243. Verbenaceae, 244. Stilbaceae, 245. Dicrastylidaceae, 246. Phrymataceae

Division 2. Herbaceae

Order 55. Ranales

Families : 247. Paeoniaceae, 248. Helleboraceae, 249. Ranunculaceae, 250. Nymphaeaceae, 251. Podophyllaceae, 252. Ceratophyllaceae, 253. Cabombaceae

Order 56. Berberidales

Families : 254. Sargentodoxaceae, 255. Lardizabalaceae, 256. Menispermaceae, 257. Nandinaceae, 258. Circaeasteraceae, 259. Berberidaceae

Order 57. Aristolochiales

Families : 260. Aristolochiaceae, 261. Hydnoraceae, 262. Rafflesiaceae, 263. Nepenthaceae

Order 58. Piperales

Families : 264. Piperaceae, 265. Saururaceae, 266. Choranthaceae

Order 59. Rhoeadales

Families : 267. Papaveraceae, 268. Fumariaceae

Order 60. Cruciales (Brassicales)

Families : 269. Cruciferae (Brassicaceae), 270. Cleomaceae

Order 61. Resedales

Family : 271. Resedaceae

Order 62. Caryophyllales

Families : 272. Elatinaceae, 273. Molluginaceae, 274. Caryophyllaceae, 275. Aizoaceae, 276. Portulacaceae

Order 63. Polygonales

Families : 277. Polygonaceae, 278. Illecebraceae

Order 64. Chenopodiales

Families : 279. Barbeuiaceae, 280. Phytolaccaceae, 281. Gyrostemonaceae, 282. Agdestidaceae, 283. Petiveriaceae, 284. Chenopodiaceae, 285. Amaranthaceae, 286. Theligonaceae, 287. Bataceae, 288. Basellaceae

Order 65. Lythrales (Onagrales)

Families : 289. Lythraceae, 290. Onagraceae, 291. Trapaceae, 292. Halorrhagidaceae, 293. Callitrichaceae

Order 66. Gentianales

Families : 294. Gentianaceae, 295. Menyanthaceae

Order 67. Primulales

Families : 296. Primulaceae, 297. Plumbaginaceae

Order 68. Plantaginales

Family : 298. Plantaginaceae

Order 69. Saxifragales

Families : 299. Crassulaceae, 300. Cephalotaceae, 301. Saxifragaceae, 302. Eremosynaceae, 303. Vahliaceae, 304. Francoaceae, 305. Donatiaceae, 306. Parnassiaceae, 307. Adoxaceae

Order 70. Sarraceniales

Families : 308. Droseraceae, 309. Sarraceniaceae

Order 71. Podostemales

Families : 310. Podostemaceae, 311. Hydrostachyaceae

Order 72. Umbellales

Family : 312. Apiaceae (Umbelliferae)

Order 73. Valerianales

Families : 313. Valerianaceae, 314. Dipsacaceae, 315. Calyceraceae

Order 74. Campanales

Families : 316. Campanulaceae, 317. Lobeliaceae

Order 75. Goodeniales

Families : 318. Goodeniaceae, 319. Brunoniaceae, 320. Stylidiaceae

Order 76. Asterales

Family : 321. Asteraceae (Compositae)

Order 77. Solanales

Families : 322. Solanaceae, 323. Convolvulaceae, 324. Nolanaceae

Order 78. Personales

Families : 325. Scrophulariaceae, 326. Acanthaceae, 327. Gesneriaceae, 328. Orobanchaceae, 329. Lentibulariaceae, 330. Columelliaceae

Order 79. Geraniales

Families : 331. Geraniaceae, 332. Limnanthaceae, 333. Oxalidaceae, 334. Tropaeolaceae, 335. Balsaminaceae

Order 80. Polemoniales

Families : 336. Polemoniaceae, 337. Hydrophyllaceae, 338. Cuscutaceae

Order 81. Boraginales

Family : 339. Boraginaceae

Order 82. Lamiales

Families : 340. Myoporaceae, 341. Selaginaceae, 342. Globulariaceae, 343. Lamiaceae (Labiatae)

Sub phylum —Monocotyledones

Division 1. Calyciferae

Order 83. Butomales

Families : 344. Butomaceae, 345. Hydrocharitaceae

Order 84. Alismatales

Families : 346. Alismataceae, 347. Scheuchzeriaceae, 348. Petrosaviaceae

Order 85. Triuridales

Families : 349. Triuridaceae

Order 86. Juncaginales

Families : 350. Juncaginaceae, 351. Lilaeaceae, 352. Posidoniaceae

Order 87. Aponogetonales

Families : 353. Aponogetonaceae, 354. Zosteraceae

Order 88. Potamogetonales

Families : 355. Potamogetonaceae, 356. Ruppiaceae

Order 89. Najadales

Families : 357. Zannichelliaceae, 358. Najadaceae

Order 90. Commelinales

Families : 359. Commelinaceae, 360. Cartonemataceae, 361. Flagellariaceae, 362. Mayacaceae

Order 91. Xyridales

Families : 363. Xyridaceae, 364. Rapateaceae

Order 92. Eriocaulales

Family : 365. Eriocaulaceae

Order 93. Bromeliales

Family : 366. Bromeliaceae

Order 94. Zingiberales

Families : 367. Musaceae, 368. Strelitziaceae, 369. Lowiaceae, 370. Zingiberaceae, 371. Cannaceae, 372. Marantaceae

Division 2. Corolliferae

Order 95. Liliales

Families : 373. Liliaceae, 374. Tecophilaeaceae, 375. Trilliaceae, 376. Potenderiaceae, 377. Smilacaceae, 378. Ruscaceae

Order 96. Alstroemeriales

Families : 379. Alstroemeriaceae, 380. Petermanniaceae, 381. Philesiaceae

Order 97. Arales

Families : 382. Araceae, 383. Lemnaceae

Order 98. Typhales

Families : 384. Sparganiaceae, 385. Typhaceae

Order 99. Amaryllidales

Family : 386. Amaryllidaceae

Order 100. Iridales

Family : 387. Iridaceae

Order 101. Dioscoreales

Families : 388. Stenomeridaceae, 389. Trichopodaceae, 390. Roxburghiaceae, 391. Dioscoreaceae

Order 102. Agavales

Families : 392. Xanthorrhoeaceae, 393. Agavaceae

Order 103. Palmales

Family : 394. Palmae (Arecaceae)

Order 104. Pandanales

Family : 395. Pandanaceae

Order 105. Cyclanthales

Family : 396. Cyclanthaceae

Order 106. Haemodorales

Families : 397. Haemodoraceae, 398. Hypoxidaceae, 399. Velloziaceae, 400. Apostasiaceae, 401. Taccaceae, 402. Philydraceae

Order 107. Burmannniales

Families : 403. Burmanniaceae, 404. Thismiaceae, 405. Corsiaceae

Order 108. Orchidales

Family : 406. Orchidaceae

Division 3. Glumiflorae

Order 109. Juncales

Families : 407. Juncaceae, 408. Thurniaceae, 409. Centrolepidaceae, 410. Retionaceae

Order 110. Cyperales

Family : 411. Cyperaceae

Order 111. Graminales

Family : 412. Gramineae (Poaceae)

7.2.6.3 Linear sequence of orders in Hutchinson's system of classification (1973)

ERBENALES (5)
BIGNONIALES (4)
RUBIALES (2)
APOCYNALES (4)
LOGANIALES (7)
SAPINDALES (11)
MELIALES (1)
RUTALES (4)
EBENALES (3)
MYRSINALES (3)
RHAMNALES (4)
SANTALALES (5)
OLACALES (6)
CELASTRALES (19)
MYRTALES (7)
GUTTIIFERALES (4)
ERICALES (8)
OCHNALES (6)
THEALES (10)
EUPHORBIALES (1)
MALPIGHIALES (12)
MALVALES (1)
TILIALES (6)
CACTALES (1)
CUCURBITALES (4)
PASSIFLORALES (3)
LOASALES (2)
POLYGALALES (4)
VIOLALES (1)
TAMARICALES (3)
CAPPARALES (3)
PITTOSPORALES (5)
PROTEALES (1)
THYMELEALES (6)
BIXALES (7)
URTICALES (6)
CASUARINALES (1)
JUGLANDALES (3)
FAGALES (3)
BALANOPSIDALES (1)
MYRICALES (1)
LEITNERIALES (1)
SALICALES (1)
HAMAMELIDALES (8)
ARALIALES (5)
STYRACALES (3)
CUNONIALES (10)
LEGUMINALES (3)
ROSALES (3)
CORIARIALES (1)
DILLENIALES (4)
LAURALES (7)
ANNONALES (2)
MAGNOLIALES (9)

Figure gives the linear sequence of orders in Hutchinson's system (1973). Figures in parentheses indicate the number of families. In the phylogenetic tree, Hutchinson did not derive any order directly from the other but from its ancestral stock. This is more safe and acceptable to believe any present-day order to be directly ancestral to any other. It is also evident from the number of families that, their delimitation was done very critically and therefore their relationships appear to be more natural.

LAMIALES (4)
BORAGINALES (1)
POLEMONIALES (3)
GERANIALES (5)
PERSONALES (6)
SOLANALES (3)
ASTERALES (1)
GOODENIALES (3)
CAMPANALES (2)
VALERIANALES (3)
UMBELLALES (1)
PODOSTEMALES (2)
SARRACENIALES (2)
SAXIFRAGALES (9)
PLANTAGINALES (1)
PRIMULALES (2)
GENTIANALES (2)
LYTHRALES (5)
CHENOPODIALES (10)
POLYGONALES (2)
CARYOPHYLLALES (5)
RESEDALES (1)
CRUCIALES (2)
RHOEADALES (2)
PIPERALES (3)
ARISTOLOCHIALES (4)
BERBERIDALES (6)
RANALES (7)

GRAMINALES (1)
CYPERALES (1)
JUNCALES (4)
ORCHIDALES (1)
BURMANNIALES (3)
HAEMODORALES (6)
CYCLANTHALES (1)
PANDANALES (1)
PALMALES (1)
AGAVALES (2)
DIOSCOREALES (4)
IRIDALES (1)
AMARYLLIDALES (1)
TYPHALES (2)
ARALES (2)
ALSTROEMERIALES (3)
LILIALES (6)
ZINGIBERALES (6)
BROMELIALES (1)
ERIOCAULALES (1)
XYRIDALES (2)
COMMELINALES (4)
NAJADALES (2)
POTAMOGETONALES (2)
APONOGETONALES (2)
JUNCAGINALES (3)
TRIURIDALES (1)
ALISMATALES (3)
BUTOMALES (2)

LIGNOSAE — HERBACEAE
DICOTYLEDONS — MONOCOTYLEDONS
HYPOTHETICAL PROANGIOSPERMS

7.2.6.4 An outline of Hutchinson's system of classification

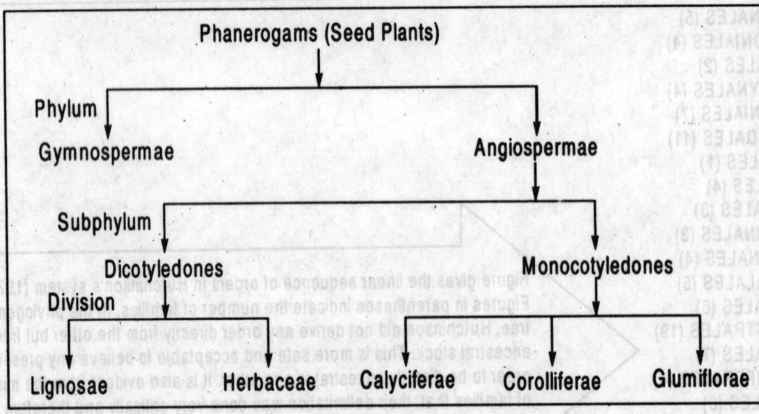

7.2.6.5 Merits of Hutchinson's system of classification

(a) Although Hutchinson's system has been followed widely, it has provided a sound basis for the later phylogenetic systems of Oswald Tippo, Cronquist, Takhtajan and others. It is in conformity with the modern views of the phylogeny of angiosperms, as it considers the Ranales and Magnoliales as the starting points among dicots.

(b) The families and orders are of small size and comprise only very closely related taxa. The arrangement of families within the monocots has been particularly appreciated the world over.

(c) In this system, the monocots have been discussed after dicots.

7.2.6.6 Demerits of Hutchinson's system of classification

(a) In this system the dicots have been divided into two major groups (Lignosae and Herbaceae) on the basis of their woody herbaceous evolutionary lines. This is strongly objectionable, as later studies by Bailey, Sinnott and others on overall resemblances and critical anatomical studies have shown that, herbaceous habit has been derived from woody habit and the reverse has also occurred occasionally. Therefore, the two evolutionary lines cannot be considered distinct.

(b) The primary division on the basis of woody or herbaceous habitat has resulted into wide separation of families, which resemble one another rather closely on the basis of floral structure. Several herbaceous families in the Ranales are clearly related to the woody members of Magnoliales. For example, herbaceous Umbelliferae is more closely related to and probably derived from the woody Cornaceae and Araliaceae. Similarly, herbaceous Cruciferae is clearly derived from the predominantly woody Capparidaceae most probably via Cleomaceae.

(c) This system is not of much utility from the point of view of plant classification.

7.2.7 Classification system of Armen Takhtajan (1910-)

Armen Takhtajan (1910-)

7.2.7.1 Armen Takhtajan (1910-)

Armen Takhtajan who is the Head of the Department of Higher plants at the Komarov Botanical Institute in Leningrad, Russia, presented a system of classification, which was first published in a very preliminary form in 1942. The actual system however appeared in 1959 in German in his book *Die Evolution der Angiospermen.* After 1966, this system has gone through a series of modifications and elaborations, but the nucleus of the system and the basic features remained essentially the same. It was published in English in 1969 in *Flowering Plants: Origin and Dispersal.* A new version of the system has been published in *Botanical Review* recently in 1980. Takhtajan was inspired by Hailier's attempt to create a synthetic evolutionary classification of flowering plants based on Darwinian philosophy and his approach is very much similar to that of Cronquist.

7.2.7.2 Armen Takhtajan's system of classification

Takhtajan's system of classification is basically of Bessey-Hallier tradition which considered all evidences from different fields including morphological, anatomical, embryological, cytological, palynological, paleobotanical, chemical and ultrastructural evidences while classifying Angiosperms. He considered Angiosperms to be monophyletic in origin and that they arose from some ancient group of Gymnosperms. The Monocotyledons are, considered to have been derived from primitive Dicotyledons

Takhtajan divided the Magnoliophyta (Angiosperms) into two Classes – Magnoliopsida (Dicotyledons) and Liliopsida (Monocotyledons), of which Magnoliopsida is considered primitive and Liliopsida to have been derived from Magnoliales under Magnoliopsida. The two Classes have been further divided into 10 Subclasses, 7 under the Magnoliopsida and 3 under Liliopsida (1980), which are as follows :

Class : Magnoliopsida (Dicots)	Class : Liliopsida (Monocots)
Magnoliidae	Alismatidae
Ranunculidae	Liliidae
Hamamelididae	Arecidae
Caryophyllidae	
Dilleniidae	
Rosidae	
Asteridae	

However according to Takhtajan's system of classification, 1997, the two Classes have been divided into 15 Subclasses, 11 under the Magnoliopsida and 5 under Liliopsida, which are as follows :

Class : Magnoliopsida (Dicots)	Class : Liliopsida (Monocots)
Magnoliidae	Liliidae
Nymphaeidae	Commelinidae
Nelumbonidae	Arecidae
Ranunculidae	Alismatidae
Caryophyllidae	Triuridae
Hamamelididae	Aridae
Dilleniidae	
Rosidae	
Cornidae	
Asteridae	
Lamiidae	

Among the Subclasses, Takhtajan considered Magnoliidae to be the most primitive, forming the basal group from which all other Subclasses have been derived (Fig. 7.2). Among the Magnoliopsida, he considered the Asteridae the most advanced. The three Subclasses under Liliopsida (Alismatadae, Liliidae and Arecidae) are considered to be primitive than the other Subclasses of Magnoliopsida, and have been derived from Magnoliidae. Among the Angiosperms, the Winteraceae family is most primitive and Poaceae the most advanced.

Takhtajan's system is based on 67 phyletic principles. Some of the important criteria used by him to evaluate the relative degree of advancement of flowering plants are as follows:

1. Woody plants are primitive than herbaceous plants.
2. Deciduous woody plants are considered evolved from evergreen plants.
3. Xylem fibres evolved from tracheids to libriform fibres, through fibre tracheids.
4. Trilacunar or pentalacunar nodes are primitive to unilacunar nodes.
5. Alternate leaf arrangement is primitive while parallel venation is most advanced.
6. Stomata with subsidiary cells are primitive than those lacking subsidiary cells.
7. Cymose inflorescence is primitive than racemose.
8. Flowers with an indefinite or variable number of floral parts are primitive.
9. Pollen with unsculptured exine is primitive to sculptured pollen.
10. Apocarpous gynoecium is primitive.
11. Bitegmic ovules are primitive than unitegmic ovules, while anatropous ovule is most primitive.
12. 8-nucleated Polygonum type female gametophyte is most primitive.
13. Mesogamic and Chalazogamic conditions have evolved from Porogamic condition.
14. Among fruits many-seeded follicle is most primitive.

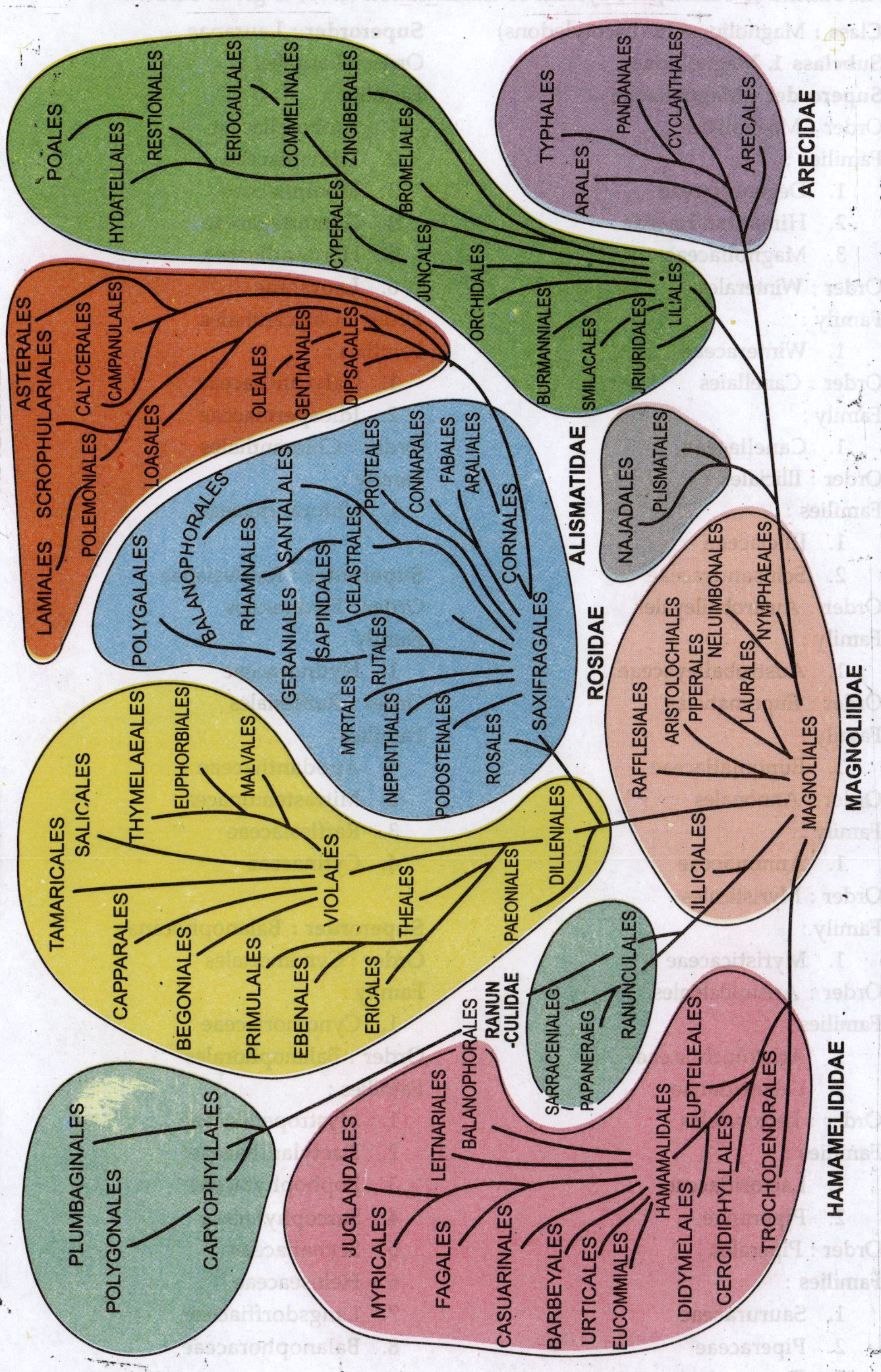

Fig. 7.2 : Takhtajan's (1980) chart showing putative relationships among the Classes, Subclasses, and Orders of flowering plants

An outline of Takhtajan's system of classification (1997) is given below :

Class : Magnoliopsida (Dicotyledons)

Subclass 1. Magnoliidae

Superorder : Magnolianae

Order : Magnoliales

Families :

1. Degeneriaceae
2. Himantandraceae
3. Magnoliaceae

Order : Winterales

Family :

1. Winteraceae

Order : Canellales

Family :

1. Canellaceae

Order : Illiciales

Families :

1. Illiciaceae
2. Schisandraceae

Order : Austrobaileyales

Family :

1. Austrobaileyaceae

Order : Eupomatiales

Family :

1. Eupomatiaceae

Order : Annonales

Family :

1. Annonaceae

Order : Myristicales

Family :

1. Myristicaceae

Order : Aristolochiales

Families :

1. Aristolochiaceae
2. Lactoridanae

Order : Lactoridales

Families :

1. Lactoridaceae
2. Piperanae

Order : Piperales

Families :

1. Saururaceae
2. Piperaceae
3. Peperomiaceae

Superorder : Lauranae

Order : Laurales

Families :

1. Amborellaceae
2. Trimeniaceae
3. Monimiaceae
4. Gomortegaceae
5. Hernandiaceae
6. Lauraceae

Order : Calycanthales

Families :

1. Calycanthaceae
2. Idiospermaceae

Order : Chloranthales

Family :

1. Chloranthaceae

Superorder : Rafflesianae

Order : Hydnorales

Family :

1. Hydnoraceae

Order : Rafflesiales

Families :

1. Apodanthaceae
2. Mitrastemonaceae
3. Rafflesiaceae
4. Cytinaceae

Superorder : Balanophoranae

Order : Cynomoriales

Family :

1. Cynomoriaceae

Order : Balanophorales

Families :

1. Mystropetalaceae
2. Dactylanthaceae
3. Lophophytaceae
4. Sarcophytaceae
5. Scybaliaceae
6. Heloseaceae
7. Langsdorffiaceae
8. Balanophoraceae

Subclass 2. Nymphaeidae

Superorder : Nymphaeanae

Order : Hydropeltidales

Families :

1. Hydropeltidaceae
2. Cabombaceae

Order : Nymphaeales

Families :

1. Nupharaceae
2. Nymphaeaceae
3. Barclayaceae

Superorder : Ceratophyllanae

Order : Ceratophyllales

Family :

1. Ceratophyllaceae

Subclass 3. Nelumbonidae

Superorder : Nelumbonanae

Order : Nelumbonales

Family :

1. Nelumbonaceae

Subclass 4. Ranunculidae

Superorder : Ranunculanae

Order : Lardizabales

Families :

1. Lardizabalaceae
2. Sargentodoxaceae

Order : Menispermales

Family :

1. Menispermaceae

Order : Berberidales

Families :

1. Nandinaceae
2. Berberidaceae
3. Ranzaniaceae
4. Podophyllaceae

Order : Ranunculales

Family :

1. Ranunculaceae

Order : Circaeasterales

Families :

1. Kingdoniaceae
2. Circaeasteraceae

Order : Hydrastidales

Family :

1. Hydrastidaceae

Order : Glaucidiales

Family :

1. Glaucidiaceae

Order : Paeoniales

Family :

1. Paeoniaceae

Order : Papaverales

Families :

1. Papaveraceae
2. Pteridophyllaceae
3. Hypecoaceae
4. Fumariaceae

Subclass 5. Caryophyllidae

Superorder : Caryophyllanae

Order : Caryophyllales

Families :

1. Phytolaccaceae
2. Gisekiaceae
3. Agdestidaceae
4. Barbeuiaceae
5. Achatocarpaceae
6. Petiveriaceae
7. Nyctaginaceae
8. Aizoaceae
9. Sesuviaceae
10. Tetragoniaceae
11. Stegnospermataceae
12. Portulacaceae
13. Hectorellaceae
14. Basellaceae
15. Halophytaceae
16. Cactaceae
17. Didiereaceae
18. Molluginaceae
19. Caryophyllaceae
20. Amaranthaceae
21. Chenopodiaceae

Superorder : Gyrostemonanae

Order : Gyrostemonales

Family :

1. Gyrostemonaceae

Superorder : Polygonanae

Order : Polygonales

Family :

1. Polygonaceae

Superorder : Plumbaginanae

Order : Plumbaginales

Family :

1. Plumbaginaceae

Subclass 6. Hamamelididae

Superorder : Trochodendranae

Order : Trochodendrales

Families :

1. Trochodendraceae
2. Tetracentraceae

Order : Cercidiphyllales

Families :

1. Cercidiphyllaceae
2. Eupteleales
3. Eupteleaceae

Superorder : Myrothamnanae

Order : Myrothamnales

Family :

1. Myrothamnaceae

Superorder : Hamamelidanae

Order : Hamamelidales

Families :

1. Hamamelidaceae
2. Altingiaceae
3. Platanaceae

Superorder : Barbeyanae

Order : Barbeyales

Family :

1. Barbeyaceae

Superorder : Daphniphyllanae

Order : Daphniphyllales

Family :

1. Daphniphyllaceae

Order : Balanopales

Family :

1. Balanopaceae

Superorder : Buxanae

Order : Didymelales

Family :

1. Didymelaceae

Order : Buxales

Family :

1. Buxaceae

Order : Simmondsiales

Family :

1. Simmondsiaceae

Superorder : Faganae

Order : Fagales

Families :

1. Fagaceae
2. Nothofagaceae

Order : Corylales

Families :

1. Betulaceae
2. Corylaceae
3. Ticodendraceae

Superorder : Casuarinanae

Order : Casuarinales

Family :

1. Casuarinaceae

Superorder : Juglandanae

Order : Myricales

Family :

1. Myricaceae

Order : Rhoipteleales

Family :

1. Rhoipteleaceae

Order : Juglandales

Family :

1. Juglandaceae

Subclass 7. Dilleniidae
Superorder : Dillenianae
Order : Dilleniales
Family :
1. Dilleniaceae
Superorder : Theanae
Order : Paracryphiales
Family :
1. Paracryphiaceae
Order : Theales
Families :
1. Stachyuraceae
2. Theaceae
3. Asteropeiaceae
4. Pentaphylacaceae
5. Tetrameristaceae
6. Oncothecaceae
7. Marcgraviaceae
8. Caryocaraceae
9. Pellicieraceae
Order : Hypericales
Families :
1. Bonnetiaceae
2. Clusiaceae
3. Hypericaceae
Order : Physenales
Family :
1. Physenaceae
Order : Medusagynales
Family :
1. Medusagynaceae
Order : Ochnales
Families :
1. Strasburgeriaceae
2. Ochnaceae
3. Sauvagesiaceae
4. Lophiraceae
5. Quiinaceae
6. Scytopetalaceae
Order : Elatinales
Family :
1. Elatinaceae

Order : Ancistrocladales
Family :
1. Ancistrocladaceae
Order : Dioncophyllales
Family :
1. Dioncophyllaceae
Order : Lecythidales
Families :
1. Barringtoniaceae
2. Lecythidaceae
3. Napoleonaeaceae
4. Foetidiaceae
5. Asteranthaceae
Superorder : Sarracenianae
Order : Sarraceniales
Family :
1. Sarraceniaceae
Superorder : Nepenthanae
Order : Nepenthales
Family :
1. Nepenthaceae
Order : Droserales
Family :
1. Droseraceae
Superorder : Ericanae
Order : Actinidiales
Family :
1. Actinidiaceae
Order : Ericales
Families :
1. Clethraceae
2. Cyrillaceae
3. Ericaceae
4. Epacridaceae
5. Empetraceae
Order : Diapensiales
Family :
1. Diapensiaceae
Order : Bruniales
Family :
1. Bruniaceae

Order : Geissolomatales
Family :
1. Geissolomataceae

Order : Fouquieriales
Family :
1. Fouquieriaceae

Superorder : Primulanae

Order : Styracales
Families :
1. Styracaceae
2. Symplocaceae
3. Ebenaceae
4. Lissocarpaceae

Order : Sapotales
Family :
1. Sapotaceae

Order : Myrsinales
Families :
1. Myrsinaceae
2. Theophrastaceae

Order : Primulales
Family :
1. Primulaceae

Superorder : Violanae

Order : Violales
Families :
1. Berberidopsidaceae
2. Aphloiaceae
3. Bembiciaceae
4. Flacourtiaceae
5. Lacistemataceae
6. Peridiscaceae
7. Violaceae
8. Dipentodontaceae
9. Scyphostegiaceae

Order : Passiflorales
Families :
1. Passifloraceae
2. Turneraceae
3. Malesherbiaceae
4. Achariaceae

Order : Caricales
Family :
1. Caricaceae

Order : Salicales
Family :
1. Salicaceae

Order : Tamaricales
Families :
1. Reaurmuriaceae
2. Tamaricaceae
3. Frankeniaceae

Order : Cucurbitales
Family :
1. Cucurbitaceae

Order : Begoniales
Families :
1. Datiscaceae
2. Tetramelaceae

Order : Capparales
Families :
1. Capparaceae
2. Pentadiplandraceae
3. Koeberliniaceae
4. Brassicaceae
5. Tovariaceae
6. Resedaceae

Order : Moringales
Family :
1. Moringaceae

Order : Batales
Family :
1. Bataceae

Superorder : Malvanae

Order : Cistales
Families :
1. Bixaceae
2. Cochlospermaceae
3. Cistaceae

Order : Elaeocarpales
Family :
1. Elaeocarpaceae

Order : Malvales

Families :

1. Tiliaceae
2. Dirachmaceae
3. Monotaceae
4. Dipterocarpaceae
5. Sarcolaenaceae
6. Plagiopteraceae
7. Huaceae
8. Sterculiaceae
9. Diegodendraceae
10. Sphaerosepalaceae
11. Bombacaceae
12. Malvaceae

Superorder : Urticanae

Order : Urticales

Families :

1. Ulmaceae
2. Moraceae
3. Cannabaceae
4. Cecropiaceae
5. Urticaceae

Superorder : Euphorbianae

Order : Euphorbiales

Families :

1. Euphorbiaceae
2. Pandaceae
3. Aextoxicaceae
4. Dichapetalaceae

Order : Thymelaeales

Families :

1. Gonystylaceae
2. Thymelaeaceae

Subclass 8. Rosidae

Superorder : Saxifraganae

Order : Cunoniales

Families :

1. Cunoniaceae
2. Davidsoniaceae
3. Eucryphiaceae
4. Brunelliaceae

Order : Saxifragales

Families :

1. Tetracarpaeaceae
2. Penthoraceae
3. Crassulaceae
4. Saxifragaceae
5. Grossulariaceae
6. Pterostemonaceae
7. Iteaceae
8. Eremosynaceae
9. Vahliaceae

Order : Cephalotales

Family :

1. Cephalotaceae

Order : Greyiales

Family :

1. Greyiaceae

Order : Francoales

Family :

1. Francoaceae

Order : Haloragales

Family :

1. Haloragaceae

Order : Podostemales

Family :

1. Podostemaceae

Order : Gunnerales

Family :

1. Gunneraceae

Superorder : Rosanae

Order : Rosales

Families :

1. Rosaceae
2. Neuradaceae

Order : Crossosomatales

Family :

1. Crossosomataceae

Order : Chrysobalanales

Family :

1. Chrysobalanaceae

Superorder : Rhizophoranae

Order : Anisophylleales

Family :

1. Anisophylleaceae

Order : Rhizophorales

Family :

1. Rhizophoraceae

Superorder : Myrtanae

Order : Myrtales

Families :

1. Alzateaceae
2. Rhynchocalycaceae
3. Penaeaceae
4. Oliniaceae
5. Combretaceae
6. Crypteroniaceae
7. Memecylaceae
8. Melastomataceae
9. Lythraceae
10. Punicaceae
11. Duabangaceae
12. Sonneratiaceae
13. Onagraceae
14. Trapaceae
15. Psiloxylaceae
16. Heteropyxidaceae
17. Myrtaceae

Superorder : Fabanae

Order : Fabales

Family :

1. Fabaceae

Superorder : Rutanae

Order : Sapindales

Families :

1. Staphyleaceae
2. Tapisciaceae
3. Melianthaceae
4. Sapindaceae
5. Hippocastanaceae
6. Aceraceae
7. Bretschneideraceae
8. Akaniaceae

Order : Tropaeolales

Family :

1. Tropaeolaceae

Order : Sabiales

Families :

1. Sabiaceae
2. Meliosmaceae

Order : Connarales

Family :

1. Connaraceae

Order : Rutales

Families :

1. Rutaceae
2. Rhabdodendraceae
3. Cneoraceae
4. Simaroubaceae
5. Surianaceae
6. Irvingiaceae
7. Kirkiaceae
8. Ptaeroxylaceae
9. Tepuianthaceae
10. Meliaceae

Order : Leitneriales

Family :

1. Leitneriaceae

Order : Coriariales

Family :

1. Coriariaceae

Order : Burserales

Families :

1. Burseraceae
2. Anacardiaceae
3. Podoaceae

Superorder : Geranianae

Order : Linales

Families :

1. Hugoniaceae
2. Linaceae
3. Ctenolophonaceae
4. Ixonanthaceae
5. Humiriaceae
6. Erythroxylaceae

Order : Oxalidales
Families :
1. Oxalidaceae
2. Lepidobotryaceae

Order : Geraniales
Families :
1. Hypseocharitaceae
2. Vivianiaceae
3. Geraniaceae
4. Ledocarpaceae
5. Rhynchothecaceae

Order : Biebersteiniales
Family :
1. Biebersteiniaceae

Order : Balsaminales
Family :
1. Balsaminaceae

Order : Zygophyllales
Families :
1. Zygophyllaceae
2. Peganaceae
3. Balanitaceae
4. Nitrariaceae
5. Tetradiclidaceae

Order : Vochysiales
Families :
1. Malpighiaceae
2. Trigoniaceae
3. Vochysiaceae
4. Tremandraceae

Order : Polygalales
Families :
1. Polygalaceae
2. Xanthophyllaceae
3. Emblingiaceae

Superorder : Corynocarpanae

Order : Corynocarpales
Family :
1. Corynocarpaceae

Superorder : Celastranae

Order : Brexiales
Families :
1. Ixerbaceae
2. Brexiaceae
3. Rousseaceae

Order : Parnassiales
Families :
1. Parnassiaceae
2. Lepuropetalaceae

Order : Celastrales
Families :
1. Goupiaceae
2. Celastraceae
3. Lophopyxidaceae
4. Stackhousiaceae

Order : Salvadorales
Family :
1. Salvadoraceae

Order : Icacinales
Families :
1. Aquifoliaceae
2. Phellinaceae
3. Icacinaceae
4. Sphenostemonaceae

Order : Metteniusales
Family :
1. Metteniusaceae

Order : Cardiopteridales
Family :
1. Cardiopteridaceae

Superorder : Santalanae

Order : Medusandrales
Family :
1. Medusandraceae

Order : Santalales
Families :
1. Olacaceae
2. Opiliaceae
3. Aptandraceae
4. Octoknemaceae
5. Santalaceae
6. Misodendraceae
7. Loranthaceae
8. Viscaceae
9. Eremolepidaceae

Superorder : Rhamnanae

Order : Rhamnales
Family :
1. Rhamnaceae

Order : Elaeagnales
Family :
1. Elaeagnaceae
Order : Proteales
Family :
1. Proteaceae
Superorder : Vitanae
Order : Vitales
Families :
1. Vitaceae
2. Leeaceae
Subclass 9. Cornidae
Superorder : Cornanae
Order : Hydrangeales
Families :
1. Escalloniaceae
2. Hydrangeaceae
3. Abrophyllaceae
4. Argophyllaceae
5. Corokiaceae
6. Alseuosmiaceae
7. Carpodetaceae
8. Phyllonomaceae
9. Pottingeriaceae
10. Tribelaceae
11. Melanophyllaceae
12. Montiniaceae
13. Kaliphoraceae
14. Columelliaceae
Order : Desfontainiales
Family :
1. Desfontainiaceae
Order : Roridulales
Family :
1. Roridulaceae
Order : Cornales
Families :
1. Davidiaceae
2. Nyssaceae
3. Mastixiaceae
4. Curtisiaceae
5. Cornaceae
6. Alangiaceae

Order : Garryales
Family :
1. Garryaceae
Order : Aucubales
Family :
1. Aucubaceae
Order : Griseliniales
Family :
1. Griseliniaceae
Order : Eucommiales
Family :
1. Eucommiaceae
Order : Aralidiales
Family :
1. Aralidiaceae
Order : Torricelliales
Family :
1. Torricelliaceae
Superorder : Aralianae
Order : Helwingiales
Family :
1. Helwingiaceae
Order : Araliales
Families :
1. Araliaceae
2. Hydrocotylaceae
3. Apiaceae
Order : Pittosporales
Family :
1. Pittosporaceae
Order : Byblidales
Family :
1. Byblidaceae
Superorder : Dipsacanae
Order : Viburnales
Family :
1. Viburnaceae
Order : Adoxales
Families :
1. Sambucaceae
2. Adoxaceae

Order : Dipsacales

Families :

1. Caprifoliaceae
2. Valerianaceae
3. Triplostegiaceae
4. Dipsacaceae
5. Morinaceae

Subclass 10. Asteridae

Superorder : Campanulanae

Order : Campanulales

Families :

1. Pentaphragmataceae
2. Sphenocleaceae
3. Campanulaceae
4. Cyphocarpaceae
5. Nemacladaceae
6. Cyphiaceae
7. Lobeliaceae

Order : Goodeniales

Families :

1. Brunoniaceae
2. Goodeniaceae

Order : Stylidales

Families :

1. Donatiaceae
2. Stylidiaceae

Order : Menyanthales

Family :

1. Menyanthaceae

Superorder : Asteranae

Order : Calycerales

Family :

1. Calyceraceae

Order : Asterales

Family :

1. Asteraceae

Subclass 11. Lamiidae

Superorder : Gentiananae

Order : Gentianales

Families :

1. Gelsemiaceae
2. Loganiaceae
3. Strychnaceae
4. Antoniaceae
5. Spigeliaceae
6. Gentianaceae
7. Saccifoliaceae
8. Geniostomaceae
9. Plocospermataceae

Order : Rubiales

Families :

1. Dialypetalanthaceae
2. Rubiaceae
3. Theligonaceae
4. Carlemanniaceae

Order : Apocynaceae

Family :

1. Apocynaceae

Superorder : Solananae

Order : Solanales

Families :

1. Solanaceae
2. Sclerophylacaceae
3. Duckeodendraceae
4. Goetzeaceae

Order : Convolvulales

Families :

1. Convolvulaceae
2. Cuscutaceae

Order : Polemoniales

Family :

1. Polemoniaceae

Order : Boraginales
Families :
1. Hydrophyllaceae
2. Boraginaceae
3. Tetrachondraceae
4. Hoplestigmataceae

Order : Limnanthales
Family :
1. Limnanthaceae

Order : Loasales
Family :
1. Loasaceae

Superorder : Oleanae
Order :Oleales
Family :
1. Oleaceae

Superorder : Lamianae
Order : Scrophulariales
Families :
1. Buddlejaceae
2. Retziaceae
3. Stilbaceae
4. Scrophulariaceae
5. Oftiaceae
6. Globulariaceae
7. Gesneriaceae
8. Plantaginaceae
9. Bignoniaceae
10. Pedaliaceae
11. Martyniaceae
12. Trapellaceae
13. Myoporaceae
14. Acanthaceae
15. Lentibulariaceae

Order : Lamiales
Families :
1. Verbenaceae
2. Phrymaceae
3. Cyclocheilaceae
4. Symphoremataceae
5. Avicenniaceae
6. Viticaceae
7. Lamiaceae

Order : Callitrichales
Family :
1. Callitrichaceae

Order : Hydrostachyales
Family :
1. Hydrostachyaceae

Order : Hippuridales
Family :
1. Hippuridaceae

Class : Liliopsida (Monocotyledons)
Subclass 1. Liliidae
Superorder : Lilianae
Order : Melanthiales
Families :
1. Tofieldiaceae
2. Melanthiaceae
3. Japonoliriaceae
4. Xerophyllaceae
5. Nartheciaceae
6. Heloniadaceae
7. Chionographidaceae
8. Colchicales
9. Tricyrtidaceae
10. Burchardiaceae
11. Uvulariaceae
12. Campynemataceae
13. Scoliopaceae
14. Colchicaceae
15. Calochortaceae

Order : Trilliales
Family :
1. Trillaceae

Order : Liliales
Families :
1. Liliaceae
2. Medeolaceae

Order : Alstroemeriales
Family :
1. Alstroemeriaceae

Order : Iridales
Families :
1. Isophysidaceae
2. Geosiridaceae
3. Iridaceae

Order : Tecophilaeales

Families :

1. Ixioliriaceae
2. Lanariaceae
3. Walleriaceae
4. Tecophilaeaceae
5. Cyanastraceae
6. Eriospermaceae

Order : Burmanniales

Families :

1. Burmanniaceae
2. Thismiaceae
3. Corsiaceae

Order : Hypoxidales

Family :

1. Hypoxidaceae

Order : Orchidales

Family :

1. Orchidaceae

Order : Amaryllidales

Families :

1. Hemerocallidaceae
2. Hyacinthaceae
3. Alliaceae
4. Hesperocallidaceae
5. Hostaceae
6. Agavaceae
7. Amaryllidaceae

Order : Asparagales

Families :

1. Convallariaceae
2. Ophiopogonaceae
3. Ruscaceae
4. Asparagaceae
5. Dracaenaceae
6. Nolinaceae
7. Blandfordiaceae
8. Herreriaceae
9. Phormiaceae
10. Dianellaceae
11. Doryanthaceae
12. Asteliaceae
13. Asphodelaceae
14. Aloaceae
15. Anthericaceae
16. Aphyllanthaceae

Order : Xanthorrhoeales

Families :

1. Baxteriaceae
2. Lomandraceae
3. Dasypogonaceae
4. Calectasiaceae
5. Xanthorrhoeaceae

Order : Hanguanales

Family :

1. Hanguanaceae

Superorder : Dioscoreanae

Order : Stemonales

Families :

1. Stemonaceae
2. Croomiaceae
3. Pentastemonaceae

Order : Smilacales

Families :

1. Luzuriagaceae
2. Philesiaceae
3. Rhipogonaceae
4. Smilacaceae
5. Petermanniaceae

Order : Dioscoreales

Families :

1. Stenomeridaceae
2. Trichopodaceae
3. Avetraceae
4. Dioscoreaceae

Order : Taccales

Family :

1. Taccaceae

Subclass 2. Commelinidae

Superorder : Bromelianae

Order : Bromeliales

Family :

1. Bromeliaceae

Order : Velloziales

Family :

1. Velloziaceae

Superorder : Pontederianae

Order : Philydrales

Family :

1. Philydraceae

Order : Pontederiales
Family :
1. Pontederiaceae

Order : Haemodorales
Families :
1. Haemodoraceae
2. Conostylidaceae

Superorder : Zingiberanae
Order : Musales
Families :
1. Strelitziaceae
2. Musaceae
3. Heliconiaceae

Order : Lowiales
Family :
1. Lowiaceae

Order : Zingiberales
Families :
1. Zingiberaceae
2. Costaceae

Order : Cannales
Families :
1. Cannaceae
2. Marantaceae

Superorder : Commelinanae
Order : Commelinales
Family :
1. Commelinaceae

Order : Mayacales
Family :
1. Mayacaceae

Order : Xyridales
Family :
1. Xyridaceae

Order : Rapateales
Family :
1. Rapateaceae

Order : Eriocaulales
Family :
1. Eriocaulaceae

Superorder : Hydatellanae
Order : Hydatellales
Family :
1. Hydatellaceae

Superorder : Juncanae
Order : Juncales
Families :
1. Juncaceae
2. Thurniaceae

Order : Cyperales
Family :
1. Cyperaceae

Superorder : Poanae
Order : Flagellariales
Family :
1. Flagellariaceae

Order : Restionales
Families :
1. Joinvilleaceae
2. Restionaceae
3. Anarthriaceae
4. Ecdeiocoleaceae

Order : Centrolepidales
Family :
1. Centrolepidaceae

Order : Poales
Family :
1. Poaceae

Subclass 3. Arecidae
Superorder : Arecanae
Order : Arecales
Family :
1. Arecaceae

Subclass 4. Alismatidae
Superorder : Alismatane
Order : Butomales
Family :
1. Butomaceae

Order : Hydrocharitales
Families :
1. Hydrocharitaceae
2. Thalassiaceae
3. Halophilaceae

Order : Najadales
Family :
1. Najadaceae

Order : Alismatales

Families :

1. Limnocharitaceae
2. Alismataceae

Order : Aponogetonales

Family :

1. Aponogetonaceae

Order : Jucaginales

Family :

1. Scheuchzeriaceae

Order : Juncaginales

Families :

1. Juncaginaceae
2. Lilaeaceae
3. Maundiaceae

Order : Potamogetonales

Families :

1. Potamogetonaceae
2. Ruppiaceae

Order : Posidoniales

Family :

1. Posidoniaceae

Order : Cymodoceales

Families :

1. Zannichelliaceae
2. Cymodoceaceae

Order : Zosterales

Family :

1. Zosteraceae

Subclass 5. Triurididae

Superorder : Triuridanae

Order : Petrosaviales

Family :

1. Petrosaviaceae

Order : Triuridales

Family :

1. Triuridaceae

Subclass 6. Aridae

Superorder : Aranae

Order : Arales

Families :

1. Araceae
2. Pistiaceae
3. Lemnaceae

Order : Acorales

Family :

1. Acoraceae

Superorder : Cyclanthanae

Order : Cyclanthales

Family :

1. Cyclanthaceae

Superorder : Pandananae

Order : Pandanales

Family :

1. Pandanaceae

Superorder : Typhanae

Order :

Families :

1. Typhales
2. Sparganiaceae
3. Typhaceae

7.2.7.3 Merits of Takhtajan's system of classification

Takhtajan's system was not much cared of when it first appeared in 1959. However ever since its publication in the revised form in English in 1969, it has attracted the attention of the entire world and is considered one of the most satisfactory phylogenetic system, which is followed in many parts of the world. This system is largely accepted mainly due to its clearly defined evolution principles and has the following plus points to its credit:

(a) The Dicots (Magnoliopsida) has been discussed prior to Monocots (Liliopsida).

(b) The families are small homogeneous units made up of closely related genera.

(c) The Dicots begin with Magnoliales, which is highly satisfactory as Magnoliales are universally considered to be the most primitive living Angiosperms.

(d) Among Monocots, the Alismatales, which are considered to be the most primitive living Monocots, have been placed at the starting point, which is, satisfactory.

(e) Engler and Prantl's division of Dicots into two traditional groups – Archichlamydae and Metachlamydae, has been abolished in this system.

(f) Problems such as monophyly or polyphyly, interrelationships of Dicots and Monocots, primitive position of Magnoliales, the secondary nature of anemophilous families with reduced unisexual flowers, etc. have been satisfactorily settled.

7.2.7.4 Demerits of Takhtajan's system of classification

(a) The main objection to Takhtajan's system is his derivation of Monocotyledons from the stocks ancestral to the Nymphaeales.

(b) The extremely narrowly defined taxa in this system has resulted in the unwarranted splitting of related groups.

7.2.8 Classification system of Arthur Cronquist (1919-1992)

Arthur Cronquist (1919-1992)
Courtesy : Science Home .... Mertz Library ... Archives and Manuscripts. The New York Botanical Garden

7.2.8.1 Arthur Cronquist (1919-1992)

Arthur Cronquist (1919-1992), an internationally renowned systematic botanist, was born on March 19, 1919, in San Jose, California. During his childhood itself, which was largely spent in Portland, Oregon and in Pocatello, Idaho, he became interested in the natural history of the mountain regions of the west. He was greatly influenced by his mentor Professor Ray J. Davis after he joined the University of Idaho (now Idaho State University). He obtained a B.S. (1938) and M.S. (1940)

degree from Utah State University where he also worked on the *Aster foliaceus* complex under the direction of Dr. Bassett Maguire who later became his colleague at The New York Botanical Garden. Cronquist obtained his Ph.D. degree in 1944 from the University of Minnesota under the supervision of Dr. C. O. Rosenthal and joined the New York Botanical Garden in 1943. In between, during the periods of 1946-1948 and 1948-1951, he left the garden to teach botany at the University of Georgia and at Washington State University respectively. He finally returned back to the Garden in 1952 where he was greatly influenced by Henry A. Gleason in floristics. Jointly they published *New Britton and Brown Illustrated Flora* (1952), to which Cronquist contributed the section on the Asteraceae and *Manual of Vascular Plants of Northeastern United States and Adjacent Canada* (1963), which is an encyclopedic reference and field book, whose intellectual genealogy can be traced to Nathaniel Lord Britton and Addison Brown's original *Illustrated Flora* of 1896-1898 and is known to botany students as the "Green Bible". He also jointly published another book entitled *The Natural Geography of Plants* (1964) with Henry Gleason. He also conducted extensive surveys throughout North America, and studied and collected plants of the intermountain regions of the western United States. These studies resulted in the publication of *Intermountain Flora* (1972, volume one), with Arthur H. Holmgren, Noel H. Holmgren, and James L. Reveal as co-authors. In the later 1950's he began a correspondence and collaboration with the Armenian botanist, Armen Takhtajan, of the Komarov Institute in Leningrad, U.S.S.R. and associate biologists, which proved to be a critical stimulus in the development of his plant classification systems. During this association, he also became proficient in Russian, and visited the (then) Soviet Union on several occasions, and promoted scientific exchanges between the two countries.

Cronquist carried out important administrative duties at the Garden and at its satellite facility, the Cary Arboretum as Director of Botany (1971-74) and Senior Scientist (1974-92). During this period, he also held faculty appointments at Columbia University and the City University of New York, where he served as part of the Executive Committee on Biology. He was also professionally affiliated to many other organizations, which included the American Society of Plant Taxonomists (president, 1962), the Botanical Society of America (president, 1973), the International Association of Plant Taxonomy (council member), and the Torrey Botanical Club (president, 1976). He was the proud recipient of many Professional awards and honors, which included the Leidy Medal of the Academy of Natural Sciences, Philadelphia (1970), honorary vice-president of the XII International Botanical Congress, Leningrad (1975), the Asa Gray Award (American Society of Plant Taxonomists, 1985), and the Medal for Botany, Linnean Society of London (1986). He was a great botanist and educator and his advancement of taxonomy, plant systematics, and floristics was of lasting significance to the science of botany. He died on March 22, 1992 while, studying plant specimens in the herbarium of Brigham Young University in Utah.

7.2.8.2 Arthur Cronquist's system of classification

Arthur Cronquist presented a system of classification for Dicotyledons in 1957, which was later modified and published in collaboration with two other eminent botanists in 1966. However the final modification appeared in 1988. The system presented in his book *Evolution and Classification of Flowering Plants* is largely based on literature, personal observations and discussions with other botanists. His revised system appeared in his book *An Integrated System of Classification of Flowering Plants* in 1981.

This system of classification is an elaboration of Bessey's system of classification and a refinement over Takhtajan's system (1964), which is based on morphological, anatomical, embryological, palynological, serological, cytological, chemical as well as ultrastructural evidences. According to Cronquist's system of classification, the angiosperms have been divided into two Classes - Magnoliopsida and Liliopsida, which have been further divided into Subclasses, Orders and Families. There are a total of 382 Families under the Angiosperms of which, 317 are under Magnoliopsida (Dicotyledons) and 65 under Liliopsida (Monocotyledons). There are 6 Subclasses under Magnoliopsida and 5 under Liliopsida as shown below :

Class : Magnoliopsida (Dicots)	Class : Liliopsida (Monocots)
1. Magnoliidae	1. Alismatidae
2. Hamamelidae	2. Arecidae
3. Caryophyllidae	3. Commelinidae
4. Dilleniidae	4. Zingiberidae
5. Rosidae	5. Liliidae
6. Asteridae	

An outline of Cronquist's system of classification (1988) upto the level of families is given below :

Class 1. Magnoliopsida (Dicots)

Subclass 1. Magnoliidae

Order 1. Magnoliales

Families :

1. Winteraceae
2. Degeneriaceae
3. Himantandraceae
4. Eupomatiaceae
5. Austrobaileyaceae
6. Magnoliaceae
7. Lactoridaceae
8. Annonaceae
9. Myristicaceae
10. Canellaceae

Order 2. Laurales

Families :

1. Amborellaceae
2. Trimeniaceae
3. Monimiaceae
4. Gomortegaceae
5. Calycanthaceae
6. Idiospermaceae
7. Lauraceae
8. Hernandiaceae

Order 3. Piperales

Families:

1. Cloranthaceae
2. Sauruaceae
3. Piperaceae

Order 4. Aristolochiales

Family :

1. Aristolochiaceae

Order 5. Illiciales

Families:

1. Illiciaceae
2. Schisandraceae

Order 6. Nymphaeales

Families:

1. Nelumbonaceae
2. Nymphaeaceae
3. Barclayaceae
4. Cabombaceae
5. Ceratophyllaceae

Order 7. Ranunculales
Families:
1. Ranunculaceae
2. Circaeasteraceae
3. Berberidaceae
4. Sargentodoxaceae
5. Lardizabalaceae
6. Menispermaceae
7. Coriariaceae
8. Sabiaceae

Order 8. Papaverales
Families:
1. Papaveraceae
2. Fumariaceae

Subclass 2. Hamamelidae

Order 1. Trochodendrales
Families:
1. Tetracentraceae
2. Trochodendraceae

Order 2. Hamamelidales
Families:
1. Cercidiphyllaceae
2. Eupteleaceae
3. Platanaceae
4. Hamamelidaceae
5. Myrothamnaceae

Order 3. Daphniphyllales
Family:
1. Daphniphyllaceae

Order 4. Didymelales
Family:
1. Didymelaceae

Order 5. Eucommiales
Family:
1. Eucommiaceae

Order 6. Urticales
Families:
1. Barbeyaceae
2. Ulmaceae
3. Cannabaceae
4. Moraceae
5. Cecropiaceae
6. Urticaceae
7. Physenaceae

Order 7. Leitneriales
Family:
1. Leitneriaceae

Order 8. Juglandales
Families:
1. Rhoipteleaceae
2. Juglandaceae

Order 9. Myricales
Family:
1. Myricaceae

Order 10. Fagales
Families:
1. Balanopaceae
2. Ticodendraceae
3. Fagaceae
4. Nothofagaceae
5. Betulaceae

Order 11. Casuarinales
Family:
1. Casuarinaceae

Subclass 3. Caryophyllidae

Order 1. Caryophyllales
Families:
1. Phytolaccaceae
2. Achatocarpaceae
3. Nyctaginaceae
4. Aizoaceae
5. Didiereaceae
6. Cactaceae
7. Chenopodiaceae
8. Amaranthaceae
9. Portulacaceae
10. Basellaceae
11. Molluginaceae
12. Caryophyllaceae

Order 2. Polygonales
Family:
1. Polygonaceae

Order 3. Plumbaginales
Family:
1. Plumbaginaceae

Subclass 4. Dilleniidae

Order 1. Dilleniales

Families:

1. Dilleniaceae
2. Paeoniaceae

Order 2. Theales

Families:

1. Ochnaceae
2. Sphaerosepalaceae
3. Sarcolaenaceae
4. Dipterocarpaceae
5. Caryocaraceae
6. Theaceae
7. Actinidiaceae
8. Scytopetalaceae
9. Pentaphylacaceae
10. Tetrameristaceae
11. Pellicieraceae
12. Oncothecaceae
13. Marcgraviaceae
14. Quiinaceae
15. Elatinaceae
16. Paracryphiaceae
17. Medussagynaceae
18. Clusiaceae (Guttiferae)

Order 3. Malvales

Families:

1. Elaeocarpaceae
2. Tiliaceae
3. Sterculiaceae
4. Bombacaceae
5. Malvaceae

Order 4. Lecythidales

Family:

1. Lecythidaceae

Order 5. Nepenthales

Families:

1. Sarraceniaceae
2. Nepenthaceae
3. Droseraceae

Order 6. Violales

Families:

1. Flacourtiaceae
2. Peridiscaceae
3. Bixaceae
4. Cistaceae
5. Huaceae
6. Lacistemataceae
7. Scyphostegiaceae
8. Stachyuraceae
9. Violaceae
10. Tamaricaceae
11. Frankeniaceae
12. Dioncophyllaceae
13. Ancistrocladaceae
14. Turneraceae
15. Malesherbiaceae
16. Passofloraceae
17. Achariaceae
18. Caricaceae
19. Fouquieriaceae
20. Hoplestigmataceae
21. Cucurbitaceae
22. Datiscaceae
23. Begoniaceae
24. Loasaceae

Order 7. Salicales

Family:

1. Salicaceae

Order 8. Capparales

Families:

1. Tovariaceae
2. Capparaceae
3. Brassicaceae (Cruciferae)
4. Moringaceae
5. Resedaceae

Order 9. Batales

Families:

1. Gyrostemonaceae
2. Bataceae

Order 10. Ericales

Families:

1. Cyrillaceae
2. Clethraceae
3. Grubbiaceae
4. Empetraceae
5. Epacridaceae
6. Ericaceae
7. Pyrolaceae
8. Monotropaceae

Order 11. Diapensiales

Family:

1. Diapensiaceae

Order 12. Ebenales

Families:

1. Sapotaceae
2. Ebenaceae
3. Styracaceae
4. Lissocarpaceae
5. Symplocaceae

Order 13. Primulales

Families:

1. Theophrastaceae
2. Myrsinaceae
3. Primulaceae

Subclass 5. Rosidae

Order 1. Rosales

Families:

1. Brunelliaceae
2. Connaraceae
3. Eucryphiaceae
4. Cunoniaceae
5. Davidsoniaceae
6. Dialypetalanthaceae
7. Pittosporaceae
8. Byblidaceae
9. Hydrangeaceae
10. Columelliaceae
11. Grossulariaceae
12. Greyiaceae
13. Bruniaceae
14. Anisophyllaceae
15. Alseuosmiaceae
16. Crassulaceae
17. Cephalotaceae
18. Saxifragaceae
19. Rosaceae
20. Neuradaceae
21. Crossosomataceae
22. Chrysobalanaceae
23. Surianaceae
24. Rhabdodendraceae

Order 2. Fabales

Families:

1. Minosaceae
2. Caesalpiniaceae
3. Fabaceae (Papilionaceae)

Order 3. Proteales

Families:

1. Elaeagnaceae
2. Proteaceae

Order 4. Podestemales

Family:

1. Podostemaceae

Order 5. Haloragales

Families:

1. Haloragaceae
2. Gunneraceae

Order 6. Myrtales

Families:

1. Sonneratiaceae
2. Lythraceae
3. Rhynchocalycaceae
4. Alzateaceae
5. Penaeaceae
6. Crypteroniaceae
7. Thymelaeaceae
8. Trapaceae
9. Myrtaceae
10. Punicaceae
11. Onagraceae
12. Oliniaceae
13. Melastomataceae
14. Combretaceae

Order 7. Rhizophorales
Family:
1. Rhizophoraceae

Order 8. Cornales
Families:
1. Alangiaceae
2. Cornaceae
3. Garryaceae

Order 9. Santalales
Families:
1. Medusandraceae
2. Dipentodontaceae
3. Olacaceae
4. Opiliaceae
5. Santalaceae
6. Missodendraceae
7. Loranthaceae
8. Viscaceae
9. Eremolepidaceae
10. Balanophoraceae

Order 10. Rafflesiales
Families:
1. Hydnoraceae
2. Mitrastemonaceae
3. Rafflesiaceae

Order 11. Celastrales
Families:
1. Geissolomataceae
2. Celastraceae
3. Hippocrateaceae
4. Stackhousiaceae
5. Salvadoraceae
6. Tepuianthaceae
7. Aquifoliaceae
8. Icacinaceae
9. Aextoxicaceae
10. Cardiopteridaceae
11. Corynocarpaceae
12. Dichapetalaceae

Order 12. Euphorbiales
Families:
1. Buxaceae
2. Simmondsiaceae
3. Pandaceae
4. Euphorbiaceae

Order 13. Rhamnales
Families:
1. Rhamnaceae
2. Leeaceae
3. Vitaceae

Order 14. Linales
Families:
1. Erythroxylaceae
2. Humiriaceae
3. Ixonanthaceae
4. Hugoniaceae
5. Linaceae

Order 15. Polygalales
Families:
1. Malpighiaceae
2. Vochysiaceae
3. Trigoniaceae
4. Tremandraceae
5. Polygalaceae
6. Xanthophyllaceae
7. Krameriaceae

Order 16. Sapindales
Families:
1. Staphyleaceae
2. Melianthaceae
3. Bretschneideraceae
4. Akaniaceae
5. Sapindaceae
6. Hippocastanaceae
7. Aceraceae
8. Burseraceae
9. Anacardiaceae
10. Julianiaceae
11. Simaroubaceae
12. Cneroraceae
13. Meliaceae
14. Rutaceae
15. Zygophyllaceae

Order 17. Geraniales
Families:
1. Oxalidaceae
2. Geraniaceae
3. Limnanthaceae
4. Tropaeolaceae
5. Balsaminaceae

Order 18. Apiales
Families:
1. Araliaceae (Thorne)
2. Apiaceae (Umbelliferae)

Subclass 6. Asteridae

Order 1. Gentianales
Families:
1. **Loganiaceae**
2. **Gentianaceae**
3. **Saccifoliaceae**
4. **Apocynaceae**
5. **Asclepiadaceae**

Order 2. Solanales
Families:
1. **Duckeodendraceae**
2. **Nolanaceae**
3. **Solanaceae**
4. **Convolvulaceae**
5. **Cuscutaceae**
6. **Retziaceae**
7. **Menyanthaceae**
8. **Polemoniaceae**
9. **Hydrophyllaceae**

Order 3. Lamiales
Families:
1. **Lennoaceae**
2. **Boraginaceae**
3. **Verbenaceae**
4. **Lamiaceae (Labiatae)**

Order 4. Callitrichales
Families:
1. **Hippuridaceae**
2. **Callitrichaceae**
3. **Hydrostachyaceae**

Order 5. Plantaginales
Family:
1. **Plantaginaceae**

Order 6. Scrophulariales
Families:
1. **Buddlejaceae**
2. **Oleaceae**
3. **Scrophulariaceae**
4. **Globulariaceae**
5. **Myoporaceae**
6. **Orobanchaceae**
7. **Gesneriaceae**
8. **Acanthaceae**
9. **Pedaliaceae**
10. **Bignoniaceae**
11. **Mendonciaceae**
12. **Lentibulariaceae**

Order 7. Campanulales
Families:
1. **Pentaphragmataceae**
2. **Sphenocleaceae**
3. Campanulaceae
4. Stylidiaceae
5. Donatiaceae
6. Brunoniaceae
7. Goodeniaceae

Order 8. Rubiales
Families:
1. Rubiaceae
2. **Theligonaceae**

Order 9. Dipsacales
Families:
1. **Caprifoliaceae**
2. **Adoxaceae**
3. **Valerianaceae**
4. **Dipsacaceae**

Order 10. Calycerales
Family:
1. **Calyceraceae**

Order 11. Asterales
Family:
1. **Asteraceae (Compositae)**

Class 2. Liliopsida (Monocots)

Subclass 1. Alismatidae

Order 1. Alismatales
Families:
1. **Butomaceae**
2. **Limnocharitaceae**
3. **Alismataceae**

Order 2. Hydrocharitales
Family:
1. Hydrocharitaceae

Order 3. Najadales
Families:
1. Aponogetonaceae
2. Scheuchzeriaceae
3. Juncaginaceae
4. Potamogetonaceae
5. Ruppiaceae
6. Najadaceae
7. Zannichelliaceae
8. Posidoniaceae
9. Cymodoceaceae
10. Zosteraceae

Order 4. Triuridales
Families:
1. Petrosaviaceae
2. Triuridaceae

Subclass 2. Arecidae

Order 1. Arecales
Family:
1. Arecaceae (Palmae)

Order 2. Cyclanthales
Family:
1. Cyclanthaceae

Order 3. Pandanales
Family:
1. Pandanaceae

Order 4. Arales
Families:
1. Acoraceae
2. Araceae
3. Lemnaceae

Subclass 3. Commelinidae

Order 1. Commelinidales
Families:
1. Rapateaceae
2. Xyridaceae
3. Mayacaceae
4. Commelinaceae

Order 2. Eriocaulales
Family:
1. Eriocaulaceae

Order 3. Restionales
Families:
1. Flagellariaceae
2. Joinvilleaceae
3. Restionaceae
4. Centrolepidaceae

Order 4. Juncales
Families:
1. Juncaceae
2. Thurniaceae

Order 5. Cyperales
Families:
1. Cyperaceae
2. Poaceae (Gramineae)

Order 6. Hydatellales
Family:
1. Hydatellaceae

Order 7. Typhales
Families:
1. Sparganiaceae
2. Typhaceae

Subclass 4. Zingiberidae

Order 1. Bromeliales
Family:
1. Bromeliaceae

Order 2. Zingiberales
Families:
1. Strelitziaceae
2. Heliconiaceae
3. Musaceae
4. Lowiaceae
5. Zingiberaceae
6. Costaceae
7. Cannaceae
8. Marantaceae

Subclass 5. Liliidae

Order 1. Liliales
Families:
1. Philydraceae
2. Pontederiaceae
3. Haemodoraceae
4. Cyanastraceae
5. Liliaceae
6. Iridaceae

7. Velloziaceae
8. Aloeaceae
9. Agavaceae
10. Xanthorrhoeaceae
11. Habguanaceae
12. Taccaceae
13. Stemonaceae
14. Smilacaceae
15. Dioscoreaceae

Order 2. Orchidales

Families:

1. Geosiridaceae
2. Burmanniaceae
3. Corsiaceae
4. Orchidaceae

7.2.8.3 The Classification of Magnoliophyta according to Cronquist's system

Classification of flowering plant is an on-going scientific activity. Classification systems are conceptual structures that reflect interpretation of available data regarding evolutionary relationships. Since interpretations differ and new informations are constantly emerging from many sources, there is no single, stable classification. Pre-Darwinian classification systems either followed a metaphysical rationale, charting the Divine pattern of creation, or grouped flowering plants on the basis of abritrary ('artifical') criteria. Later classification systems have largely focused on the determination of phylogenetic or evolutionary patterns, which consider a 'model' archaic or primitive type as a foundation from which modern groups have evolved. The model and perspectives (primitive vs. derived or specialized) of Charles Bessey (1915) is followed by most current classification systems. Regardless of perspective, all current systems treat the flowering plants as a two-parted entity, composed of the Dicots (Magnoliopsida) and Monocots (Liliopsida), both of which share a complex suite of features, and all data available indicate that these features were present in an ancestral, dicot-like plant. Thus, the flowering plants appear to be monophyletic (both monocots and dicots evolved from a common ancestor) and the extant angiosperms most similar to the ancestral type are those that are placed in the basal dicot groups.

All taxonomic circumscriptions are based on common or 'typical' key characters. At the higher taxonomic ranks, critical evaluation of multiple key characters is very important. Since the two elements (Magnoliopsida and Liliopsida) of the Magnoliophyta diverged early in the evolutionary history of the group, they are fairly distinct. Key characters used to circumscribe the two Classes in the Cronquist's system are as follows:

Characters	Magnoliopsida	Liliopsida
Number of cotyledons	Usually 2	Usually 1
Leaves	Usually veined reticulately	Usually veined parallely
Vascular Cambium	Often present (ca. 50% woody)	Absent
Primary Vascular Bundles	Arranged in a ring	Often scattered
Pollen	Various types	Monosulcate
Floral Parts	Usually in sets of 4s or 5s	Usually in sets of 3
Root System	Primary and adventitious	Adventitious only

Class : Magnoliopsida

In this system, the subclass Magnoliidae under the class Magnoliopsida, which, consists mainly of the families of the traditional Ranales, is regarded as the basal complex from which other angiosperms are considered to have been derived. The Hamamelidae is considered to consist of a group of wind-pollinated families with reduced, chiefly apetalous flowers that are often borne in catkins (the traditional Amentiferae). The families under Caryophyllidae have a strong tendency towards free-central placentation and have betalains (nitrogenous anthocyanins and anthoxanthins) in many of them. The Rosidae and Dilleniidae, which are morphologically very much similar, are regarded to be parallel groups. The Asteridae consists of the higher sympetalous families, which are mostly derived from the Rosidae (Fig. 7.3).

The key characters used to circumscribe the various Subclasses under the Class Magnoliopsida in the Cronquist's system are as follows:

Subclass	Characteristic features	Number of Orders	Number of Families	Number of Species
Magnoliidae	Archaic, flowers many-parted	8	39	11,000
Hamamelidae	Ancient, floral parts reduced, mostly wind pollinated	11	24	3,400
Caryophyllidae	Herbs with betalains and 'centered' placentation types	3	14	11,000
Dillenidae	Some sympetalous, few apocarpous	13	78	25,000
Rosidae	Polypetalous, often numerous stamens	18	114	58,000
Asteridae	Mostly sympetalous	11	49	56,000

Subclass 1. Magnoliidae

Subclass Magnoliidae includes 8 Orders, 39 Families and about 11,000 Species. According to Cronquist's system of classification, Families included under this Subclass show a tendency of retention of primitive features and, regardless of the classification system, most are placed at the base of the dicot lineage.

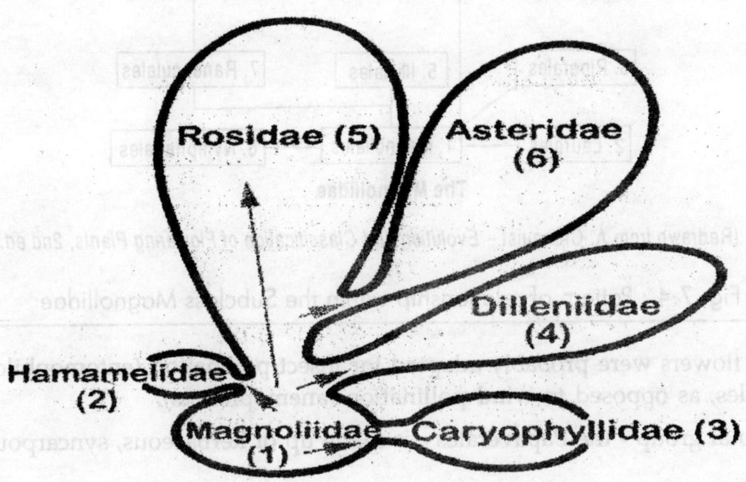

⇘ Fig. 7.3 : The figure above (redrawn from Cronquist, 1988,) depicts phyletic relationships among subclasses of the Class Magnoliopsida, as envisioned by Cronquist. This alignment features the Magnoliidae as including extant dicot elements that carry the greatest number of archaic features (similar to the 'original' flowering plant) and the Asteridae (Sunflowers and relatives) as the most 'derived' or specialized element of the Class. The size of the 'balloons' is roughly proportional to the number of species per subclass.
(*Courtesy* : Professor Hugh D. Wilson, Professor of Biology, Curator, Departmental Herbarium, Department of Biology, College of Science, Texas A&M University, College Station, Texas 77843)

In this Subclass (Fig. 7.4), the most archaic element i.e., families of the largest Order, the Magnoliales tend to express the highest level of primitive features, as originally defined by Charles Bessey in his description what the first angiosperm might have been like:

(a) A woody, dicot-like plant - maybe arborescent, but more likely a fruticose, shrubby plant of seasonally dry, sub-tropical habitat.

(b) Leaves were persisent (evergreen), simple, alternate, entire, pinnately veined, glabrous, and probably stipulate.

(c) Flowers were solitary, terminal, perfect, actinomorphic, with no connation or adnation. The numerous floral parts of this flower were spirally arranged on an elongate floral axis.

(d) Perianth were many parted with poor differentiation between appendages of the corolla and calyx (Tepals).

(e) Androecium included many stamens that were broad and laminar with no distinct filament.

(f) Gynoecium was apocarpous and produced a follicle.

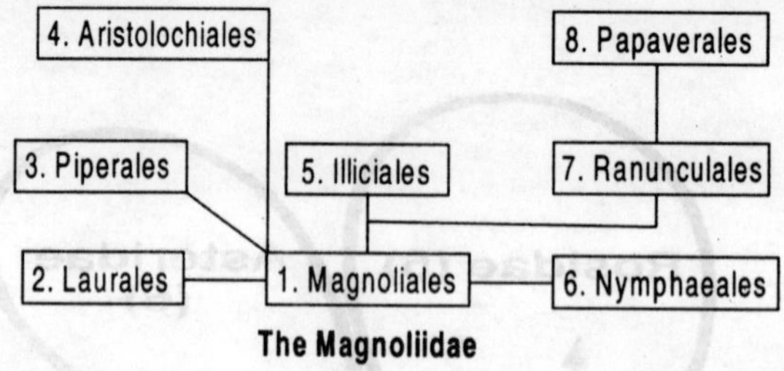

(Redrawn from A. Cronquist – *Evolution and Classification of Flowering Plants, 2nd ed.*)

Fig. 7.4 : Pattern of relationship within the Subclass Magnollidae

(g) The flowers were probably adapted for insect pollination (entomophilous), most likely beetles, as opposed to wind pollination (anemophilous).

The terminal group - the Papaverales - is made up of herbaceous, syncarpous plants.

Subclass 2. Hamamelidae

The smallest of the dicot Subclasses is Hamamelidae, which includes 11 Orders, 24 Families and about 3,400 Species with the Urticales and Fagales including about 90% of the Species. This group is weighted in taxa at the higher ranks (Orders with few Families and Families with few Species). Its fossil record extends to the lower Cretaceous (100 million years ago) and it is believed that the group is either marked by high levels of extinction, thereby producing an array of well separated, more or less relictual taxa, or, perhaps, polyphyletic (an artifical assemblage with elements drawn from several ancestral types). Families included under this Subclass are typically woody (Urticales an exception) and characterized by a suite of structural adaptations for wind pollination or anemophily. These features include :

(a) Unisexual (imperfect) flowers with either monoecious or dioecious taxa.

(b) Perianth reduced, weakly developed or absent (apetalous).

(c) Inflorescences with numerous, reduced, often apetalous flowers with the most specialized type - the catkin or ament (older name for the group is Amentiferae).

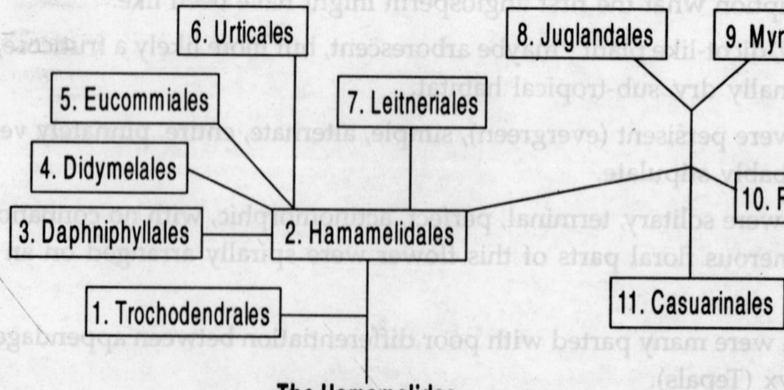

(Redrawn from A. Cronquist – *Evolution and Classification of Flowering Plants, 2nd ed.*)

Fig. 7.5 : Pattern of relationship within the Subclass Hamamelidae

In the Subclass Hamamelidae, the Order Trochodendrales is placed at the base and the Genus *Casuarina* (Australian Pine - Casuarinaceae, Casuarinales) takes the most derived or specialized position (Fig. 7.5). The two monotypic families included under the Trochodendrales lack vessel elements and tend toward apocarpy whereas *Casuarina* shows multiple (fruiting pistillate inflorescence) fruiting structures.

Subclass 3. Caryophyllidae

Subclass Caryophyllidae includes only 3 Orders, 14 Families and about 11,000 Species. About 90% of the Species are included under the large, diverse Order Caryophyllales and also carries the relatively distinct set of features that mark the Subclass (Fig. 7.7). These include:

(a) Most families of the Caryophyllales lack the anthocyanin pigments typical of most angiosperms and are replaced by betalains, a class of pigments, that are found only in this Order (and some fungi).

(b) The perianth tends to be uniseriate via apetaly but often showy via petaloid sepals.

(c) The placentation pattern if often basal or free central thereby tending to place the ovules in a central position within the ovary - the archaic term 'Centrospermae' denotes this tendency.

(d) This Order is marked by a suite of unusual embryological features. For example, the nutritive tissue found in the seeds of most Caryophyllales is perisperm, derived from sporophytic (diploid) tissue and not endosperm. Further, the embryo tends to occupy a peripheral position in the seed, which often produces a characteristic 'beaked' (protrusion of the radical) asymmetry (Fig. 7.6).

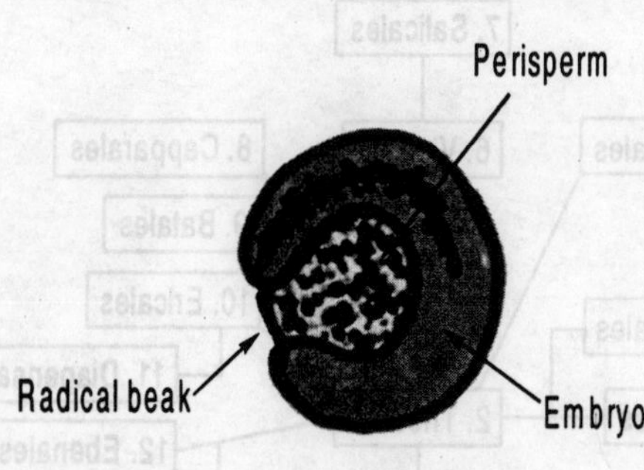

Fig. 7.6 : Embryological features of most Caryophyllales

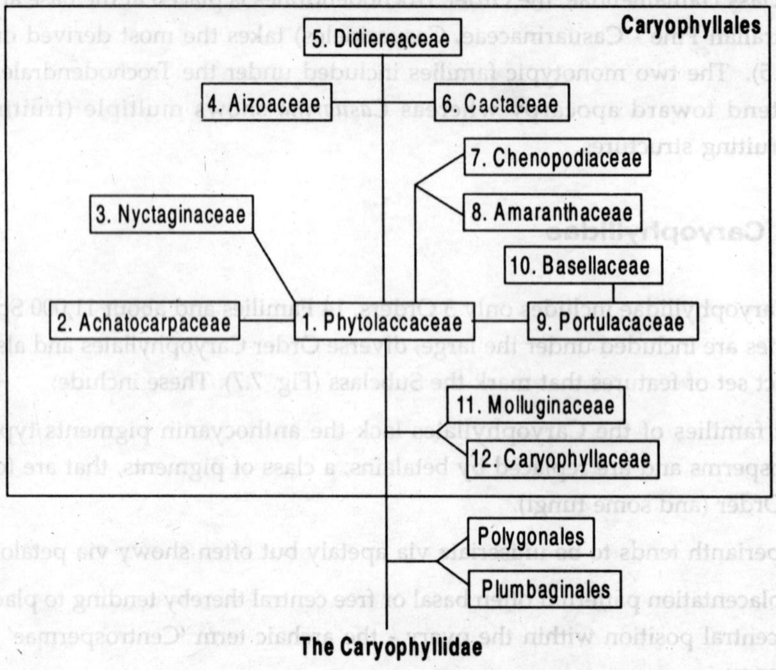

(Redrawn from A. Cronquist – *Evolution and Classification of Flowering Plants, 2nd ed.*)

Fig. 7.7 : Pattern of relationship within the Subclass Caryophyllidae

Subclass 4. Dilleniidae

This Subclass includes 13 Orders, 78 Families and about 25,000 Species. Five Orders - the Violales (5,000) Capparales (4,000), Ericales (4,000), Theales (3,500) and Malvales (3,000 to 3,500) include about 75% of the Species under this Subclass.

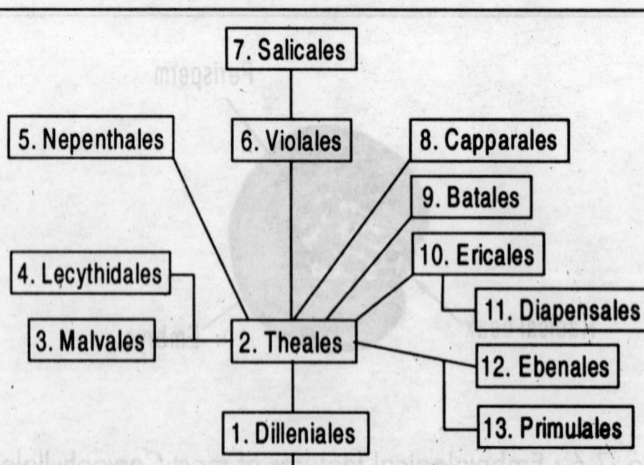

(Redrawn from A. Cronquist – *Evolution and Classification of Flowering Plants, 2nd ed.*)

Fig. 7.8 : Pattern of relationship within the Subclass Dilleniidae

Dilleniidae include a diverse set of flowering plant Orders that lacks a distinct set of key characters. General characteristic features of this Subclass include:

(a) All the members are syncarpous except Dilleniales, which stands as a structural link to the Magnoliidae.

(b) More than 1/3 of the species show parietal placentation.

(c) About 1/3 of the species are sympetalous.

Due to lack of connation in the gynoecium, as expressed by the peony (*Paeonia* - Paeoniaceae), the Order Dilleniales, including two Families, is placed at the base of this Subclass and forms a structural link to the basal flowering plant Subcalss, the Magnoliidae (Fig. 7.8).

Subclass 5. Rosidae

It includes 18 Orders, 114 Families and over 60,000 Species. This is the largest Subclass with regard to the number of Families and is more or less tied with the Asteridae with regard to species diversity. Here also five Orders - the Fabales (18,000), Myrtales (9,000), Euphorbiales (8,000), Rosales (6,600) and Sapindales (5,400) include about 75% of the Species (Fig. 7.9).

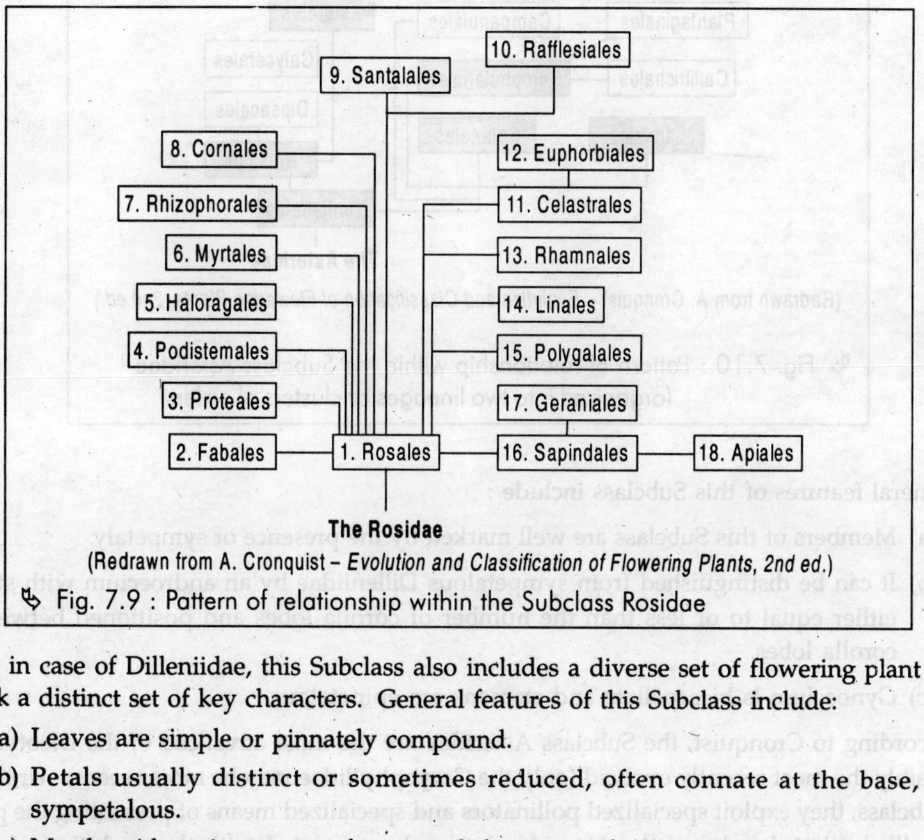

Fig. 7.9 : Pattern of relationship within the Subclass Rosidae

As in case of Dilleniidae, this Subclass also includes a diverse set of flowering plant orders that lack a distinct set of key characters. General features of this Subclass include:

(a) Leaves are simple or pinnately compound.

(b) Petals usually distinct or sometimes reduced, often connate at the base, rarely sympetalous.

(c) Mostly with various types of nectary disks, many of which are stamodial in origin and positioned at the base of the ovary.

(d) Generally syncarpous, except in the Rosales, where it is apocarpous and the Fabales and Proteales, where it is monocarpous.

(e) Placentation is of various types, but most often axile with, generally, fewer ovules per locule than found in the Dilleniidae.

Cronquist has placed the Subclasses Rosidae and Dilleniidae apart because each seems to constitute a natural group separately derived from the ancentral Magnoliidae, rather than because of any definitive distinguishing characters. Hence it is conceptually more useful to hold the two Subclasses separate rather than to combine them into one or to abandon any attempt at organization of the Magnoliopsida into Subclasses.

Subclass 6. Asteridae

This Subclass, includes 11 Orders, 49 Families and nearly 60,000 Species. Species diversity is comparable to that of the Rosidae, but unlike Rosidae or Dilleniidae, the family level is fewer in this Subclass. The single Family Asteraceae includes about a 1/3rd of the species in this Subclass. In terms of class-level phylogenetic position, level of unique adaptive modification, and species diversity, Asteraceae is comparable to the monocot family Orchidaceae.

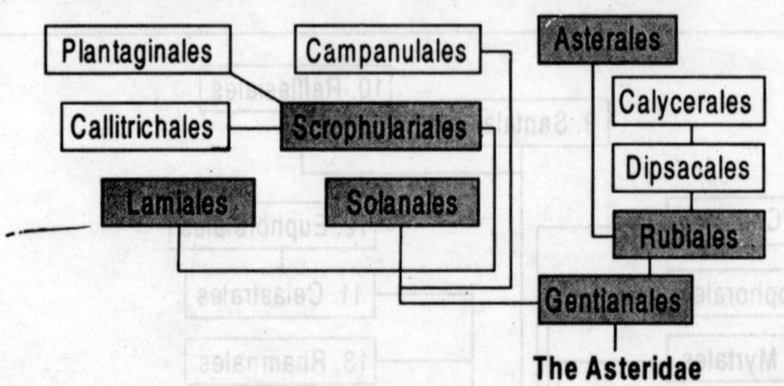

(Redrawn from A. Cronquist – *Evolution and Classification of Flowering Plants, 2nd ed.*)

Fig. 7.10 : Pattern of relationship within the Subclass Asteridae (organized into two lineages or clusters of orders)

General features of this Subclass include :

(a) Members of this Subclass are well marked by the presence of sympetaly.

(b) It can be distinguished from sympetalous Dilleniidae by an androecium with stamens either equal to or less than the number of corolla lobes and positioned between the corolla lobes.

(c) Gynoecium is bicarpellate and stamens are epipetalous.

According to Cronquist, the Subclass Asteridae are the most advanced of the Dicotyledons, and possibly the most recently evolved (only the Caryophyllidae may be more recent). Unlike any other Subclass, they exploit specialized pollinators and specialized means of presenting the pollen. It seems likely that the rise of the Asteridae is closely correlated with the evolution of insects capable of recognizing complex floral patterns. Cronquist has roughly organized this Subclass into two lineages or clusters of Orders, one of which is marked by a dominance of hypogyny and the other, terminated by the Asterales, by epigyny.

Class : Liliopsida

The two Classes Magnoliopsida and Liliopsida under Magnoliophyta have diverged early in the evolutionary history of the group and hence they are fairly distinct and, in terms of flowering plant identification, they represent the first level of recognition. In the Class Liliopsida, the Alismatidae chiefly includes the aquatic members, whereas the other Subclasses includes those Families which typically have large often petiolate leaves, and usually are either arborescent or have flowers crowded into a spadix. The Families under Commelinidae are characterized by floral reduction and are mostly wind-pollinated. The Liliidae includes the insect-pollinated members (Fig. 7.11).

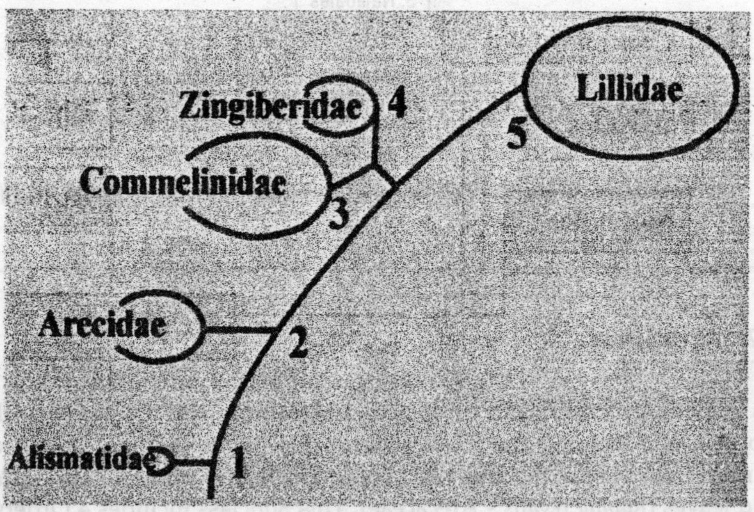

Fig. 7.11 : The figure above (redrawn from Cronquist, 1988,) depicts phyletic relationships among Subclasses of the Class Liliopsida, as envisioned by Cronquist. This alignment features the Alismatidae as the extant Monocot element that carries the greatest number of archaic features (similar to the 'original' Monocot) and the Liliidae (Orchids and relatives) as the most 'derived' or specialized element of the Class. The size of the 'balloons' is roughly proportional to the number of Species per Subclass. (*Courtesy* : Professor Hugh D. Wilson, Professor of Biology, Curator, Departmental Herbarium, Department of Biology, College of Science, Texas A&M University, College Station, Texas 77843)

The key characters used to circumscribe the two Classes include :

Subclass	Characteristic features	Number of Orders	Number of Families	Number of Species
Alismatidae	Aquatics with archaic, many-parted flowers	4	16	500
Arecidae	Inflorescences specialized	4	5	5,600
Commelinidae	Mostly herbs with reduced flowers	7	16	16.200
Zingiberidae	Tropical herbs, epigynous	2	9	3,800
Liliidae	Flowers showy with sepals petaloid	2	19	25,000

Subclass 1. Alismatidae

The Subclass Alismatidae, includes 4 Orders, 16 Families and about 500 Species (Fig. 7.12). It is the smallest Subclass of the Magnoliophyta in terms of species diversity. Cronquist has considered the characteristic features of the gynoecium and apocarpy as a central, unifying feature for this group. Aquatic adaptation of all the Orders except the Triuridales, also provides a tentative link to the archaic dicot element (Magnoliidae - Nymphaeales) placed by Cronquist as a possible extant link to the monocots.

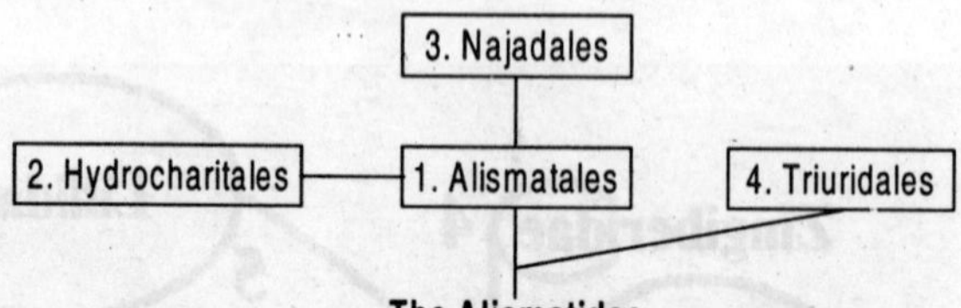

The Alismatidae

(Redrawn from A. Cronquist – *Evolution and Classification of Flowering Plants, 2nd ed.*)

Fig. 7.12 : Pattern of relationship within the Subclass Alismatidae

Subclass 2. Arecidae

In terms of both structural and ecological adaptations, this is, perhaps, the most diverse flowering Subclass as it includes the smallest flowering plants (Lemnaceae - Arales), the largest Monocots (Arecaceae - Arecales) and the largest flowering plant seeds (*Lodoicea maldivica* - Arecaceae) and leaves (*Raphia regalis* - Arecaceae). This Subclass includes 4 Orders, 5 Families and about 5,600 Species and more than half of the Species of this Subclass belong to the monotypic Order Arecales (Fig. 7.13).

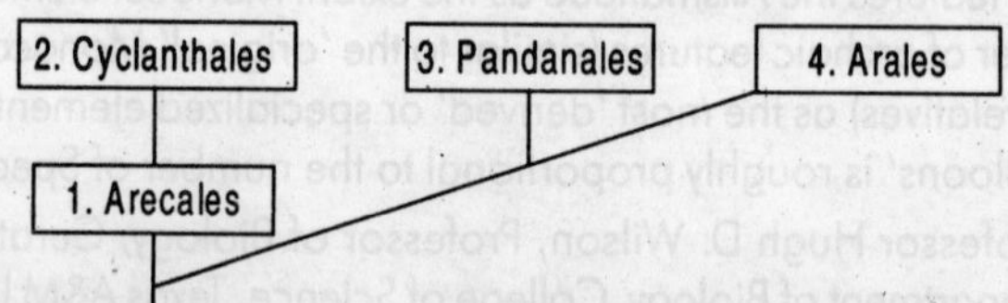

The Arecidae

(Redrawn from A. Cronquist – *Evolution and Classification of Flowering Plants, 2nd ed.*)

Fig. 7.13 : Pattern of relationship within the Subclass Arecidae

Subclass 3. Commelinidae

It includes 7 Orders, 16 Families and 16,200 Species (Fig. 7.14) that are phylogenetically aligned according to pollination system. More than half of the Species i.e. about 15, 000 species of the Commelinidae are included within a single Family, the Grass Family or Poaceae (Gramineae) which, is the most important Family of flowering plants from the point of view of both economic and ecological importance, at least from the perspective of mammals and, more specifically, humans.

Commelinales, which forms the basal or archaic group, shows floral adaptations for entomophilous mode of pollination with showy flowers and well-developed corolla, whereas the more derived Orders are mostly anemophilous. Thus, according to Cronquist, the trend or phyletic pattern within this group, is towards floral reduction and other adaptive changes for wind-pollination.

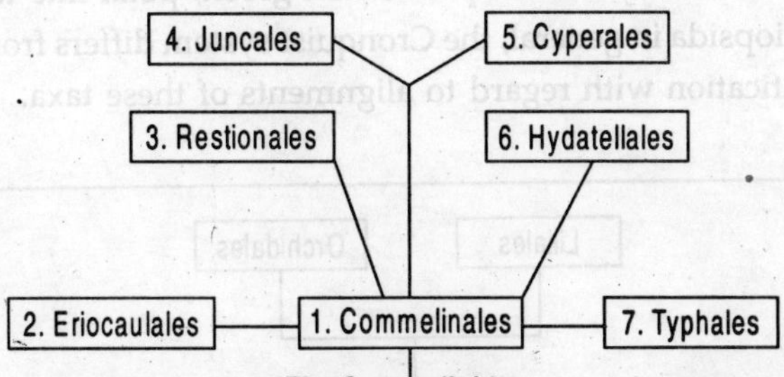

The Commelinidae

(Redrawn from A. Cronquist – *Evolution and Classification of Flowering Plants, 2nd ed.*)

↳ Fig. 7.14 : Pattern of relationship within the Subclass Commelinidae

Subclass 4. Zingiberidae

This Subclass includes 2 Orders, 9 Families and 3,800 Species (Fig. 7.15). Members of Zingiberidae are often associated with Orders of the Commelinidae and Liliidae. Cronquist was of the view that these plants are "discordant" when so treated, i.e., the Bromeliales does not 'fit' well within the Commelinidae and the Zingiberales are a bit anomalous within the context of the Liliidae. If both Orders are excluded from these Subclasses, it serves to "sharpen" the definition or circumscription of both. It however does create a unit that is not recognized by other current systems of flowering plant classification.

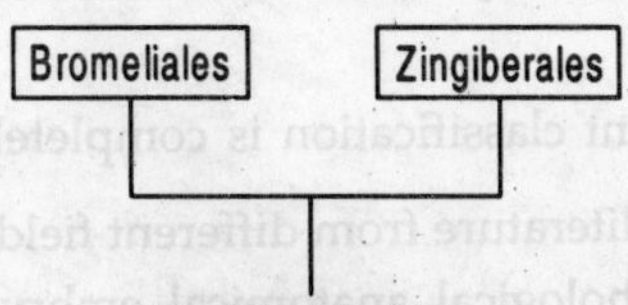

The Zingiberidae

(Redrawn from A. Cronquist – *Evolution and Classification of Flowering Plants, 2nd ed.*)

↳ Fig. 7.15 : Pattern of relationship within the Subclass Zingiberidae

Within this Subclass the order Bromeliales (monotypic) includes the epiphytes or xerophytes, which are characterized by narrow, parallel-veined leaves that are often leathery and not petiolate and the flowers are usually actinomorphic with 6 stamens. The other Order Zingiberales include mesophytes of the forest floor or emergent aquatics with broadly expanded, petiolate leaves that carry a distinctive pattern of veination and flowers are usually zygomorphic with 1 or 5 (rarely 6) stamens.

Subclass 5. Liliidae

This Subclass includes 2 Orders, 19 Families, and about 25,000 Species (Fig. 7.16). The members of the Liliidae are roughly characterized by large, showy flowers that tend to show petaloid sepals, i.e., the perianth which is usually made up of 6 non-green, petal-like tepals. As is the case of members of the Class Liliopsida in general, the Cronquist system differs from other current systems of flowering plant classification with regard to alignments of these taxa.

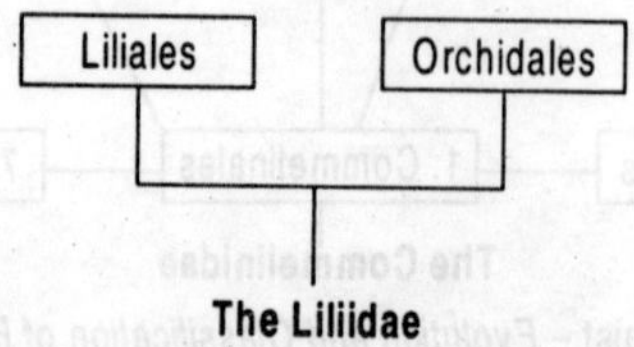

(Redrawn from A. Cronquist – *Evolution and Classification of Flowering Plants, 2nd ed.*)

Fig. 7.16 : Pattern of relationship within the Subclass Liliidae

The Order Liliales under this Subclass is characterized by plants which are usually not mycotrophic (associated with fungi), with mostly actinomorphic flowers having inferior or superior ovary and seeds of ordinary size and structure, usually with endosperm and a well-developed embryo. The other Order Orchidales is characterized by plants which are usually mycotrophic, sometimes without chlorophyll, flowers strongly zygomorphic, ovary inferior and seeds very numerous and minute with undifferentiated embryo lacking endosperm.

(*Courtesy* : Professor Hugh D. Wilson, Professor of Biology, Curator, Departmental Herbarium, Department of Biology, College of Science, Texas A&M University, College Station, Texas 77843)

7.2.8.4 Merits of Cronquist's system of classification

(a) Cronquist system of plant classification is completely a phylogenetic classification.

(b) Conquist has taken vast literature from different fields of science into consideration. His system is based on morphological, anatomical, embryological, palynological, serological, cytological, chemical as well as ultrastructural evidences.

(c) Synoptical keys for each group up to the family level are very much useful.

7.2.8.5 Demerits of Cronquist's system of classification

(a) Cronquist system of classification at some level has shown too much reliance on single characters like centrifugal stamens, free-central placentation, etc.

(b) The arrangement of certain Families in Liliales has been strongly criticized, particularly the submergence of Amaryllidaceae into Liliaceae

7.2.9 Rolf M.T. Dahlgren's system of classification (1932-1987)

Professor Rolf M.T. Dahlgren (1932-1987)

7.2.9.1 Rolf M.T. Dahlgren (1932-1987)

Rolf M.T. Dahlgren, Professor at the Botanical Museum of the University of Copenhagen, Denmark, proposed a new phylogenetic system of classification and presented a method of illustrating an angiosperm system as an imaginary phylogenetic shrub in transection (Fig. 7.18), which first appeared in *A Textbook of Angiosperm Taxonomy* in 1974 (Dahlgren et al. 1974). This system was however later revised and a revised system with a diagram, was presented in 1975. A complete revision again appeared in 1980 wherein the author has stated that the new classification is still provisional.

7.2.9.2 Rolf M.T. Dahlgren's system of classification

Dahlgren's system of classification is based on characters drawn from several branches of botany including anatomy, embryology and chemistry. Dahlgren believed that none of the extant groups of flowering plants is ancestral to any present-day group. He was of the view that the Angiosperms are of monophyletic origin and have evolved from one particular line of Gymnosperms. Among Dicotyledons, Magnoliiflorae was considered most primitive and among Monocotyledons, Alismatiflorae has been regarded as most primitive.

The Angiosperms (Magnoliopsida) represent a Class, which has been divided into two Subclasses, Magnoliidae and Liliidae, corresponding to the Dicotyledons and Monocotyledons. Magnoliidae has been further divided into 24 Superorders, 80 Orders and 346 Families, while Liliidae has been divided into 7 Superorders, 26 Orders and 92 Families, as shown in the following table (Fig. 7.17) :

Synopsis of Dahlgren's system of classification (1980)

	Class : Magnoliopsida	
	Subclass : Magnoliidae (Dicotyledons)	Subclass : Liliidae (Monocotyledons)
Number of Superorders	24	6
Number of Orders	80	26
Number of Families	346	92

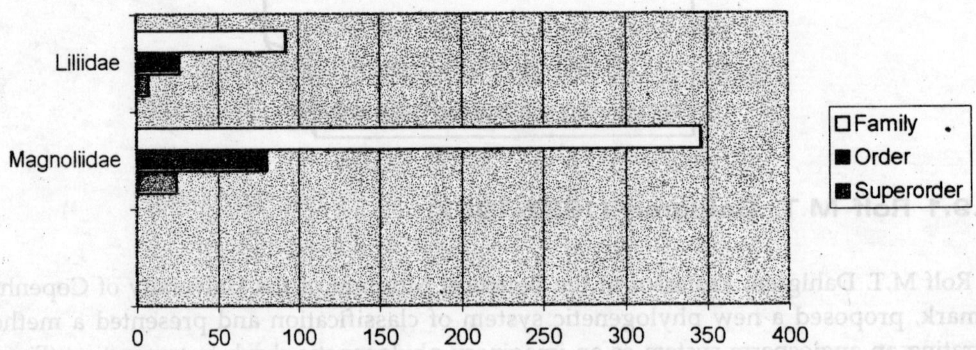

Fig. 7.17 : Statistical analysis of the two Subclasses of Class Magnoliopsida according to Dahlgren's system of classification

The division of the Class Magnoliopsida into two main Subclasses, Magnoliidae and Liliidae has been considered somewhat arbitrary and would probably not stand a test according to cladistic methods. However, Liliidae stands out as a possibly monophyletic group by virtue of their single cotyledon and characteristic triangular protein bodies in the sieve tube plastids (Behnke, 1969), in addition to the less constant differences from Magnoliidae.

Dahlgren illustrated the angiosperm system as an imaginary phylogenetic shrub (Fig. 7.18) to depict the placement of variously shaped figures representing Orders close to one another which are similar in several characters, thus reflecting their presumably phylogenetic relationships. The size of these figures correspond to the number of species in the Orders and the extensions of the figures to various directions indicate diversity of characters within them.

PG183CH7FIG7.18

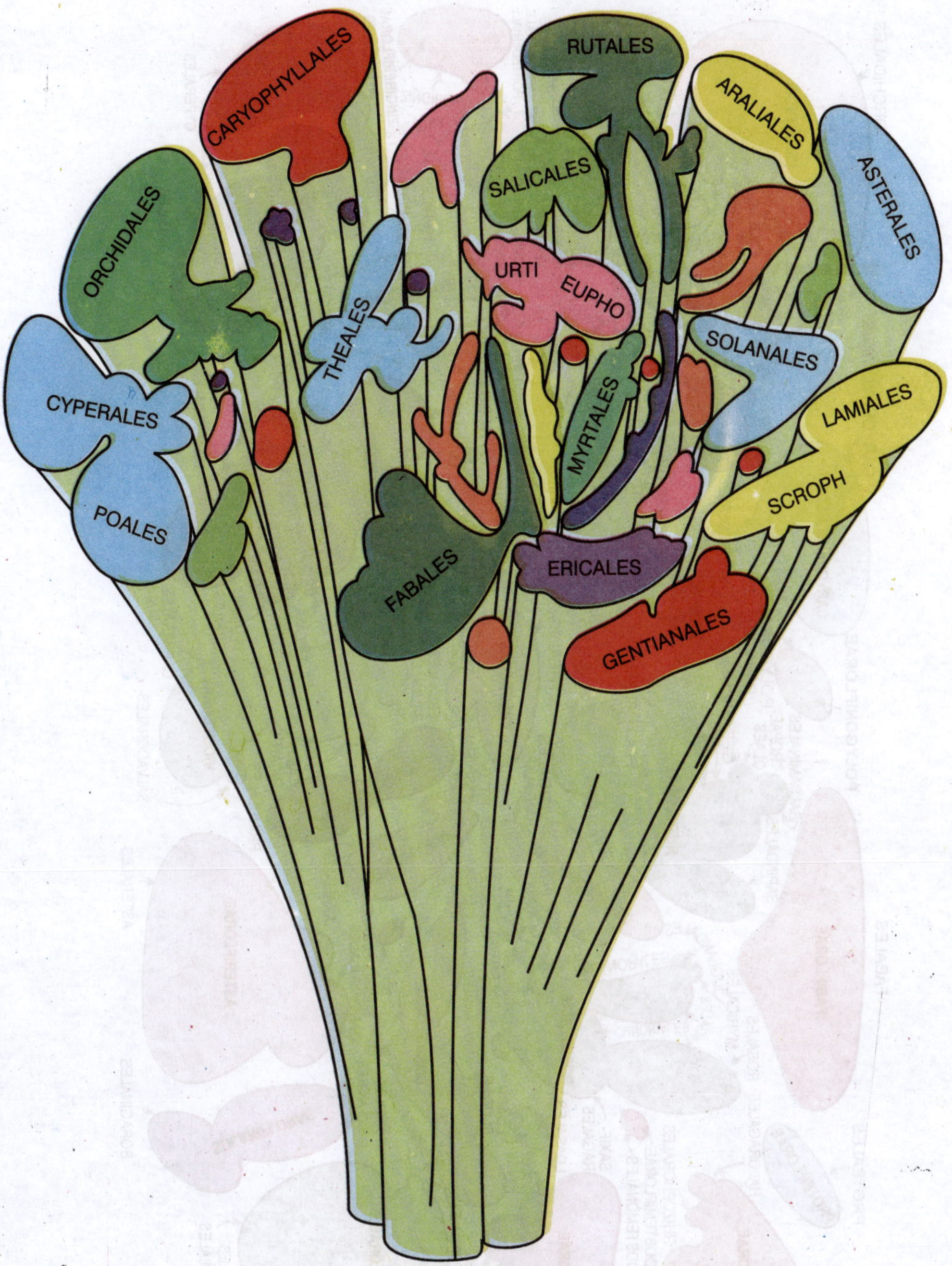

PG183CH7FIG7.18

Fig. 7.18 : Phylogenetic shrub (in cross section) by Dahlgren's system of classification (Adopted from Prof. K. Kubitzki, 1977)

Fig. 7.19 : The Superorders of Angiosperms illustrated as clusters of Orders, as seen by Dahlgren (1980)

An outline of Dahlgren's system of classification (1980)

Class : Magnoliopsida
Subclass : Magnoliidae (Dicotyledons)
Superorder 1. Magnoliiflorae
Orders :
1. Annonales (has 4 families)
2. Illiciales (has 2 families
3. Laurales (has 8 families)
4. Nelumbonales (has 1 family)
5. Aristolochiales (has 1 family)
6. Rafflesiales (has 2 families)
7. Magnoliales (has 6 families)

Superorder 2. Nymphaeaflorae
Orders :
1. Nymphaeales (has 3 families)
2. Piperales (has 2 families)

Superorder 3. Ranunculiflorae
Orders :
1. Papaverales (has 2 families)
2. Ranunculales (has 8 families)

Superorder 4. Caryophylliflorae
Order :
1. Caryopyllales (has 9 families)

Superorder 5. Polygoniflorae
Order :
1. Polygonales (has 1 family)

Superorder 6. Malviflorae
Orders :
1. Paeoniales (has 1 family)
2. Elaeagnales (has 1 family)
3. Dilleniales (has 1 family)
4. Malvales (has 13 families)
5. Plumbaginales (has 2 families)
6. Thymelaeales (has 1 family)
7. Urticales (has 6 families)
8. Euphorbiales (has 6 families)
9. Rhamnales (has 1 family)

Superorder 7. Violiflorae
Orders :
1. Violiflorae (has 13 families)
2. Salicales (has 1 family)
3. Tamaricales (has 2 families)
4. Capparales (has 7 families)
5. Salvadorales (has 1 family)

Superorder 8. Theiflorae

Orders :

1. Theales (has 17 families)
2. Droserales (has 3 families)

Superorder 9. Primuliflorae

Orders :

1. Ebenales (has 4 families)
2. Primulales (has 5 families)

Superorder 10. Rosiflorae

Orders :

1. Trochodendrales (has 4 families)
2. Hamamelidales (has 4 families
3. Fagales (has 3 families)
4. Balanopales (has 1 family)
5. Juglandales (has 2 families)
6. Myricales (has 1 family)
7. Casuarinales (has 1 family)
8. Buxales (has 2 families)
9. Cunoniales (has 8 families)
10. Saxifragales (has 7 families)
11. Gunnerales (has 1 family)
12. Rosales (has 3 families)
13. Malales (has 2 families)

Superorder 11. Podostemiflorae

Order :

1. Podostemales (has 1 family)

Superorder 12. Fabiflorae

Order :

1. Fabales (has 3 families)

Superorder 13. Proteiflorae

Order :

1. Proteales (has 1 family)

Superorder 14. Myrtiflorae

Orders :

1. Myrtales (has 12 families)
2. Haloragales (has 1 family)
3. Rhizophorales (has 1 family)

Superorder 15. Rutiflorae

Orders :

1. Rutales (has 6 families)
2. Sapindales (has 15 families)
3. Balsaminales (has 1 family)
4. Polygalales (has 5 families)
5. Geraniales (has 16 families)
6. Tropaeolales (has 2 families)

Superorder 16. Santaliflorae

Orders :

1. Celastrales (has 5 families)
2. Vitales (has 1 family)
3. Santalales (has 7 families)

Superorder 17. Balanoflorae

Order :

1. Balanophorales (has 2 families)

Superorder 18. Araliiflorae

Orders :

1. Pittosporales (has 3 families)
2. Araliales (has 3 families)

Superorder 19. Asteriflorae

Orders :

1. Campanulales (has 3 families)
2. Asterales (has 1 family)

Superorder 20. Solaniflorae

Orders :

1. Solanales (has 7 families)
2. Boraginales (has 6 families)

Superorder 21. Corniflorae

Orders :

1. Fouquieriales (has 1 family)
2. Serraceniales (has 1 family)
3. Ericales (has 10 families)
4. Eucomiales (has 1 family)
5. Cornales (has 29 families)
6. Dipsacales (has 7 families)

Superorder 22. Loasiflorae

Order :

1. Loasales (has 1 family)

Superorder 23. Gentianiflorae

Orders :

1. Goodeniales (has 1 family)
2. Gentianales (has 6 families)
3. Oleales (has 1 family)

Superorder 24. Lamiflorae

Orders :

1. Scrophulariales (has 18 families)
2. Hydrostachyales (has 1 family)
3. Hippuridales (has 1 family)
4. Lamiales (has 3 families)

Subclass : Liliidae

Suporder 1. Alismatiflorae

Orders :

1. Hydrocharitales (has 3 families)
2. Zosterales (has 8 families)
3. Alismatales (has 1 family)

Superorder 2. Triuridiflorae

Order :

1. Triuridales (has 1 family)

Superorder 3. Ariflorae

Order :

1. Arales (2 families)

Superorder 4. Liliifloræ

Orders :

1. Dioscoreales (has 5 families)
2. Asparagales (has 31 families)
3. Hydatellales (has 1 family)
4. Liliales (has 8 families)
5. Burmanniales (has 3 families)
6. Philydrales (has 1 family)
7. Orchidales (has 3 families)
8. Velloziales (has 1 family)
9. Bromeliales (has 1 family)
10. Haemodorales (has 1 family)
11. Pontederiales (has 1 family)
12. Typhales (has 2 families)

Superorder 5. Zingiberiflorae

Order :

1. Zingiberales (has 8 families)

Superorder 6. Commeliniflorae

Orders :

1. Commelinales (has 2 families)
2. Eriocaulales (has 2 families)
3. Juncales (has 1 family)
4. Cyperales (has 1 family)
5. Poales (has 5 families)

Superorder 6. Areciflorae

Orders :

1. Arecales (has 1 family)
2. Cyclanthales (has 1 family)
3. Pandanales (has 1 family)

7.2.9.3 Merits of Dahlgren's system of classification

(a) It is a modification of Takhtajan's original plan taking into account additional data based on anatomy, embryology and chemistry.

(b) The Dicots (Magnoliopsida) have been discussed prior to Monocots (Liliopsida).

(c) The Angiosperm system has been illustrated as a phylogenetic shrub depicting the relation between Orders. The closely related Orders have been placed close to each other.

(d) Dahlgren (1980) has attempted to clarify most of the doubts and objections raised by Corner. There will continue to be two distinct groups of investigators working in two widely different areas of taxonomic research i.e., one group which follows the basic philosophy of intensive investigation in limited groups like Prof. Corner, and the second group attempting extensive general purpose investigation. Since their approaches are vastly different, there is bound to be controversial assessment of the results achieved by them.

7.2.9.4 Demerits of Dahlgren's system of classification

(a) This classification includes unnatural Orders and Superorders.

(b) The nomenclature of various groups adopted by Dahlgren has been strongly criticized and the name "Pseudon" has been suggested for a conjectural taxon.

(c) Dahlgren did not consider the polyphyletic origin of Angiosperms from different Gymnosperms. He rather believed that the combination of several very characteristic attributes like companion cells in the phloem, 8-nucleate embryo sac, secondary endosperm, etc. would hardly have evolved independently in different gymnospermous groups.

7.2.10 Robert Thorne's (1983) system of classification

Robert F. Throne

7.2.10.1 Robert Thorne's (1983) system of classification

Robert Thorne (1983) proposed a pure phylogenetic system of classification on the basis of characters of comparative morphology, cytology, embryology, paleobotany, plant geography, pollen and seed morphology, ultrastructure, and host-parasite relationships. He also believed in the monophyletic origin of angiosperms.

He treats the flowering plants at the rank of Class with an initial bifurcation into two Subclasses (Fig. 7.21). Related Orders are placed within each Subclass as Superorders and within these, clusters of related Families are grouped into Suborders. He divided the Class Angiospermae (Annonopsida) into 2 Subclasses – Dicotyledonae (Magnoliidae) and Monocotyledonae (Liliidae), which were further subdivided into 28 Superorders, 54 Orders, 73 Suborders, 350 Families, 12,255 genera and 2,25,490 Species. A statistical summary of Thorne's system of classification is as follows (Fig. 7.20) :

	Subclass : Dicotyledoneae	Subclass : Monocotyledoneae	Total number of Angiosperms
Superorders	19	9	28
Orders	41	12	53
Suborders	56	17	73
Families	297	53	350
Subfamilies	350	102	452
Genera	9,640	2,615	12,255
Species	1,73,370	52,120	2,25,490

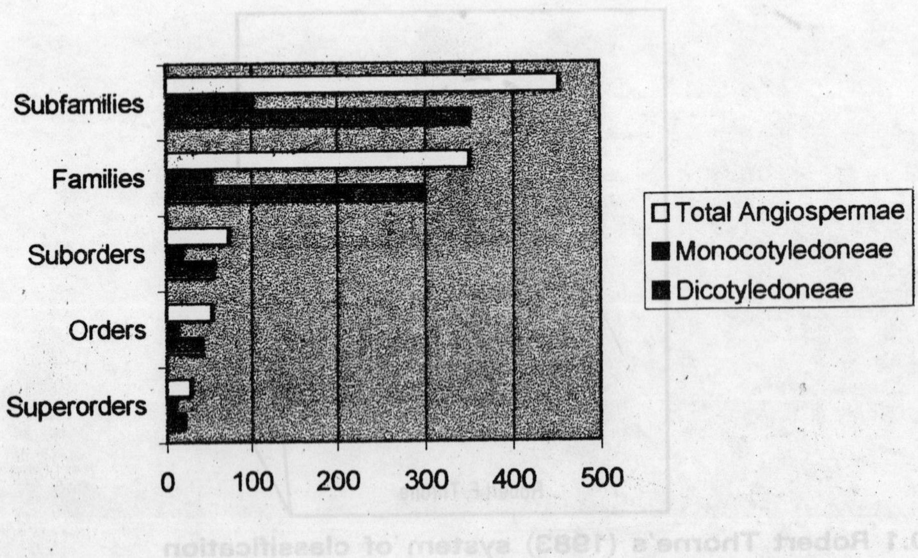

Fig. 7.20 : A statistical synopsis of various taxa of Thorne's system of plant classification

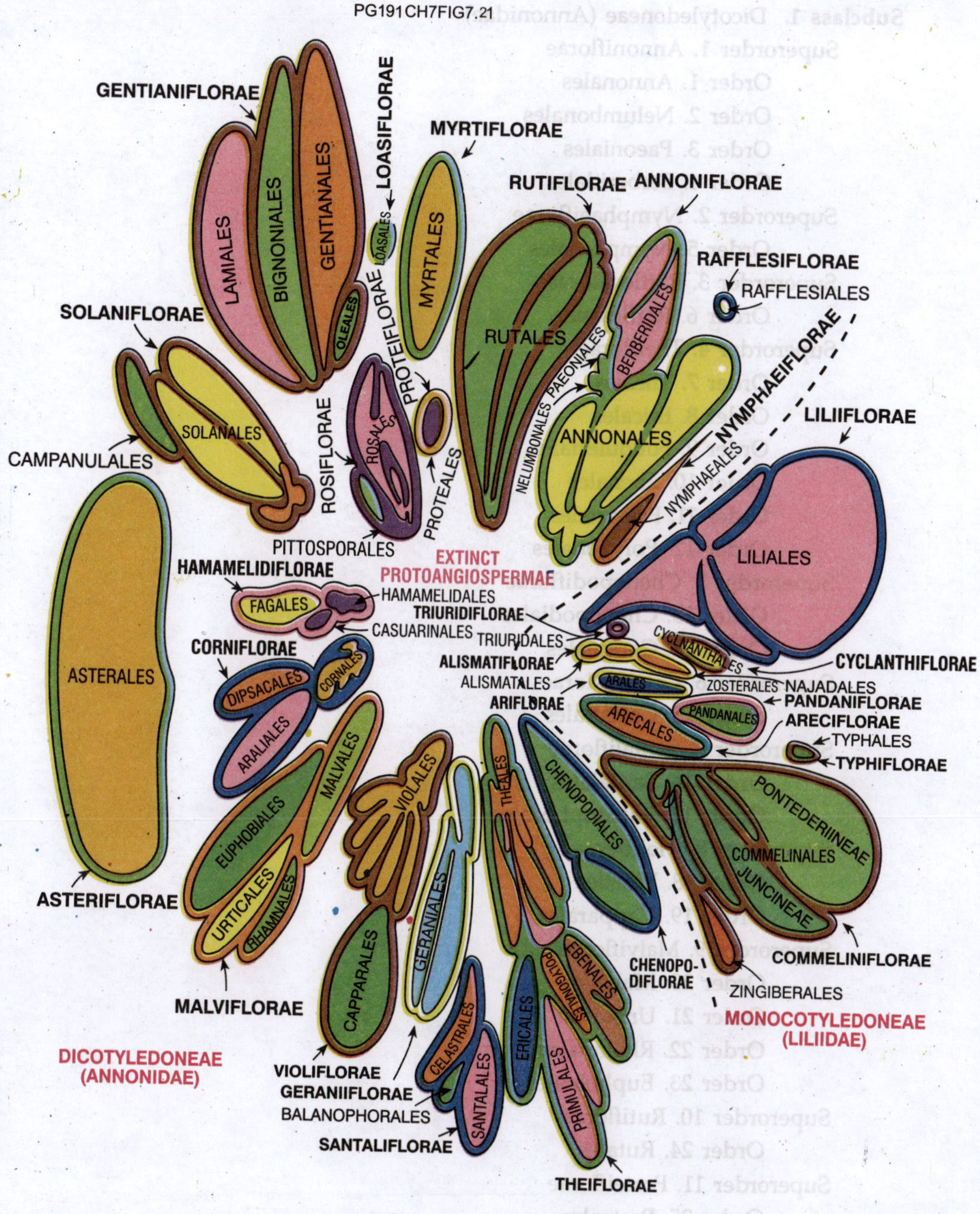

Fig. 7.21 : A diagrammatic representation of the evolutionary interrelationships of the extinct angiosperms, as seen by Thorne (1983)

An outline of Thorne's system of classification (1983) upto Orders

Class : Angiospermae (Annonopsida)

- **Subclass 1.** Dicotyledoneae (Annonidae)
  - Superorder 1. Annoniflorae
    - Order 1. Annonales
    - Order 2. Nelumbonales
    - Order 3. Paeoniales
    - Order 4. Berberidales
  - Superorder 2. Nymphaeiflorae
    - Order 5. Nymphaeales
  - Superorder 3. Rafflesiflorae
    - Order 6. Rafflesiales
  - Superorder 4. Theiflorae
    - Order 7. Theales
    - Order 8. Ericales
    - Order 9. Fouquieriales
    - Order 10. Ebenales
    - Order 11. Primulales
    - Order 12. Polygonales
  - Superorder 5. Chenopodiflorae
    - Order 13. Chenopodiales
    - Order 14. Geraniales
  - Superorder 6. Celastriflorae
    - Order 15. Celastrales
  - Superorder 7. Santaliflorae
    - Order 16. Santalales
    - Order 17. Balanophorales
  - Superorder 8. Violiflorae
    - Order 18. Violales
    - Order 19. Capparales
  - Superorder 9. Malviflorae
    - Order 20. Malvales
    - Order 21. Urticales
    - Order 22. Rhamnales
    - Order 23. Euphobiales
  - Superorder 10. Rutiflorae
    - Order 24. Rutales
  - Superorder 11. Proteiflorae
    - Order 25. Proteales
  - Superorder 12. Hamamelidiflorae
    - Order 26. Hamamelidales
    - Order 27. Casuarinales
    - Order 28. Fagales

Superorder 13. Rosiflorae
- Order 29. Rosales
- Order 30. Pittosporales

Superorder 14. Loasiflorae
- Order 31. Loasales

Superorder 15. Myrtiflorae
- Order 32. Myrtales

Superorder 16. Gentianiflorae
- Order 33. Oleales
- Order 34. Gentianales
- Order 35. Bignoniales
- Order 36. Lamiales

Superorder 17. Solaniflorae
- Order 37. Solanales
- Order 38. Campanulales

Superorder 18. Corniflorae
- Order 39. Cornales
- Order 40. Araliales
- Order 41. Dipsacales

Superorder 19. Asteriflorae
- Order 42. Asterales

Subclass 2. Monocotyledoneae (Liliidae)

Superorder 1. Liliiflorae
- Order 1. Liliales

Superorder 2. Triuridiflorae
- Order 2. Triuridales

Superorder 3. Alismatiflorae
- Order 3. Alismatales
- Order 4. Zosterales
- Order 5. Najadales

Superorder 4. Ariflorae
- Order 6. Arales

Superorder 5. Cyclanthiflorae
- Order 7. Cyclanthales

Superorder 6. Pandaniflorae
- Order 8. Pandanales

Superorder 7. Areciflorae
- Order 9. Arecales

Superorder 8. Typhiflorae
- Order 10. Typhales

Superorder 9. Commeliniflorae
- Order 11. Commelinales
- Order 12. Zingiberales

7.2.10.2 Merits of Thorne's system of classification

1. Annonales which forms the starting point of this system is universally accepted as the most primitive living angiosperms.
2. In this system, closely related taxa are by and large placed nearer to one another.
3. The Orders Malvales, Urticales, Rhamnales and Euphorbiales have been included under one Superorder Malviflorae, which is highly satisfactory.
4. The related Orders Cornales and Dipsacales have also been placed under one Superorder Corniflorae, which is also the merit of this system
5. The families of Amentiferae (a group of unrelated families) have been distributed into different Orders and the group Amentiferae has been abolished.

7.2.10.3 Demerits of Thorne's system of classification

(a) It is not of much practical utility in the identification of plants.

(b) Thorne's view that angiosperms might have originated from some Pteridospermous members in early Cretaceous times is not accepted by several taxonomists.

7.2.10.4 An outline of the revised Robert F. Thorne system of classification (*Classification and Geography of Flowering Plants*. Botanical Review 58: 225-348 as updated December, 1992 and March, 1999)

Subclass : Magnoliidae (Dicots)	Subclass : Liliidae (Monocots)
1. Magnolianae	1. Triuridanae
2. Nymphaeanae	2. Acoranae
3. Rafflesianae	3. Aranae
4. Caryophyllanae	4. Alismatanae
5. Dillenianae	5. Lilianae
6. Celastranae	6. Pandanaranae
7. Malvanae	7. Arecanae
8. Violanae	8. Hydatellanae
9. Santalanae	9. Commelinanae
10. Geranianae	
11. Rutanae	
12. Proteanae	
13. Rosanae	
14. Aralianae	
15. Asteranae	
16. Solananae	
17. Loasanae	
18. Myrtanae	
19. Lamianae	

DICOTS

Superorder 1. Magnolianae
Order 1. Magnoliales
Suborder 1. Illiciineae
Families:
1. Illiceae
2. Schisandraceae

Suborder 2. Magnoliineae
Families:
1. Magnoliaceae
   Subfamilies:
   Magnolioideae
   Liriodendroideae
2. Degeneriaceae
3. Himantandraceae
4. Eupomatiaceae
5. Annonaceae
6. Aristolochiaceae
   Subfamilies:
   Asaroideae
   Aristolochioideae
7. Myristicaceae
8. Canellaceae

Suborder 3. Austrobaileyinea
Family :
1. Austrobaileyaceae

Suborder 4. Laurineae
Families:
1. Amborellaceae
2. Trimeniaceae
3. Chloranthaceae
4. Monimiaceae
   Subfamilies:
   Hortonioideae
   Monimioideae
   Mollinedioideae
   Atherospermatoidea
   Siparunoideae
   Glossocalycoideae
5. Gomortegaceae
6. Calycanthaceae
   Subfamilies:
   Idiospermoideae
   Calycanthoideae
7. Lauraceae
   Subfamilies:
   Lauroideae
   Cassythoideae
8. Hernandiaceae
   Subfamilies :
   Hernandioideae
   Gyrocarpoideae
   Sparattanthelioide

Suborder 5. Piperineae
Families:
1. Lactoridaceae
2. Saururaceae
3. Piperaceae
   Subfamilies:
   Piperoideae
   Peperomioideae

Order 2. Ceratophyllales
Family:
1. Ceratophyllaceae

Order 3. Nelumbonales
Family:
1. Nelumbonaceae

Order 4. Paeoniales
Families:
1. Paeoniaceae
2. Glaucidiaceae

Order 5. Berberidales
Suborder 1. Berberidineae
Families:
1. Menispermaceae
2. Lardizabalaceae
   Subfamilies:
   Decaisneoideae
   Lardizabaloideae
3. Sargentodoxaceae
4. Berberidaceae
   Subfamilies:
   Nandinoideae
   Berberidoideae
   Leonticoideae
   Epimedioideae
5. Hydrastidaceae

6. Ranunculaceae
   Subfamilies:
   Isopyroideae
   Ranumculoideae
7. Circaeasteraceae

Suborder 2. Papaverineae

Family:

1. Papaveraceae
   Subfamilies:
   Platystemonoideae
   Papaveroideae
   Eschscholzioideae
   Pteridophylloideae
   Hypecoideae
   Fumarioideae

Superorder 2. Nymphaeanae

Order 6. Nymphaeales

Families:

1. Cabombaceae
2. Nymphaeaceae
   Subfamilies:
   Nymphaeoideae
   Euryaloideae
   Barclayoideae

Superorder 3. Rafflesianae

Order 7. Rafflesiales

Families:

1. Hydnoraceae
2. Rafflesiaceae
   Subfamilies:
   Mitrastemonoideae
   Cytinoideae
   Apodanthoideae
   Rafflesioideae

Superorder 4. Caryophyllanae

Order 8. Caryophyllales

Suborder 1. Caryophyllineae

Families:

1. Caryophyllaceae
   Subfamilies:
   Alsinoideae
   Paronychioideae
   Paronychioideae
   Caryophylloideae
2. Molluginaceae

Suborder 2. Achatocarpineae

Families:

1. Achatocarpaceae

Suborder 3. Portulacineae

Families:

1. Portulacaceae
2. Hectorellaceae
3. Basellaceae
4. Didiereaceae
5. Cactaceae

Suborder 4. Phytolaccineae

Families:

1. Phytolaccaceae
   Subfamilies:
   Phytolaccoideae
   Gisekioideae
2. Nyctaginaceae
3. Aizoaceae
   Subfamilies:
   Aizooideae
   Aptenioideae
   Ruschioideae
   Sesuvioideae
   Tetragonioideae
4. Halophytaceae
5. Stegnospermataceae

Suborder 5. Chenopodiineae

Families:

1. Chenopodiaceae
   Subfamilies:
   Chenopodioideae
   Salicornioideae
   Salsoloideae
2. Amaranthaceae
   Subfamilies:
   Amaranthoideae
   Gomphrenoideae

Superorder 5. Dillenianae

Order 9. Dilleniales

Suborder 1. Dilleniineae

Family:

1. Dilleniaceae
   Subfamilies:
   Dillenioideae
   Tetraceroideae

Suborder 2. Theineae

Families:

1. Actinidiaceae
   Subfamilies:
   Actinidioideae
   Saurauioideae
   Clematoclethroideae
2. Paracryphiaceae
3. Stachyuraceae
4. Theaceae
   Subfamilies:
   Ternstroemioideae
   Theoideae
   Sladenioideae
5. Asteropeiaceae
6. Tetrameristaceae
7. Pellicieraceae
8. Chrysobalanaceae
9. Symplocaceae
10. Caryocaraceae
11. Marcgraviaceae
12. Oncothecaceae
13. Aquifoliaceae
14. Phellinaceae
15. Sphenostemonaceae

Suborder 3. Sarraceniineae

Family:

1. Sarraceniaceae

Suborder 4. Clethrineae

Families:

1. Pentaphylacaceae
2. Clethraceae
3. Cyrillaceae

Suborder 5. Scytopetalineae

Families:

1. Ochnaceae
   Subfamilies:
   Ochnoideae
   Sauvagesioideae
2. Quiinaceae
3. Scytopetalaceae
4. Medusagynaceae
5. Strasburgeriaceae

Suborder 5. Nepenthineae

Families:

1. Ancistrocladaceae
2. Dioncophyllaceae
3. Droseraceae
4. Nepenthaceae

Suborder 6. Hypericineae

Families:

1. Bonnetiaceae
2. Clusiaceae
   Subfamilies:
   Kielmeyeroideae
   Calophylloideae
   Clusioideae
   Moronoboideae
   Hypericoideae
3. Elatinaceae

Suborder 7. Lecythidineae

Family:

1. Lecythidaceae
   Subfamilies:
   Planchonioideae
   Foetidoideae
   Lecythidoideae
   Napoleonoideae
   Asteranthoideae

Order 10. Ericales
Families:
1. Ericaceae
   Subfamilies:
   Rhododendroideae
   Ericoideae
   Vaccinioideae
   Pyroloideae
   Monotropoideae
2. Epacridaceae
3. Empetraceae

Order 11. Fouquieriales
Family:
1. Fouquieriaceae

Order 12. Styracales
Suborder 1. Ebenineae
Families:
1. Ebenaceae
2. Lissocarpaceae
3. Sapotaceae

Suborder 2. Styracineae
Family:
1. Styracaceae

Order 13. Primulales
Suborder 1. Primulineae
Families:
1. Theophrastaceae
2. Myrsinaceae
   Subfamilies:
   Myrsinoideae
   Maesoideae
3. Primulaceae

Suborder 2. Plumbaginineae
Family:
1. Plumbaginaceae
   Subfamilies:
   Plumbaginoideae
   Staticoideae

Order 14. Polygonales
Family:
1. Polygonaceae
   Subfamilies:
   Eriogonoideae
   Polygonoideae
   Coccoloboideae

Superorder 6. Celastranae
Order 15. Celastrales
Families:
1. Celastraceae
   Subfamilies:
   Celastroideae
   Tripterygioideae
   Cassinoideae
   Hippocrateoideae
   Siphonodontoideae
2. Goupiaceae
3. Lophopyxidaceae
4. Stackhousiaceae
   Subfamilies:
   Macgregorioideae
   Stackhousioideae

Superorder 7. Malvanae
Order 16. Malvales
Suborder 1. Sterculiineae
Families:
1. Sterculiaceae
   Subfamilies:
   Sterculioideae
   Byttnerioideae
2. Huaceae
3. Elaeocarpaceae
4. Plagiopteraceae
5. Tiliaceae
   Subfamilies:
   Brownlowioideae
   Tetralicoideae
   Tilioideae
   Neotessmannioideae
6. Monotaceae
   Subfamilies:
   Pakaramaeoideae
   Monotoideae
7. Dipterocarpaceae
8. Sarcolaenaceae
9. Sphaerosepalaceae
   Subfamilies:
   Sphaerosepaloideae
   Diegodendroideae

Suborder 2. Malvineae

Families:

1. Bombacaceae
2. Malvaceae

Order 17. Urticales

Families:

1. Ulmaceae
   Subfamilies:
   Celtidoideae
   Ulmoideae
2. Moraceae
3. Cecropiaceae
4. Urticaceae
5. Cannabaceae

Order 18. Rhamnales

Families:

1. Rhamnaceae
2. Elaeagnaceae

Order 19. Euphorbiales

Families:

1. Euphorbiaceae
   Subfamilies:
   Phyllanthoideae
   Oldfieldioideae
   Acalyphoideae
   Crotonoideae
   Euphorbioideae
   Pandoideae
2. Simmondsiaceae
3. Dichapetalaceae
4. Gonystylaceae
5. Thymelaeaceae
   Subfamilies:
   Aquilarioideae
   Gilgiodaphnoideae
   Thymelaeoideae

Superorder 8. Violanae

Order 20. Violales

Suborder 1. Cistineae

Families:

1. Bixaceae
2. Cochlospermaceae
3. Cistaceae
4. Violaceae
   Subfamilies:
   Violoideae
   Leonioideae
   Fusispermoideae
5. Flacourtiaceae
6. Lacistemataceae
7. Salicaceae
8. Dipentodontaceae
9. Peridiscaceae
10. Scyphostegiaceae
11. Passifloraceae
12. Turneraceae
13. Malesherbiaceae
14. Achariaceae
15. Caricaceae

Suborder 2. Tamaricineae

Families:

1. Tamaricaceae
2. Frankeniaceae

Suborder 3. Begoniineae

Families:

1. Cucurbitaceae
   Subfamilies:
   Zanonioideae
   Cucurbitoideae
2. Begoniaceae
3. Datiscaceae

Order 21. Brassicales

Families:

1. Resedaceae
2. Capparaceae
   Subfamilies:
   Tovarioideae
   Pentadiplandroideae
   Koeberlinioideae
   Capparoideae
   Cleomoideae
3. Brassicaceae
4. Salvadoraceae
5. Gyrostemonaceae

Order 22. Batales

Family:

1. Bataceae

Superorder 9. Santalanae

Order 23. Santalales

Families:

1. Olacaceae
   Subfamilies:
   Olacoideae
   Octoknemoideae
   Erythropaloideae
2. Opiliaceae
3. Medusandraceae
4. Santalaceae
5. Misodendraceae
6. Loranthaceae
7. Eremolepidaceae
8. Viscaceae

Order 24. Balanophorales

Families:

1. Balanophoraceae
   Subfamilies:
   Mystropetaloideae
   Dactylanthoideae
   Sarcophytoideae
   Helosidoideae
   Lophophytoideae
   Balanophoroideae
2. Cynomoriaceae

Superorder 10. Geranianae

Order 25. Linales

Families:

1. Humiriaceae
2. Ctenolophonaceae
3. Hugoniaceae
4. Ixonanthaceae
5. Linaceae
6. Erythroxylaceae
7. Zygophyllaceae
   Subfamilies:
   Peganoideae
   Morkillioideae
   Tetradiclidoideae
   Tribuloideae
   Neoluederitzioideae
   Zygophylloideae
   Augeoideae
   Nitrarioideae
8. Balanitaceae

Order 26. Rhizophorales

Family:

1. Rhizophoraceae

Order 27. Geraniales

Families:

1. Oxalidaceae
2. Geraniaceae
   Subfamilies:
   Geranioideae
   Biebersteinioideae
   Dirachmoideae
   Vivianioideae
   Ledocarpoideae
3. Balsaminaceae
4. Tropaeolaceae
5. Limnanthaceae

Order 28. Polygalales

Families:

1. Malpighiaceae
   Subfamilies:
   Byrsonimoideae
   Gaudichaudioideae
   Malpighioideae
2. Trigoniaceae
3. Vochysiaceae
4. Polygalaceae
5. Krameriaceae

Superorder 11. Rutanae
Order 29. Rutales
Suborder 1. Rutineae
Families:
1. Rutaceae
   Subfamilies:
   Rutoideae
   Citroideae
   Flindersioideae
   Spathelioideae
   Dictyolomatoideae
2. Rhabdodendraceae
3. Cneoraceae
4. Simaroubaceae
   Subfamilies:
   Simarouboideae
   Kirkioideae
   Irvingioideae
   Picramnioideae
   Alvaradoideae
5. Ptaeroxylaceae
6. Meliaceae
   Subfamilies:
   Melioideae
   Quivisianthoideae
   Capuronianthoideae
   Swietenioideae
7. Burseraceae
8. Anacardiaceae
9. Leitneriaceae
10. Tepuianthaceae

Suborder 2. Coriariineae
Family:
1. Coriariaceae

Suborder 3. Sapindineae
Families:
1. Sapindaceae
   Subfamilies:
   Dodonaeoideae
   Koelreuterioideae
   Stylobasioideae
   Emblingioideae
   Sapindoideae
   Hippocastanoideae
   Aceroideae
2. Sabiaceae
   Subfamilies:
   Meliosmoideae
   Sabioideae
3. Melianthaceae
4. Akaniaceae
5. Bretschneideraceae
6. Moringaceae

Suborder 4. Fabineae
Families:
1. Surianaceae
2. Connaraceae
   Subfamilies:
   Connaroideae
   Jollydoroideae
3. Fabaceae
   Subfamilies:
   Caesalpinoideae
   Mimosoideae
   Swartzioideae
   Faboideae

Superorder 12. Proteanae
Order 30. Proteales
Family:
1. Proteaceae
   Subfamilies:
   Persoonioideae
   Proteoideae
   Sphalmioideae
   Carnarvonioideae
   Grevilleoideae

Superorder 13. Rosanae
Order 31. Hamamelidales
Suborder 1. Trochodendrineae
Families:
1. Trochodendraceae
   Subfamilies:
   Trochodendroideae
   Tetracentroideae
2. Eupteleaceae
3. Cercidiphyllaceae

Suborder 2. Hamamelidineae

Families:

1. Platanaceae
2. Hamamelidaceae
   Subfamilies:
   Hamamelidoideae
   Rhodoleioideae
   Exbucklandioideae
   Altingioideae

Order 32. Casuarinales

Family:

1. Casuarinaceae

Order 33. Balanopales

Suborder 1. Buxineae

Families:

1. Buxaceae
   Subfamilies:
   Buxiodeae
   Styloceratoideae
2. Didymelaceae

Suborder 2. Balanopineae

Families:

1. Daphniphyllaceae
2. Balanopaceae

Order 34. Bruniales

Families:

1. Roridulaceae
2. Bruniaceae
3. Geissolomataceae
4. Grubbiaceae
5. Myrothamnaceae
6. Hydrostachyaceae

Order 35. Juglandales

Suborder 1. Jugandineae

Families:

1. Rhoipteleaceae
2. Juglandaceae

Suborder 2. Myricineae

Family:

1. Myricaceae

Order 36. Betulales

Families:

1. Ticodendraceae
2. Betulaceae
   Subfamilies:
   Betuloideae
   Coryloideae
3. Nothofagaceae
4. Fagaceae
   Subfamilies:
   Castanoideae
   Fagoideae

Order 37. Rosales

Families:

1. Rosaceae
   Subfamilies:
   Spiraeoideae
   Quillajoideae
   Rosoideae
   Amygdaloideae
   Maloideae
2. Neuradaceae
3. Crossosomataceae
4. Anisophylleaceae

Order 38. Saxifragales

Families:

1. Tetracarpaeaceae
2. Crassulaceae
   Subfamilies:
   Sedoideae
   Cotyledonoideae
   Crassuloideae
3. Cephalotaceae
4. Penthoraceae
5. Saxifragaceae
   Subfamilies:
   Astilboideae
   Saxifragoideae
6. Francoaceae
7. Grossulariaceae

8. Vahliaceae
9. Eremosynaceae
10. Lepuropetalaceae
11. Parnassiaceae
12. Greyiaceae
13. Diapensiaceae

Order 39. Podostemales

Family:

1. Podostemaceae
   Subfamilies:
   Tristichoideae
   Podostemoideae

Order 40. Cunoniales

Families:

1. Cunoniaceae
   Subfamilies:
   Cunonioideae
   Baueroideae
   Eucryphioideae
   Brunellioideae
2. Davidsoniaceae
3. Staphyleaceae
   Subfamilies:
   Staphyleoideae
   Tapiscioideae

Superorder 14. Aralianae

Order 41. Brexiales

Families:

1. Hydrangeaceae
   Subfamilies:
   Philadelphoideae
   Kirengeshomoideae
   Hydrangeoideae
2. Escalloniaceae
   Subfamilies:
   Escallonioideae
   Iteoideae
   Tribeloideae
   Phyllonomoideae
   Pterostemonoideae
3. Griseliniaceae
4. Alseuosmiaceae
5. Montiniaceae
6. Brexiaceae
7. Columelliaceae
8. Desfontainiaceae

Order 42. Cornales

Suborder 1. Vitineae

Family:

1. Vitaceae
   Subfamilies:
   Vitoideae
   Leeoideae

Suborder 2. Gunnerineae

Family:

1. Gunneraceae

Suborder 3. Halogagineae

Family:

1. Haloragaceae

Suborder 4. Cornineae

Families:

1. Cornaceae
   Subfamilies:
   Davidioideae
   Nyssoideae
   Mastixioideae
   Cornoideae
2. Curtisiaceae
3. Alangiaceae
4. Garryaceae
5. Aucubaceae
6. Aralidiaceae
7. Eucommiaceae
8. Icacinaceae
9. Cardiopteridaceae

Order 43. Pittosporales

Families:

1. Pittosporaceae
2. Byblidaceae
3. Tremandraceae

Order 44. Araliales

Families:

1. Helwingiaceae
2. Torricelliaceae
3. Araliaceae
4. Hydrocotylaceae
5. Apiaceae
   Subfamilies:
   Saniculoideae
   Apioideae

Order 45. Dipsacales

Families:

1. Caprifoliaceae
2. Adoxaceae

   Subfamilies:

   Adoxoideae

   Viburnoideae
3. Valerianaceae
4. Triplostegiaceae
5. Dipsacaceae
6. Morinaceae

Superorder 15. Asteranae

Order 46. Asterales

Families:

1. Calyceraceae
2. Asteraceae

   Subfamilies:

   Barnadesioideae

   Lactucoideae

   Asteroideae

Order 47. Campanulales

Families:

1. Menyanthaceae
2. Pentaphragmataceae
3. Sphenocleaceae
4. Campanulaceae

   Subfamilies:

   Campanuloideae

   Cyphioideae

   Lobelioideae
5. Cyphiaceae
6. Lobeliaceae
7. Donatiaceae
8. Stylidiaceae
9. Goodeniaceae

   Subfamilies:

   Goodenioideae

   Dampieroideae

Superorder 16. Solananae

Order 48. Solanales

Suborder 1. Solanineae

Families:

1. Solanaceae

   Subfamilies:

   Solanoideae

   Cestroideae

   Sclerophylacoideae
2. Duckeodendraceae
3. Goetzeaceae
4. Convolvulaceae

   Subfamilies:

   Humbertioideae

   Dichondroideae

   Convolvuloideae

   Cuscutoideae

Suborder 2. Boraginineae

Families:

1. Hydrophyllaceae
2. Boraginaceae

   Subfamilies:

   Ehretioideae

   Cordioideae

   Heliotropoideae

   Boraginoideae

   Wellstedioideae
3. Hoplestigmataceae
4. Lennoaceae
5. Tetrachondraceae

Suborder 3. Polemoniineae

Family:

1. Polemoniaceae

Superorder 17. Loasanae

Order 49. Loasales

Family:

1. Loasaceae

   Subfamilies:

   Mentzelioideae

   Loasoideae

   Gronovioideae

Superorder 18. Myrtanae
Order 50. Myrtales
Suborder 1. Lythrineae
Families:
1. Lythraceae
   Subfamilies:
   Lythroideae
   Duabangoideae
   Sonneratioideae
   Punicoideae
2. Alzateaceae
3. Rhynchocalycaceae
4. Penaeaceae
5. Oliniaceae
6. Trapaceae
7. Crypteroniaceae
8. Melastomataceae
   Subfamilies:
   Melastomatoideae
   Memecyloideae
9. Combretaceae
   Subfamilies:
   Strephonematoideae
   Combretoideae

Suborder 2. Onagraneae
Family:
1. Onagraceae

Suborder 3. Myrtineae
Family:
1. Myrtaceae
   Subfamilies:
   Heteropyxidoideae
   Myrtoideae

Superorder 19. Lamianae
Order 51. Rubiales
Families:
1. Loganiaceae
   Subfamilies:
   Loganioideae
   Potalioideae
   Plocospermatoideae
2. Rubiaceae
   Subfamilies:
   Cinchonoideae
   Antirrheoideae
   Ixoroideae
   Rubioideae
3. Dialypetalanthaceae
4. Apocynaceae
   Subfamilies:
   Plumerioideae
   Apocynoideae
   Periplocoideae
   Secamonoideae
   Asclepiadoideae
5. Gentianaceae
6. Saccifoliaceae

Order 52. Lamiales
Suborder 1. Scrophulariineae
Families:
1. Oleaceae
   Subfamilies:
   Jasminoideae
   Oleoideae
2. Buddlejaceae
3. Stilbaceae
   Subfamilies:
   Retzioideae
   Stilboideae
4. Pedaliaceae
5. Martyniaceae
6. Myoporaceae
7. Scrophulariaceae
   Subfamilies:
   Scrophularioideae
   Rhinanthoideae
   Orobanchoideae
8. Plocospermataceae
9. Gesneriaceae
   Subfamilies:
   Gesnerioideae
   Coronantheroideae
   Cyrtandroideae
10. Globulariaceae
11. Plantaginaceae
12. Lentibulariaceae
13. Acanthaceae
    Subfamilies:
    Nelsonioideae
    Thunbergioideae
    Mendoncioideae
    Acanthoideae
    Ruellioideae
14. Callitrichaceae
15. Hippuridaceae

Suborder 2. Lamiineae
Families:
1. Verbenaceae
2. Symphoremataceae
3. Nesogenaceae
4. Avicenniaceae
5. Lamiaceae
Subfamilies:
Chloanthoideae
Teucrioideae
Ajugoideae
Scutellarioideae
Pogostemonoideae
Lamioideae
Nepetoideae

MONOCOTS

Superorder 1. Triuridanae
Order 1. Triuridales
Families:
1. Petrosaviaceae
2. Triuridaceae

Superorder 2. Acoranae
Order 2. Nartheciales
Families:
1. Tofieldiaceae
2. Nartheciaceae

Order 3. Acorales
Family:
1. Acoraceae

Superorder 3. Aranae
Order 4. Arales
Families:
1. Araceae
Subfamilies:
Gymnostachyoideae
Orontioideae
Pothoideae
Monsteroideae
Lasioideae
Calloideae
Aroideae
2. Lemnaceae
Subfamilies:
Lemnoideae
Wolffioideae

Superorder 4. Alismatanae
Order 5. Alismatales
Families:
1. Butomaceae
2. Limnocharitaceae
3. Alismataceae
4. Hydrocharitaceae
Subfamilies:
Hydrocharitoideae
Thalassioideae
Halophiloideae

Order 6. Potamogetonales
Suborder 1. Aponogetonineae
Family:
1. Aponogetonaceae

Suborder 2. Potamogetonineae
Families:
1. Scheuchzeriaceae
2. Posidoniaceae
3. Cymodoceaceae
4. Ruppiaceae
5. Juncaginaceae
Subfamilies:
Juncaginoideae
Lilaeoideae
6. Potamogetonaceae
7. Zosteraceae
8. Zannichelliaceae

Superorder 5. Lilianae
Order 7. Liliales
Families:
1. Campynemataceae
2. Melanthiaceae
3. Trilliaceae
4. Alstroemeriaceae
5. Colchicaceae
6. Rhipogonaceae
7. Smilacaceae
8. Philesiaceae
9. Luzuriagaceae
10. Calochortaceae
11. Liliaceae
Subfamilies:
Medeoloideae
Lilioideae
12. Tricyrtidaceae

Order 8. Taccales
Families:
1. Dioscoreaceae
2. Trichopodaceae
3. Pentastemonaceae
4. Stemonaceae
5. Taccaceae

Order 9. Burmanniales
Families:
1. Burmanniaceae
2. Thismiaceae
3. Corsiaceae

Order 10. Orchidales
Suborder 1. Orchidineae
Families:
1. Boryaceae
2. Hypoxidaceae
3. Orchidaceae
   Subfamilies:
   Apostasioideae
   Vanilloideae
   Cypripedioideae
   Spiranthoideae
   Orchidoideae
   Epidendroideae
4. Lanariaceae
5. Asteliaceae
6. Blandfordiaceae

Suborder 2. Iridineae
Families :
1. Tecophilaeaceae
2. Doryanthaceae
3. Ixioliriaceae
4. Iridaceae
   Subfamilies:
   Isophysidoideae
   Nivenioideae
   Iridoideae
   Ixioideae

Suborder 3. Asphodelineae
Families:
1. Xeronemaceae
2. Xanthorrhoeaceae
3. Asphodelaceae
   Subfamilies:
   Asphodeloideae
   Alooideae
4. Hemerocallidaceae
   Subfamilies:
   Hemerocallidoideae
   Phormioideae
5. Johnsoniaceae

Suborder 4. Amaryllidineae
Families:
1. Anthericaceae
2. Laxmanniaceae
   Subfamilies:
   Lomandroideae
   Laxmannioideae
3. Alliaceae
   Subfamilies:
   Allioideae
   Tulbaghioideae
   Gilliesioideae
4. Agapanthaceae
5. Amaryllidaceae
6. Hyacinthaceae
7. Themidaceae
8. Aphyllanthaceae
9. Behniaceae
10. Anemarrhenaceae
11. Herreriaceae
12. Agavaceae
    Subfamilies:
    Yuccoideae
    Agavoideae
13. Hesperocallidaceae

Suborder 5. Asparagineae
Families:
1. Asparagaceae
2. Dracaenaceae
3. Nolinaceae
4. Ruscaceae
5. Convallariaceae
6. Eriospermaceae

Superorder 6. Pandananae
Order 11. Pandanales
Family:
1. Pandanaceae
   Subfamilies:
   Pandanoideae
   Freycinetioideae

Order 12. Cyclanthales
Family:
1. Cyclanthaceae
   Subfamilies:
   Carludovicoideae
   Cyclanthoideae

Order 13. Velloziales
Families:
1. Velloziaceae
   Subfamilies:
   Vellozoideae
   Xerophytoideae
2. Acanthochlamydaceae

Superorder 7. Arecanae

Order 14. Arecales
Family:
1. Arecaceae
   Subfamilies:
   Coryphoideae
   Nypoideae
   Calamoideae
   Ceroxyloideae
   Arecoideae
   Phytelephoideae

Order 15. Dasypogonales
Families:
1. Dasypogonaceae
2. Xanthorrhoeaceae
   Subfamilies:
   Dasypogonoideae
   Kingioideae

Order 16. Hanguanales
Family:
1. Hanguanaceae

Superorder 8. Hydatellanae

Order 17. Hydatellales
Family:
1. Hydatellaceae

Superorder 9. Commelinanae

Order 18. Commelinales
Families:
1. Commelinaceae
   Subfamilies:
   Cartonematoideae
   Commelinoideae
2. Haemodoraceae
   Subfamilies:
   Haemodoroideae
   Conostylidoideae
3. Pontederiaceae
4. Philydraceae

Order 19. Cannales
Suborder 1. Musineae
Family:
1. Musaceae

Suborder 2. Strelitzineae
Family:
1. Strelitziaceae

Suborder 3. Lowiineae
Family:
1. Lowiaceae

Suborder 4. Heliconiineae
Family:
1. Heliconiaceae

Suborder 5. Zingiberineae
Families:
1. Zingiberaceae
2. Costaceae

Suborder 6. Cannineae
Families:
1. Cannaceae
2. Marantaceae

Order 20. Bromeliales
Families:
1. Bromeliaceae
   Subfamilies:
   Brocchinioideae
   Tillandsioideae
   Bromelioideae
   Pitcairnioideae
2. Rapateaceae
   Subfamilies:
   Rapateoideae
   Saxofridericioideae

Order 21. Xyridales
Suborder 1. Xyridineae
Family:
1. Xyridaceae
   Subfamilies:
   Xyridoideae
   Abolbodoideae

Suborder 2. Eriocaulineae

Family:

1. Eriocaulaceae

Order 22. Juncales

Families:

1. Prioniaceae
2. Thurniaceae
3. Juncaceae
4. Cyperaceae

 Subfamilies:

 Mapanioideae

 Cyperoideae

 Sclerioideae

 Caricoideae

Order 23. Typhales

Family:

1. Typhaceae

 Subfamilies:

 Sparganioideae

 Typhoideae

Order 24. Restionales

Suborder 1. Restionineae

Families:

1. Anarthriaceae
2. Restionaceae
3. Ecdeiocoleaceae
4. Hopkinsiaceae
5. Lyginiaceae
6. Centrolepidaceae

Suborder 2. Poineae

Families:

1. Flagellariaceae
2. Joinvilleaceae
3. Poaceae

 Subfamilies:

 Anomochlooideae

 Pharoideae

 Bambusoideae

 Ehrhartoideae

 Pooideae

 Arundinoideae

 Danthonioideae

 Aristidoideae

 Chloridoideae

 Centothecoideae

 Panicoideae

7.2.10.5 Update of the 1992 Thorne's system of classification of the Angiospermae [as published in Botanical Review (22 October, 1992)]

The following changes have been suggested in Thorne's system of classification :

Caryophyllanae

(a) Removal of Molluginaceae from Phytolaccineae to Caryophyllaneae.

(b) Recognition of Achatocarpineae for Achatocarpaceae, placed between Caryophyllaneae and Portulacineae.

(c) Placement of Stegnospermataceae as the least specialized family of the Phytolaccineae.

(d) Reduction of Petiveriaceae, Agdestidaceae, and Barbeuiaceae to subfamily status under Phytolaccaccae.

Dillenianae (1966) has priority over Theanae (1992)

(a) Acccording to the principle of autonomy, Dilleniales must be used for Theales.

(b) Expansion of Nepenthineae to include Ancistrocladaceae, Dioncopyllaceae, and Droseraceae as well as Nepenthaceae,

Malvanae

(a) Sphaerosepalaceae (1959) has priority over Diegodendraceae (1964) and Rhopalocarpaceae (1987).

Rosanae

(a) In Rosales transfer of *Lindleya, Lyonothamnus,* and *Vauquelinia* from Spiraeoideae to Quillajoideae of Rosaceae.

(b) Removal of Donatiaceae (elevated from subfamily status) and Stylidiaceae from Saxifragales to Campanulales near Goodeniaceae.

(c) Removal of Droseraceae from Saxifragales to Dilleniales.

Aralianae

Aralianae (1966) has priority over Cornanae (1992)

(a) Brexiales (1833) has priority over Hydrangeales (1943).

Asteranae

(a) Elevation of Barnadesieae to subfamily status, Barnadesioideae and recognition of tribal status for Helenieae and Eremothamneae in Lactucoideae.

(b) In Campanulales return of *Brunonia australis* as Brunonioideae to Goodeniaceae.

(c) Also in Campanulales placement of Donatiaceae and Stylidiaceae near Campanulaceae.

(d) Elevation of Cyphioideae and Lobelioideae to family status in Campanulales as Cyphiaceae and Lobeliaceae, resulting in deletion of Campanuloideae.

Solananae

(a) Reduction of Nolanaceae to tribal status as Nolaneae in Solanoideae.

Lamianae

Lamianae (1966) has priority over Gentiananae (1992)

(a) Rubiales (1829) has priority over Gentianales (1833).

(b) In Rubiales (Gentianales) removal from Loganiaceae of *Polypremum* to Scrophulariaceae of Lamiales (Scrophulariales) and Plocospermatoideae as Plocospermataceae to position near Scrophulariaceae; removal of Potalioideae to Gentianaceae, forcing recognition of other gentians as Gentianoideae and eliminating need for subfamilies in Loganiaceae.

(c) Lamiales (1838) must replace Scrophulariales (1933) because it is an autonym of Lamianae.

(d) In Lamianae Phrymaceae is the conserved spelling over Phrymataceae, and Phil Cantino in Lamiaceae recommends division of the Teucrioideae into two subfamilies, new Teucrioideae with 23 genera and c. 870 species (incl. *Aegiphila, Spartothemnella,* Monochileae, most Clerodendreae and Caryopterideae and some Ajugeae) and Viticoideae with 16 genera and c. 685 species (incl. *Teijsmanniodendron, Hymenopyramis,* and most genera of Viticeae, Callicarpeae, and Tectoneae).

Lilianae

(a) Smilacaceae, based on morphological and molecular evidence, should be transferred to Liliales from Taccales (Dioscoreales) in Amaryllidineae of Asparagales return of Velloziaceae to Bromeliales, especially on basis of epicuticular wax crystalloids.

(b) Taccales (1829) has priority over Dioscoreales (1876).

(c) Taccales (Dioscoreales) erection of *Pentastemona* to family rank, and removal of Smilacaceae to Liliales.

Cannales (1829) has priority over Zingiberales (1854), and Cannineae must replace Marantineae because it is the autonym of Cannales.

7.2.11 The Angiosperm Phylogeny Group (APG) system of flowering plant classification

7.2.11.1 The Angiosperm Phylogeny Group (APG) system of flowering plant classification

The phylogeny of flowering plants are being revealed in increasing detail by recent cladistic analysis and the morphology of many major groups above the family level are gaining stronger support through these studies. With many elements of the major branching scheme of phylogeny of these groups being established by now, a revised suprafamilial classification of flowering plants becomes both feasible and desirable. Thus, the "Angiosperm Phylogeny Group" system of flowering plant classification, which organizes flowering plants into a "selected number of monophyletic suprafamilial groups" was proposed, and was published as a paper entitled *"An Ordinal Classification for the Families of Flowering Plants"* in 1998 in the Annals of the Missouri Botanical Garden 85 : 531-553. This paper was compiled by Kare Bremer, Mark W. Chase and Peter F. Stevens, with contributions from many other taxonomists.

The groupings present in this system are viewed by the authors as monophyletic Clades. This classification includes 462 flowering plant Families in 40 putatively monophyletic Orders and a small number of monophyletic, informal higher groups. The latter are the Monocots, Commelinoids, Eudicots, Core Eudicots, Rosids including Euroids I and II, and Asterids including Euasterids I and II. Under these informal groups, there are also listed a number of Families without assignment to Order. At the end of the system is an additional list of Families of uncertain position for which no firm data exist regarding placement anywhere within the system. Thus, the three most inclusive groups are given below, for comparative purposes, as Classes :

Eudicots	Neither	Monocots
Unassigned	Unassigned	Not Commelinoids
Core Eudicots		Commelinoids
Rosids		
Asterids		
Probably Eudicots		

A comparison between this system and functional (complete) classification systems produced by Cronquist, Thorne and Takhtajan is difficult. However the APG system is now included as a comparative element of this system, with informal higher-level groups roughly placed in concordance with those employed by the other systems (Fig. 7.25).

An overview of the APG classification (Fig. 7.26) system is given below which has been derived from a copy of the published document made available by the Royal Botanical Gardens at Kew with due permission from Professor Amy Mc Pherson, Managing Editor, Missouri Botanical Garden Press, St. Louis.

7.2.11.2 Phylogenetic interrelationships of the Orders of flowering plants (AGP, 1998)

In the following list Cr, Th and Ta refer to the other systems of classification i.e., Cr for Cronquist, Th for Thorne and Ta for Takhtajan's placement. In cases where the APG family is treated as a synonym, the accepted Family name for a given system is preceded by '='. A '+' preceding an APG listed Family denotes those small Families that may be reduced to synonymy of their sister group, the latter consists of a single Family. The listing indents these below their sister Family.

I. APG Group : Eudicots

Within the eudicots there is increasing support for a large subgroup with predominantly pentamerous and isomerous flowers, the Core Eudicots, mainly comprising Caryophyllales, Santalales, Saxifragales, Rosids and Asterids. Rosids and Asterids each comprise two large subgroups, Eurosids I and II and Euasterids I and II, also receiving increasing support as monophyletic (Fig. 7.22).

1. APG Group : Eudicot - unassigned

1. Not Assigned to Order

Families :

Buxaceae — Cr: Rosidae, Euphorbiales; Th: Rosanae, Balanopales; Ta: Hamamelididae, Buxales, Didymelaceae - Cr : Hamamelidae, Didymelales; Th : Rosanae, Balanopales; Ta : Hamamelididae, Didymelales

Sabiaceae — Cr: Magnoliidae, Ranunculales; Th: Rutanae, Rutales; Ta: Rosidae, Sabiales

Trochodendraceae — Cr: Hamamelidae, Trochodendrales; Th: Rosanae, Hamamelidales; Ta: Hamamelididae, Trochodendrales

+Tetracentraceae — Cr: Hamamelidae, Trochodendrales; Ta : Hamamelididae, Trochodendrales

Order 1. Proteales

Families :

Nelumbonaceae — Cr: Magnoliidae, Nymphaeales; Th: Magnolianae, Nelumbonales; Ta: Nelumbonidae, Nelumbonales

Platanaceae — Cr: Hamamelidae, Hamamelidales; Th: Rosanae, Hamamelidales; Ta: Hamamelididae, Hamamelidales

Proteaceae — Cr: Rosidae, Proteales; Th: Proteanae, Proteales; Ta: Rosidae, Proteales

Order 2. Ranunculales

Families:

Berberidaceae — Cr: Magnoliidae, Ranunculales; Th: Magnolianae, Berberidales; Ta: Ranunculidae, Berberidales

Circaeasteraceae — Cr: Magnoliidae, Ranunculales; Th: Magnolianae, Berberidales; Ta: Ranunculidae, Circaeasterales

Kingdoniaceae — Cr = Circaeasteraceae; Ta: Ranunculidae, Circaeasterales

+Eupteleaceae — Cr: Hamamelidae, Hamamelidales; Th: Rosanae, Hamamelidales; Ta: Hamamelididae, Eupteleales

Lardizabalaceae — Cr: Magnoliidae, Ranunculales; Th: Magnolianae, Berberidales; Ta: Ranunculidae, Lardizabales

Menispermaceae — Cr: Magnoliidae, Ranunculales; Th: Magnolianae, Berberidales; Ta: Ranunculidae, Menispermales

Papaveraceae — Cr: Magnoliidae, Papaverales; Th: Magnolianae, Berberidales; Ta : Ranunculidae, Papaverales

+Fumariaceae — Cr: Magnoliidae, Papaverales; Ta: Ranunculidae, Papaverales

+Pteridophyllaceae — Cr = Fumariaceae; Ta: Ranunculidae, Papaverales

Ranunculaceae — Cr: Magnoliidae, Ranunculales; Th: Magnolianae, Berberidales; Ta: Ranunculidae, Ranunculales

2. APG Group : Core Eudicots

1. Not Assigned to Order

Families:

Aextoxicaceae — Cr: Rosidae, Celastrales; Ta: Dilleniidae, Euphorbiales

Berberidopsidaceae — Cr = Flacourtiaceae; Ta: Dilleniidae, Violales

Dilleniaceae — Cr: Dilleniidae, Dilleniales; Th: Dillenianae, Dilleniales; Ta: Dilleniidae, Dilleniales

Gunneraceae — Cr: Rosidae, Haloragales; Th: Aralianae, Cornales; Ta: Rosidae, Gunnerales

Myrothamnaceae — Cr: Hamamelidae, Hamamelidales; Th: Rosanae, Bruniales; Ta: Hamamelididae, Myrothamnales

Vitaceae — Cr: Rosidae, Rhamnales; Th: Aralianae, Cornales; Ta: Rosidae, Vitales

Order 1. Caryophyllales

Families:

Achatocarpaceae — Cr: Caryophyllidae, Caryophyllales; Th: Caryophyllana, Caryophyllales; Ta: Caryophyllidae, Caryophyllales

Aizoaceae — Cr: Caryophyllidae, Caryophyllales; Th: Caryophyllana, Caryophyllales; Ta: Caryophyllidae, Caryophyllales

Amaranthaceae — Cr: Caryophyllidae, Caryophyllales; Th: Caryophyllana, Caryophyllales; Ta: Caryophyllidae, Caryophyllales

Ancistrocladaceae — Cr: Dilleniidae, Violales; Th: Dillenianae, Dilleniales; Ta: Dilleniidae, Ancistrocladales

Asteropeiaceae — Cr = Theaceae; Th: Dillenianae, Dilleniales; Ta: Dilleniidae, Theales

Basellaceae — Cr: Caryophyllidae, Caryophyllales; Th: Caryophyllana, Caryophyllales; Ta: Caryophyllidae, Caryophyllales

Cactaceae — Cr: Caryophyllidae, Caryophyllales; Th: Caryophyllana, Caryophyllales; Ta: Caryophyllidae, Caryophyllales

Caryophyllaceae — Cr: Caryophyllidae, Caryophyllales; Th: Caryophyllana, Caryophyllales; Ta: Caryophyllidae, Caryophyllales

Didiereaceae — Cr: Caryophyllidae, Caryophyllales; Th: Caryophyllana, Caryophyllales; Ta: Caryophyllidae, Caryophyllales

Dioncophyllaceae — Cr: Dilleniidae, Violales; Th: Dillenianae, Dilleniales; Ta: Dilleniidae, Dioncophyllales

Droseraceae — Cr: Dilleniidae, Nepenthales; Th: Dillenianae, Dilleniales; Ta: Dilleniidae, Droserales

Drosophyllaceae

Frankeniaceae — Cr: Dilleniidae, Violales; Th: Violanae, Violales; Ta: Dilleniidae, Tamaricales

Molluginaceae — Cr: Caryophyllidae, Caryophyllales; Th: Caryophyllana, Caryophyllales; Ta: Caryophyllidae, Caryophyllales

Nepenthaceae — Cr: Dilleniidae, Nepenthales; Th: Dillenianae, Dilleniales; Ta: Dilleniidae, Nepenthales

Nyctaginaceae — Cr: Caryophyllidae, Caryophyllales; Th: Caryophyllana, Caryophyllales; Ta: Caryophyllidae, Caryophyllales

Physenaceae — Cr: Hamamelidae, Urticales; Ta: Dilleniidae, Physenales

Phytolaccaceae — Cr: Caryophyllidae, Caryophyllales; Th: Caryophyllana, Caryophyllales; Ta: Caryophyllidae, Caryophyllales

Plumbaginaceae — Cr: Caryophyllidae, Plumbaginales; Th: Dillenianae, Primulales; Ta: Caryophyllidae, Plumbaginales

Polygonaceae — Cr: Caryophyllidae, Polygonales; Th: Dillenianae, Polygonales; Ta: Caryophyllidae, Polygonales

Portulacaceae — Cr: Caryophyllidae, Caryophyllales; Th: Caryophyllana, Caryophyllales; Ta: Caryophyllidae, Caryophyllales

Rhabdodendraceae — Cr: Rosidae, Rosales; Th: Rutanae, Rutales; Ta: Rosidae, Rutales

Sarcobataceae

Simmondsiaceae — Cr: Rosidae, Euphorbiales; Th: Malvanae, Euphorbiales; Ta: Hamamelididae, Simmondsiales

Stegnospermataceae

Tamaricaceae — Cr: Dilleniidae, Violales; Th: Violanae, Violales; Ta: Dilleniidae, Tamaricales

Order 2. Santalales

Families:

Olacaceae — Cr: Rosidae, Santalales; Th: Santalanae, Santalales; Ta: Rosidae, Santalales

Opiliaceae — Cr: Rosidae, Santalales; Th: Santalanae, Santalales; Ta: Rosidae, Santalales

Loranthaceae — Cr: Rosidae, Santalales; Th: Santalanae, Santalales; Ta: Rosidae, Santalales

Misodendraceae — Cr: Rosidae, Santalales; Th: Santalanae, Santalales; Ta: Rosidae, Santalales

Santalaceae — Cr: Rosidae, Santalales; Th: Santalanae, Santalales; Ta: Rosidae, Santalales

Order 3. Saxifragales

Families:

Altingiaceae — Cr = Hamamelidaceae; Ta: Hamamelididae, Hamamelidales

Cercidiphyllaceae — Cr: Hamamelidae, Hamamelidales; Th: Rosanae, Hamamelidales; Ta: Hamamelididae, Cercidiphyllales

Crassulaceae — Cr: Rosidae, Rosales; Th: Rosanae, Saxifragales; Ta: Rosidae, Saxifragales

Daphniphyllaceae — Cr: Hamamelidae, Daphniphyllales; Th: Rosanae, Balanopales; Ta: Hamamelididae, Daphniphllales

Grossulariaceae — Cr: Rosidae, Rosales; Th: Rosanae, Saxifragales; Ta: Rosidae, Saxifragales

Haloragaceae — Cr: Rosidae, Haloragales; Th: Aralianae, Cornales; Ta: Rosidae, Haloragales

Hamamelidaceae — Cr: Hamamelidae, Hamamelidales; Th: Rosanae, Hamamelidales; Ta: Hamamelididae, Hamamelidales

Iteaceae — Cr = Grossulariaceae; Ta: Rosidae, Saxifragales

Paeoniaceae — Cr: Dilleniidae, Dilleniales; Th: Magnolianae, Paeoniales; Ta: Ranunculidae, Paeoniales

Penthoraceae — Cr = Saxifragaceae; Th: Rosanae, Saxifragales; Ta: Rosidae, Saxifragales

Pterostemonaceae — Cr = Grossulariaceae; Ta: Rosidae, Saxifragales

Saxifragaceae — Cr: Rosidae, Rosales; Th: Rosanae, Saxifragales; Ta: Rosidae, Saxifragales

Tetracarpaeaceae — Cr = Grossulariaceae; Th: Rosanae, Saxifragales; Ta: Rosidae, Saxifragales

3. APG Group: Eudicots - Rosids

1. Not Assigned to Order

Families:

Aphloiaceae

Crossosomataceae — Cr: Rosidae, Rosales; Th: Rosanae, Rosales; Ta: Rosidae, Crossosomatales

Ixerbaceae

Krameriaceae — Cr: Rosidae, Polygalales; Th: Geranianae, Polygalales; Ta: Rosidae, Vochysiales

Picramniaceae

Podostemaceae — Cr: Rosidae, Podestemales; Th: Rosanae, Podostemales; Ta: Rosidae, Podostemales

Stachyuraceae — Cr: Dilleniidae, Violales; Th: Dillenianae, Dilleniales; Ta: Dilleniidae, Theales

Staphyleaceae — Cr: Rosidae, Sapindales; Th: Rosanae, Cunoniales; Ta: Rosidae, Sapindales

Tristichaceae — Cr = Podostemaceae; Ta = Podostemaceae

Zygophyllaceae — Cr: Rosidae, Sapindales; Th: Geranianae, Linales; Ta: Rosidae, Zygophyllales

Order 1. Geraniales

Families:

Francoaceae — Cr = Saxifragaceae; Th: Rosanae, Saxifragales; Ta: Rosidae, Francoales

Geraniaceae — Cr: Rosidae, Geraniales; Th: Geranianae, Geraniales; Ta: Rosidae, Geraniales

+Hypseocharitaceae — Cr = Oxalidaceae; Ta: Rosidae, Geraniales

Greyiaceae — Cr: Rosidae, Rosales; Th: Rosanae, Saxifragales; Ta: Rosidae, Greyiales

Ledocarpaceae — Cr = Geraniaceae; Ta: Rosidae, Geraniales

Melianthaceae — Cr: Rosidae, Sapindales; Th: Rutanae, Rutales; Ta: Rosidae, Sapindales

Vivianiaceae — Cr = Geraniaceae; Ta: Rosidae, Geraniales

Subgroup 1. Eurosids I

1. Not Assigned to Order

Families:

Celastraceae — Cr: Rosidae, Celastrales; Th: Celastranae, Celastrales; Ta: Rosidae, Celastrales

Huaceae — Cr: Dilleniidae, Violales; Th: Malvanae, Malvales; Ta: Dilleniidae, Malvales

Parnassiaceae — Cr = Saxifragaceae; Th: Rosanae, Saxifragales; Ta: Rosidae, Parnassiales

+Lepuropetalaceae — Cr = Saxifragaceae; Th: Rosanae, Saxifragales; Ta: Rosidae, Parnassiales

Stackhousiaceae — Cr: Rosidae, Celastrales; Th: Celastranae, Celastrales; Ta: Rosidae, Celastrales

Order 1. Cucurbitales

Families:

Anisophylleaceae — Cr: Rosidae, Rosales; Th: Rosanae, Rosales; Ta: Rosidae, Anisophylleales

Begoniaceae — Cr: Dilleniidae, Violales; Th: Violanae, Violales; Ta: Dilleniidae, Begoniales

Coriariaceae — Cr: Magnoliidae, Ranunculales; Th: Rutanae, Rutales; Ta: Rosidae, Coriariales

Corynocarpaceae — Cr: Rosidae, Celastrales; Ta: Rosidae, Corynocarpales

Cucurbitaceae — Cr: Dilleniidae, Violales; Th: Violanae, Violales; Ta: Dilleniidae, Cucurbitales

Datiscaceae — Cr: Dilleniidae, Violales; Th: Violanae, Violales; Ta: Dilleniidae, Begoniales

Tetramelaceae — Cr = Datiscaceae; Ta: Dilleniidae, Begoniales

Order 2. Fabales

Families:

Fabaceae — Cr: Rosidae, Fabales; Th: Rutanae, Rutales; Ta: Rosidae, Fabales

Polygalaceae — Cr: Rosidae, Polygalales; Th: Geranianae, Polygalales; Ta: Rosidae, Polygalales

Quillajaceae

Surianaceae — Cr: Rosidae, Rosales; Th: Rutanae, Rutales; Ta: Rosidae, Rutales

Order 3. Fagales

Families:

Betulaceae — Cr: Hamamelidae, Fagales; Th: Rosanae, Betulales; Ta: Hamamelididae, Corylales

Casuarinaceae — Cr: Hamamelidae, Casuarinales; Th: Rosanae, Casuarinales; Ta: Hamamelididae, Casuarinales

Fagaceae — Cr: Hamamelidae, Fagales; Th: Rosanae, Betulales; Ta: Hamamelididae, Fagales

Juglandaceae — Cr: Hamamelidae, Juglandales; Th: Rosanae, Juglandales; Ta: Hamamelididae, Juglandales

Myricaceae — Cr: Hamamelidae, Myricales; Th: Rosanae, Juglandales; Ta: Hamamelididae, Myricales

Nothofagaceae — Cr: Hamamelidae, Fagales; Th: Rosanae, Betulales; Ta: Hamamelididae, Fagales

Rhoipteleaceae — Cr: Hamamelidae, Juglandales; Th: Rosanae, Juglandales; Ta: Hamamelididae, Rhoipteleales

Ticodendraceae — Cr: Hamamelidae, Fagales; Th: Rosanae, Betulales; Ta: Hamamelididae, Corylales

Order 4. Malpighiales

Families:

Achariaceae — Cr: Dilleniidae, Violales; Th: Violanae, Violales; Ta: Dilleniidae, Passiflorales

Balanopaceae — Cr: Hamamelidae, Fagales; Th: Rosanae, Balanopales; Ta: Hamamelididae, Balanopales

Caryocaraceae — Cr: Dilleniidae, Theales; Th: Dillenianae, Dilleniales; Ta: Dilleniidae, Theales

Chrysobalanaceae — Cr: Rosidae, Rosales; Th: Dillenianae, Dilleniales; Ta: Rosidae, Chrysobalanales

Clusiaceae — Cr: Dilleniidae, Theales; Th: Dillenianae, Dilleniales; Ta: Dilleniidae, Hypericales

Dichapetalaceae — Cr: Rosidae, Celastrales; Th: Malvanae, Euphorbiales; Ta: Dilleniidae, Euphorbiales

Erythroxylaceae — Cr: Rosidae, Linales; Th: Geranianae, Linales; Ta: Rosidae, Linales

Euphorbiaceae — Cr: Rosidae, Euphorbiales; Th: Malvanae, Euphorbiales; Ta: Dilleniidae, Euphorbiales

Euphroniaceae

Flacourtiaceae — Cr: Dilleniidae, Violales; Th: Violanae, Violales; Ta: Dilleniidae, Violales

Goupiaceae — Cr = Celastraceae; Th: Celastranae, Celastrales; Ta: Rosidae, Celastrales

Hugoniaceae — Cr: Rosidae, Linales; Th: Geranianae, Linales; Ta: Rosidae, Linales

Humiriaceae — Cr: Rosidae, Linales; Th: Geranianae, Linales; Ta: Rosidae, Linales

Irvingiaceae — Cr = Simaroubaceae; Ta: Rosidae, Rutales

Ixonanthaceae — Cr: Rosidae, Linales; Th: Geranianae, Linales; Ta: Rosidae, Linales

Lacistemataceae — Cr: Dilleniidae, Violales; Th: Violanae, Violales; Ta: Dilleniidae, Violales

Linaceae — Cr: Rosidae, Linales; Th: Geranianae, Linales; Ta: Rosidae, Linales

Malesherbiaceae — Cr: Dilleniidae, Violales; Th: Violanae, Violales; Ta: Dilleniidae, Passiflorales

Malpighiaceae — Cr: Rosidae, Polygalales; Th: Geranianae, Polygalales; Ta: Rosidae, Vochysiales

Medusagynaceae — Cr: Dilleniidae, Theales; Th: Dillenianae, Dilleniales; Ta: Dilleniidae, Medusagynales

Ochnaceae — Cr: Dilleniidae, Theales; Th: Dillenianae, Dilleniales; Ta: Dilleniidae, Ochnales

Pandaceae — Cr: Rosidae, Euphorbiales; Ta: Dilleniidae, Euphorbiales

Passifloraceae — Cr: Dilleniidae, Violales; Th: Violanae, Violales; Ta: Dilleniidae, Passiflorales

Putranjivaceae — Cr = Euphorbiaceae; Ta = Euphorbiaceae

Quiinaceae — Cr: Dilleniidae, Theales; Th: Dillenianae, Dilleniales; Ta: Dilleniidae, Ochnales

Rhizophoraceae — Cr: Rosidae, Rhizophorales; Th: Geranianae, Rhizophorales; Ta: Rosidae, Rhizophorales

Salicaceae — Cr: Dilleniidae, Salicales; Th: Violanae, Violales; Ta: Dilleniidae, Salicales

Scyphostegiaceae — Cr: Dilleniidae, Violales; Th: Violanae, Violales; Ta: Dilleniidae, Violales

Trigoniaceae — Cr: Rosidae, Polygalales; Th: Geranianae, Polygalales; Ta: Rosidae, Vochysiales

Turneraceae — Cr: Dilleniidae, Violales; Th: Violanae, Violales; Ta: Dilleniidae, Passiflorales

Violaceae — Cr: Dilleniidae, Violales; Th: Voilanae, Violales; Ta: Dilleniidae, Violales

Order 5. Oxalidales

Families:

Cephalotaceae — Cr: Rosidae, Rosales; Th: Rosanae, Saxifragales; Ta: Rosidae, Cephalotales

Connaraceae — Cr: Rosidae, Rosales; Th: Rutanae, Rutales; Ta: Rosidae, Connarales

Cunoniaceae — Cr: Rosidae, Rosales; Th: Rosanae, Cunoniales; Ta: Rosidae, Cunoniales

Elaeocarpaceae — Cr: Dilleniidae, Malvales; Th: Malvanae, Malvales; Ta: Dilleniidae, Elaeocarpales

Oxalidaceae — Cr: Rosidae, Geraniales; Th: Geranianae, Geraniales; Ta: Rosidae, Oxalidales

Tremandraceae — Cr: Rosidae, Polygalales; Th: Aralianae, Pittosporales; Ta: Rosidae, Vochysiales

Order 6. Rosales

Families:

Barbeyaceae — Cr: Hamamelidae, Urticales; Ta: Hamamelididae, Barbeyales

Cannabaceae — Cr: Hamamelidae, Urticales; Th: Malvanae, Urticales; Ta: Dilleniidae Urticales

Cecropiaceae — Cr: Hamamelidae, Urticales; Th: Malvanae, Urticales; Ta: Dilleniidae, Urticales

Celtidaceae — Cr = Ulmaceae; Ta = Ulmaceae

Dirachmaceae — Cr = Geraniaceae; Ta: Dilleniidae, Malvales

Elaeagnaceae — Cr: Rosidae, Proteales; Th: Malvanae, Rhamnales; Ta: Rosidae, Elaeagnales

Moraceae — Cr: Hamamelidae, Urticales; Th: Malvanae, Urticales; Ta: Dilleniidae, Urticales

Rhamnaceae — Cr: Rosidae, Rhamnales; Th: Malvanae, Rhamnales; Ta: Rosidae, Rhamnales

Rosaceae — Cr: Rosidae, Rosales; Th: Rosanae, Rosales; Ta: Rosidae, Rosales

Ulmaceae — Cr: Hamamelidae, Urticales; Th: Malvanae, Urticales; Ta: Dilleniidae, Urticales

Urticaceae — Cr: Hamamelidae, Urticales; Th: Malvanae, Urticales; Ta: Dilleniidae, Urticales

Subgroup 2. Eurosids II

1. Not Assigned to Order

Family:

Tapisciaceae — Cr = Staphyleaceae; Ta: Rosidae, Sapindales

Order 1. Brassicales

Families:

Akaniaceae — Cr: Rosidae, Sapindales; Th: Rutanae, Rutales; Ta: Rosidae, Sapindales

+Bretschneideraceae — Cr: Rosidae, Sapindales Th: Rutanae, Rutales; Ta: Rosidae, Sapindales

Bataceae — Cr: Dilleniidae, Batales; Th: Violanae, Batales; Ta: Dilleniidae, Batales

Brassicaceae — Cr: Dilleniidae, Capparales; Th: Violanae, Brassicales; Ta: Dilleniidae, Capparales

Caricaceae — Cr: Dilleniidae, Violales; Th: Violanae, Violales; Ta: Dilleniidae, Caricales

Emblingiaceae — Cr = Polygalaceae; Ta: Rosidae, Polygalales

Gyrostemonaceae — Cr: Dilleniidae, Batales; Th: Violanae, Brassicales; Ta: Caryophyllidae, Gyrostemonales

Koeberliniaceae

Limnanthaceae — Cr: Rosidae, Geraniales; Th: Geranianae, Geraniales; Ta: Lamiidae, Limnanthales

Moringaceae — Cr: Dilleniidae, Capparales; Th: Rutanae, Rutales; Ta: Dilleniidae, Moringales

Pentadiplandraceae - Cr = Capparaceae; Ta: Dilleniidae, Capparales

Resedaceae — Cr: Dilleniidae, Capparales; Th: Violanae, Brassicales; Ta: Dilleniidae, Capparales

Salvadoraceae — Cr: Rosidae, Celastrales; Th: Violanae, Brassicales; Ta: Rosidae, Salvadorales

Setchellanthaceae

Tovariaceae — Cr: Dilleniidae, Capparales; Ta: Dilleniidae, Capparales

Tropaeolaceae — Cr: Rosidae, Geraniales; Th: Geranianae, Geraniales; Ta: Rosidae, Tropaeolales

Order 2. Malvales

Families:

Bixaceae — Cr: Dilleniidae, Violales; Th: Violanae, Violales; Ta: Dilleniidae, Cistales

+Diegodendraceae

Cistaceae — Cr: Dilleniidae, Violales; Th: Violanae, Violales; Ta: Dilleniidae, Cistales

Cochlospermaceae — Cr = Bixaceae; Th: Violanae, Violales; Ta: Dilleniidae, Cistales

Dipterocarpaceae — Cr: Dilleniidae, Theales; Th: Malvanae, Malvales; Ta: Dilleniidae, Malvales

Malvaceae — Cr: Dilleniidae, Malvales; Th: Malvanae, Malvales; Ta: Dilleniidae, Malvales

Muntingiaceae

Neuradaceae — Cr: Rosidae, Rosales; Th: Rosanae, Rosales; Ta: Rosidae, Rosales

Sarcolaenaceae — Cr: Dilleniidae, Theales; Th: Malvanae, Malvales; Ta: Dilleniidae, Malvales

Sphaerosepalaceae — Cr: Dilleniidae, Theales; Th: Malvanae, Malvales; Ta: Dilleniidae, Malvales

Thymelaeaceae — Cr: Rosidae, Myrtales; Th: Malvanae, Euphorbiales; Ta: Dilleniidae, Thymelaeales

Order 3. Myrtales

Families:

Alzateaceae — Cr: Rosidae, Myrtales; Th: Myrtanae, Myrtales; Ta: Rosidae, Myrtales

Combretaceae — Cr: Rosidae, Myrtales; Th: Myrtanae, Myrtales; Ta: Rosidae, Myrtales

Crypteroniaceae — Cr: Rosidae, Myrtales; Th: Myrtanae, Myrtales; Ta: Rosidae, Myrtales

Heteropyxidaceae

Lythraceae — Cr: Rosidae, Myrtales; Th: Myrtanae, Myrtales; Ta: Rosidae, Myrtales

Melastomataceae — Cr: Rosidae, Myrtales; Th: Myrtanae, Myrtales; Ta: Rosidae, Myrtales

Memecylaceae — Cr = Melastomataceae; Ta: Rosidae, Myrtales

Myrtaceae — Cr: Rosidae, Myrtales; Th: Myrtanae, Myrtales; Ta: Rosidae, Myrtales

Oliniaceae — Cr: Rosidae, Myrtales; Th: Myrtanae, Myrtales; Ta: Rosidae, Myrtales

Onagraceae — Cr: Rosidae, Myrtales; Th: Myrtanae, Myrtales; Ta: Rosidae, Myrtales

Penaeaceae — Cr: Rosidae, Myrtales; Th: Myrtanae, Myrtales; Ta: Rosidae, Myrtales

Psiloxylaceae — Cr = Myrtaceae; Ta: Rosidae, Myrtales

Rhynchocalycaceae — Cr: Rosidae, Myrtales; Th: Myrtanae, Myrtales; Ta: Rosidae, Myrtales

Vochysiaceae — Cr: Rosidae, Polygalales; Th: Geranianae, Polygalales; Ta: Rosidae, Vochysiales

Order 4. Sapindales

Families:

Anacardiaceae — Cr: Rosidae, Sapindales; Th: Rutanae, Rutales; Ta: Rosidae, Burserales

Biebersteiniaceae — Cr = Geraniaceae; Ta: Rosidae, Biebersteiniales

Burseraceae — Cr: Rosidae, Sapindales; Th: Rutanae, Rutales; Ta: Rosidae, Burserales

Kirkiaceae — Cr = Simaroubaceae; Ta: Rosidae, Rutales

Meliaceae — Cr: Rosidae, Sapindales; Th: Rutanae, Rutales; Ta: Rosidae, Rutales

Nitrariaceae — Cr = Zygophyllaceae; Ta: Rosidae, Zygophyllales

+Peganaceae — Cr = Zygophyllaceae; Ta: Rosidae, Zygophyllales

Rutaceae — Cr: Rosidae, Sapindales; Th: Rutanae, Rutales; Ta: Rosidae, Rutales

Sapindaceae — Cr: Rosidae, Sapindales; Th: Rutanae, Rutales; Ta: Rosidae, Sapindales

Simaroubaceae — Cr: Rosidae, Sapindales; Th: Rutanae, Rutales; Ta: Rosidae, Rutales

4. APG Group: Eudicots - Asterids

Order 1. Cornales

Families:

Cornaceae — Cr: Rosidae, Cornales; Th: Aralianae, Cornales; Ta: Cornidae, Cornales

+Nyssaceae — Cr = Cornaceae; Ta: Cornidae, Cornales

Grubbiaceae — Cr: Dilleniidae, Ericales; Th: Rosanae, Bruniales; Ta: Dilleniidae, Bruniales

Hydrangeaceae — Cr: Rosidae, Rosales; Th: Aralianae, Brexiales; Ta: Cornidae, Hydrangeales

Hydrostachyaceae — Cr: Asteridae, Callitrichales; Th: Rosanae, Bruniales; Ta: Lamiidae, Hydrostachyales

Loasaceae — Cr: Dilleniidae, Violales; Th: Loasanae, Loasales; Ta: Lamiidae, Loasales

Order 2. Ericales

Families:

Actinidiaceae — Cr: Dilleniidae, Theales; Th: Dillenianae, Dilleniales; Ta: Dilleniidae, Actinidiales

Balsaminaceae — Cr: Rosidae, Geraniales; Th: Geranianae, Geraniales; Ta: Rosidae, Balsaminales

Clethraceae — Cr: Dilleniidae, Ericales; Th: Dillenianae, Dilleniales; Ta: Dilleniidae, Ericales

Cyrillaceae — Cr: Dilleniidae, Ericales; Th: Dillenianae, Dilleniales; Ta: Dilleniidae, Ericales

Diapensiaceae — Cr: Dilleniidae, Diapensiales; Th: Rósanae, Saxifragales; Ta: Dilleniidae, Diapensiales

Ebenaceae — Cr: Dilleniidae, Ebenales; Th: Dillenianae, Styracales; Ta: Dilleniidae, Styracales

Ericaceae — Cr: Dilleniidae, Ericales; Th: Dillenianae, Ericales; Ta: Dilleniidae, Ericales

Fouquieriaceae — Cr: Dilleniidae, Violales; Th: Dillenianae, Fouquieriales; Ta: Dilleniidae, Fouquieriales

Halesiaceae

Lecythidaceae — Cr: Dilleniidae, Lecythidales; Th: Dillenianae, Dilleniales; Ta: Dilleniidae, Lecythidales

Marcgraviaceae — Cr: Dilleniidae, Theales; Th: Dillenianae, Dilleniales; Ta: Dilleniidae, Theales

Myrsinaceae — Cr: Dilleniidae, Primulales; Th: Dillenianae, Primulales; Ta: Dilleniidae, Myrsinales

Pellicieraceae — Cr: Dilleniidae, Theales; Th: Dillenianae, Dilleniales; Ta: Dilleniidae, Theales

Polemoniaceae — Cr: Asteridae, Solanales; Th: Solananae, Solanales; Ta: Lamiidae, Polemoniales

Primulaceae — Cr: Dilleniidae, Primulales; Th: Dillenianae, Primulales; Ta: Dilleniidae, Primulales

Roridulaceae — Cr = Byblidaceae; Th: Rosanae, Bruniales; Ta: Cornidae, Roridulales

Sapotaceae — Cr: Dilleniidae, Ebenales; Th: Dillenianae, Styracales; Ta: Dilleniidae, Sapotales

Sarraceniaceae — Cr: Dilleniidae, Nepenthales; Th: Dillenianae, Dilleniales; Ta: Dilleniidae, Sarraceniales

Styracaceae — Cr: Dilleniidae, Ebenales; Th: Dillenianae, Styracales; Ta: Dilleniidae, Styracales

Symplocaceae — Cr: Dilleniidae, Ebenales; Th: Dillenianae, Dilleniales; Ta: Dilleniidae, Styracales

Ternstroemiaceae — Cr = Theaceae; Ta = Theaceae

Tetrameristaceae — Cr: Dilleniidae, Theales; Th: Dillenianae, Dilleniales; Ta: Dilleniidae, Theales

Theaceae — Cr: Dilleniidae, Theales; Th: Dillenianae, Dilleniales; Ta: Dilleniidae, Theales

Theophrastaceae — Cr: Dilleniidae, Primulales; Th: Dillenianae, Primulales; Ta: Dilleniidae, Myrsinales

Subgroup 1. Euasterids I

1. Not Assigned to Order

Families:

Boraginaceae — Cr: Asteridae, Lamiales; Th: Solananae, Solanales; Ta: Lamiidae, Boraginales

Plocospermataceae — Cr = Apocynaceae; Th: Lamianae, Lamiales; Ta: Lamiidae, Gentianales

Vahliaceae — Cr = Saxifragaceae; Th: Rosanae, Saxifragales; Ta: Rosidae, Saxifragales

Order 1. Garryales

Families:

Aucubaceae

Eucommiaceae — Cr: Hamamelidae, Eucommiales; Th: Aralianae, Cornales; Ta: Cornidae, Eucommiales

Garryaceae — Cr: Rosidae, Cornales; Th: Aralianae, Cornales; Ta: Cornidae, Garryales

Oncothecaceae — Cr: Dilleniidae, Theales; Th: Dillenianae, Dilleniales; Ta: Dilleniidae, Theales

Order 2. Gentianales

Families:

Apocynaceae — Cr: Asteridae, Gentianales; Th: Lamianae, Rubiales; Ta: Lamiidae, Apocynaceae

Gelsemiaceae

Gentianaceae — Cr: Asteridae, Gentianales; Th: Lamianae, Rubiales; Ta: Lamiidae, Gentianales

Loganiaceae — Cr: Asteridae, Gentianales; Th: Lamianae, Rubiales; Ta: Lamiidae, Gentianales

Rubiaceae — Cr: Asteridae, Rubiales; Th: Lamianae, Rubiales; Ta: Lamiidae, Rubiales

Order 3. Lamiales

Families:

Acanthaceae — Cr: Asteridae, Scrophulariales; Th: Lamianae, Lamiales; Ta: Lamiidae, Scrophulariales

Avicenniaceae — Cr = Verbenaceae; Th: Lamianae, Lamiales; Ta: Lamiidae, Lamiales

Bignoniaceae — Cr: Asteridae, Scrophulariales; Ta: Lamiidae, Scrophulariales

Buddlejaceae — Cr: Asteridae, Scrophulariales; Th: Lamianae, Lamiales; Ta: Lamiidae, Scrophulariales

Byblidaceae — Cr: Rosidae, Rosales; Th: Aralianae, Pittosporales; Ta: Cornidae, Byblidales

Cyclocheilaceae

Gesneriaceae — Cr: Asteridae, Scrophulariales; Th: Lamianae, Lamiales; Ta: Lamiidae, Scrophulariales

Lamiaceae — Cr: Asteridae, Lamiales; Th: Lamianae, Lamiales; Ta: Lamiidae, Lamiales

Lentibulariaceae — Cr: Asteridae, Scrophulariales; Th: Lamianae, Lamiales; Ta: Lamiidae, Scrophulariales

Myoporaceae — Cr: Asteridae, Scrophulariales; Th: Lamianae, Lamiales; Ta: Lamiidae, Scrophulariales

Oleaceae — Cr: Asteridae, Scrophulariales; Th: Lamianae, Lamiales; Ta: Lamiidae, Oleales

Orobanchaceae — Cr: Asteridae, Scrophulariales; Ta = Scrophulariaceae

Paulowniaceae

Pedaliaceae — Cr: Asteridae, Scrophulariales; Th: Lamianae, Lamiales; Ta: Lamiidae, Scrophulariales

+Martyniaceae — Cr = Pedaliaceae; Th: Lamianae, Lamiales; Ta: Lamiidae, Scrophulariales

Phrymaceae — Cr = Verbenaceae; Ta: Lamiidae, Lamiales

Plantaginaceae — Cr: Asteridae, Plantaginales; Th: Lamianae, Lamiales; Ta: Lamiidae, Scrophulariales

Schlegeliaceae

Scrophulariaceae — Cr: Asteridae, Scrophulariales; Th: Lamianae, Lamiales; Ta: Lamiidae, Scrophulariales

Stilbaceae — Cr = Verbenaceae; Th: Lamianae, Lamiales; Ta: Lamiidae, Scrophulariales

Tetrachondraceae — Cr = Lamiaceae; Th: Solananae, Solanales; Ta: Lamiidae, Boraginales

Verbenaceae — Cr: Asteridae, Lamiales; Th: Lamianae, Lamiales; Ta: Lamiidae, Lamiales

Order 4. Solanales

Families:

Convolvulaceae — Cr: Asteridae, Solanales; Th: Solananae, Solanales; Ta: Lamiidae, Convolvulales

Hydroleaceae

Montiniaceae

Solanaceae — Cr: Asteridae, Solanales; Th: Solananae, Solanales; Ta: Lamiidae, Solanales

Sphenocleaceae — Cr: Asteridae, Campanulales; Th: Asteranae, Campanulales; Ta: Asteridae, Campanulales

Subgroup 2. Euasterids II

1. Not Assigned to Order

Families:

Adoxaceae — Cr: Asteridae, Dipsacales; Th: Aralianae, Dipsacales; Ta: Cornidae, Adoxales

Bruniaceae — Cr: Rosidae, Rosales; Th: Rosanae, Bruniales; Ta: Dilleniidae, Bruniales

Carlemanniaceae — Cr = Caprifoliaceae; Ta: Lamiidae, Rubiales

Columelliaceae — Cr: Rosidae, Rosales; Th: Aralianae, Brexiales; Ta: Cornidae, Hydrangeales

+Desfontainiaceae — Cr = Loganiaceae; Th: Aralianae, Brexiales; Ta: Cornidae, Desfontainiales

Eremosynaceae — Cr = Saxifragaceae; Th: Rosanae, Saxifragales; Ta: Rosidae, Saxifragales

Escalloniaceae — Cr = Grossulariaceae; Th: Aralianae, Brexiales; Ta: Cornidae, Hydrangeales

Icacinaceae — Cr: Rosidae, Celastrales; Th: Aralianae, Cornales; Ta: Rosidae, Icacinales

Polyosmaceae

Sphenostemonaceae — Cr = Aquifoliaceae; Th: Dillenianae, Dilleniales; Ta: Rosidae, Icacinales

Tribelaceae — Cr = Grossulariaceae; Ta: Cornidae, Hydrangeales

Order 1. Apiales

Families:

Apiaceae — Cr: Rosidae, Apiales; Th: Aralianae, Araliales; Ta: Cornidae, Araliales

Araliaceae — Cr: Rosidae, Apiales; Th: Aralianae, Araliales; Ta: Cornidae, Araliales

Aralidiaceae — Cr = Cornaceae; Th: Aralianae, Cornales; Ta: Cornidae, Aralidiales

Griseliniaceae — Cr = Cornaceae; Th: Aralianae, Brexiales; Ta: Cornidae, Griseliniales

Melanophyllaceae — Cr = Cornaceae; Ta: Cornidae, Hydrangeales

Pittosporaceae — Cr: Rosidae, Rosales; Th: Aralianae, Pittosporales; Ta: Cornidae, Pittosporales

Torricelliaceae

Order 2. Aquifoliales

Families:

Aquifoliaceae — Cr: Rosidae, Celastrales; Th: Dillenianae, Dilleniales; Ta: Rosidae, Icacinales

Helwingiaceae — Cr = Cornaceae; Th: Aralianae, Araliales; Ta: Cornidae, Helwingiales

Phyllonomaceae — Cr = Grossulariaceae; Ta: Cornidae, Hydrangeales

Order 3. Asterales

Families:

Alseuosmiaceae — Cr: Rosidae, Rosales; Th: Aralianae, Brexiales; Ta: Cornidae, Hydrangeales

Argophyllaceae — Cr = Grossulariaceae; Ta: Cornidae, Hydrangeales

Asteraceae — Cr: Asteridae, Asterales; Th: Asteranae, Asterales; Ta: Asteridae, Asterales

Calyceraceae — Cr: Asteridae, Calycerales; Th: Asteranae, Asterales; Ta: Asteridae, Calycerales

Campanulaceae — Cr: Asteridae, Campanulales; Th: Asteranae, Campanulales; Ta: Asteridae, Campanulales

+Lobeliaceae — Cr = Campanulaceae; Th: Asteranae, Campanulales; Ta: Asteridae, Campanulales

Carpodetaceae — Cr = Grossulariaceae; Ta: Cornidae, Hydrangeales

Donatiaceae — Cr: Asteridae, Campanulales; Th: Asteranae, Campanulales; Ta: Asteridae, Stylidales

Goodeniaceae — Cr: Asteridae, Campanulales; Th: Asteranae, Campanulales; Ta: Asteridae, Goodeniales

Menyanthaceae — Cr: Asteridae, Solanales; Th: Asteranae, Campanulales; Ta: Asteridae, Menyanthales

Pentaphragmataceae — Cr: Asteridae, Campanulales; Th: Asteranae, Campanulales; Ta: Asteridae, Campanulales

Phellinaceae — Cr = Aquifoliaceae; Th: Dillenianae, Dilleniales; Ta: Rosidae, Icacinales

Rousseaceae — Cr = Grossulariaceae; Ta: Rosidae, Brexiales

Stylidiaceae — Cr: Asteridae, Campanulales; Th: Asteranae, Campanulales; Ta: Asteridae, Stylidales

Order 4. Dipsacales

Families:

Caprifoliaceae — Cr: Asteridae, Dipsacales; Th: Aralianae, Dipsacales; Ta: Cornidae, Dipsacales

Diervillaceae

Dipsacaceae — Cr: Asteridae, Dipsacales; Th: Aralianae, Dipsacales; Ta: Cornidae, Dipsacales

Linnaeaceae

Morinaceae — Cr = Dipsacaceae; Th: Aralianae, Dipsacales; Ta: Cornidae, Dipsacales

Valerianaceae — Cr: Asteridae, Dipsacales; Th: Aralianae, Dipsacales; Ta: Cornidae, Dipsacales

5. APG Group: Probably Eudicot

1. Not Assigned to Order

Families :

Balanophoraceae — Cr: Rosidae, Santalales; Th: Santalanae, Balanophorales; Ta: Magnoliidae, Balanophorales

Bonnetiaceae — Cr = Theaceae; Th: Dillenianae, Dilleniales; Ta: Dilleniidae, Hypericales

Cardiopteridaceae — Cr: Rosidae, Celastrales; Th: Aralianae, Cornales; Ta: Rosidae, Cardiopteridales

Ctenolophonaceae

Cynomoriaceae — Cr = Balanophoraceae; Th: Santalanae, Balanophorales; Ta: Magnoliidae, Cynomoriales

Cytinaceae — Cr = Rafflesiaceae; Ta: Magnoliidae, Rafflesiales

Dipentodontaceae — Cr: Rosidae, Santalales; Th: Violanae, Violales; Ta: Dilleniidae, Violales

Elatinaceae — Cr: Dilleniidae, Theales; Th: Dillenianae, Dilleniales; Ta: Dilleniidae, Elatinales

Geissolomataceae — Cr: Rosidae, Celastrales; Th: Rosanae, Bruniales; Ta: Dilleniidae, Geissolomatales

Hoplestigmataceae — Cr: Dilleniidae, Violales; Th: Solananae, Solanales; Ta: Lamiidae, Boraginales

Kaliphoraceae

Lepidobotryaceae — Cr = Oxalidaceae; Ta: Rosidae, Oxalidales

Lissocarpaceae — Cr: Dilleniidae, Ebenales; Th: Dillenianae, Styracales; Ta: Dilleniidae, Styracales

Lophopyxidaceae — Cr = Celastraceae; Th: Celastranae, Celastrales; Ta: Rosidae, Celastrales

Medusandraceae — Cr: Rosidae, Santalales; Th: Santalanae, Santalales; Ta: Rosidae, Medusandrales

Metteniusaceae

Mitrastemonaceae — Cr: Rosidae, Rafflesiales; Ta: Magnoliidae, Rafflesiales

Paracryphiaceae — Cr: Dilleniidae, Theales; Th: Dillenianae, Dilleniales; Ta: Dilleniidae, Paracryphiales

Pentaphylacaceae — Cr: Dilleniidae, Theales; Th: Dillenianae, Dilleniales; Ta: Dilleniidae, Theales

Peridiscaceae — Cr: Dilleniidae, Violales; Th: Violanae, Violales; Ta: Dilleniidae, Violales

Plagiopteraceae

Pottingeriaceae — Cr = Hydrangeaceae; Ta: Cornidae, Hydrangeales

Sladeniaceae — Cr = Theaceae; Ta = Theaceae

Strasburgeriaceae

Tepuianthaceae — Cr: Rosidae, Celastrales; Th: Rutanae, Rutales; Ta: Rosidae, Rutales

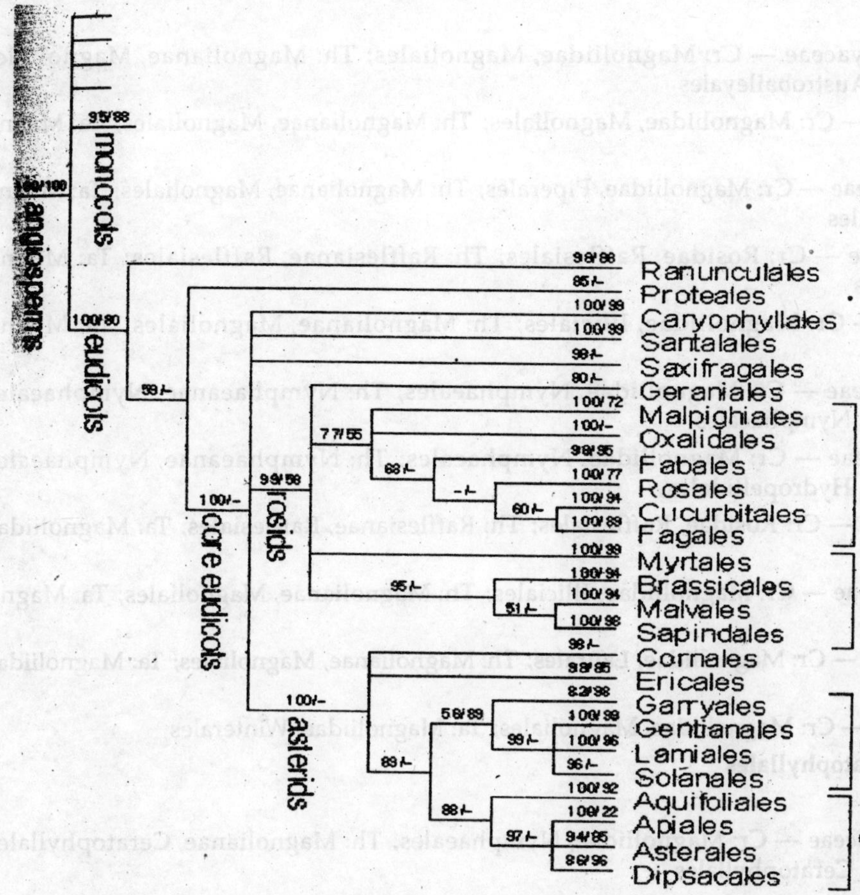

↬ Fig. 7.22 : Phylogenetic interrelationships of the orders of flowering plants, compiled from recent cladistic analyses. Jackknife support is given on the branches (a dash for values <50%), first jackknife values from analysis of 545 sequences of the rbcL, atpB, and 18S rDNA genes (D. E. Soltis, M. W. Chase, P. S. Soltis, D. Albach, M. E. Mort, V. Savolainen, M. Zanis, and J. S. Farris, unpublished, in preparation) and second jackknife values from analysis of 2538 rbcL sequences (Källersjö et al., 1998).

II. APG Group: Neither Monocots nor Dicots

Under each of the supraordinal groups of Monocots, Commelinoids, Core Eudicots, Rosids, etc., there are a number of Families listed without assignment to Order. These Families are known to belong within the major group under which they are listed, but their ordinal position is still uncertain. Similarly, Amborellaceae, Austrobaileyaceae, Canellaceae, etc., are listed at the beginning because they belong neither in any of the phylogenetically "basal" Orders at the beginning nor in the Monocots or Eudicots (Fig. 7.23).

1. Not Assigned to Order

Families:

Austrobaileyaceae — Cr: Magnoliidae, Magnoliales; Th: Magnolianae, Magnoliales; Ta: Magnoliidae, Austrobaileyales

Canellaceae — Cr: Magnoliidae, Magnoliales; Th: Magnolianae, Magnoliales; Ta: Magnoliidae, Canellales

Chloranthaceae — Cr: Magnoliidae, Piperales; Th: Magnolianae, Magnoliales; Ta: Magnoliidae, Chloranthales

Hydnoraceae — Cr: Rosidae, Rafflesiales; Th: Rafflesianae, Rafflesiales; Ta: Magnoliidae, Hydnorales

Illiciaceae — Cr: Magnoliidae, Illiciales; Th: Magnolianae, Magnoliales; Ta: Magnoliidae, Illiciales

Nymphaeaceae — Cr: Magnoliidae, Nymphaeales; Th: Nymphaeanae, Nymphaeales; Ta: Nymphaeidae, Nymphaeales

+Cabombaceae — Cr: Magnoliidae, Nymphaeales; Th: Nymphaeanae, Nymphaeales; Ta: Nymphaeidae, Hydropeltidales

Rafflesiaceae — Cr: Rosidae, Rafflesiales; Th: Rafflesianae, Rafflesiales; Ta: Magnoliidae, Rafflesiales

Schisandraceae — Cr: Magnoliidae, Illiciales; Th: Magnolianae, Magnoliales; Ta: Magnoliidae, Illiciales

Trimeniaceae — Cr: Magnoliidae, Laurales; Th: Magnolianae, Magnoliales; Ta: Magnoliidae, Laurales

Winteraceae — Cr: Magnoliidae, Magnoliales; Ta: Magnoliidae, Winterales

Order 1. Ceratophyllales

Family:

Ceratophyllaceae — Cr: Magnoliidae, Nymphaeales; Th: Magnolianae, Ceratophyllales; Ta: Nymphaeidae, Ceratophyllales

Order 2. Laurales

Families:

Atherospermataceae — Cr = Monimiaceae; Ta = Monimiaceae

Calycanthaceae — Cr: Magnoliidae, Laurales; Th: Magnolianae, Magnoliales; Ta: Magnoliidae, Calycanthales

Gomortegaceae — Cr: Magnoliidae, Laurales; Th: Magnolianae, Magnoliales; Ta: Magnoliidae, Laurales

Hernandiaceae — Cr: Magnoliidae, Laurales; Th: Magnolianae, Magnoliales; Ta: Magnoliidae, Laurales

Lauraceae — Cr: Magnoliidae, Laurales; Th: Magnolianae, Magnoliales; Ta: Magnoliidae, Laurales

Monimiaceae — Cr: Magnoliidae, Laurales; Th: Magnolianae, Magnoliales; Ta: Magnoliidae, Laurales

Siparunaceae — Cr = Monimiaceae; Ta = Monimiaceae

Order 3. Magnoliales

Families:

Annonaceae — Cr: Magnoliidae, Magnoliales; Th: Magnolianae, Magnoliales; Ta: Magnoliidae, Annonales

Degeneriaceae — Cr: Magnoliidae, Magnoliales; Th: Magnolianae, Magnoliales; Ta: Magnoliidae, Magnoliales

Eupomatiaceae — Cr: Magnoliidae, Magnoliales; Th: Magnolianae, Magnoliales; Ta: Magnoliidae, Eupomatiales

Himantandraceae — Cr: Magnoliidae, Magnoliales; Th: Magnolianae, Magnoliales; Ta: Magnoliidae, Magnoliales

Magnoliaceae — Cr: Magnoliidae, Magnoliales; Th: Magnolianae, Magnoliales; Ta: Magnoliidae, Magnoliales

Myristicaceae — Cr: Magnoliidae, Magnoliales; Th: Magnolianae, Magnoliales; Ta: Magnoliidae, Myristicales

Order 4. Piperales

Families:

Aristolochiaceae — Cr: Magnoliidae, Aristolochiales; Th: Magnolianae, Magnoliales; Ta: Magnoliidae, Aristolochiales

Lactoridaceae — Cr: Magnoliidae, Magnoliales; Th: Magnolianae, Magnoliales; Ta: Magnoliidae, Lactoridales

Piperaceae — Cr: Magnoliidae, Piperales; Th: Magnolianae, Magnoliales; Ta: Magnoliidae, Piperales

Saururaceae — Cr: Magnoliidae, Piperales; Th: Magnolianae, Magnoliales; Ta: Magnoliidae, Piperales

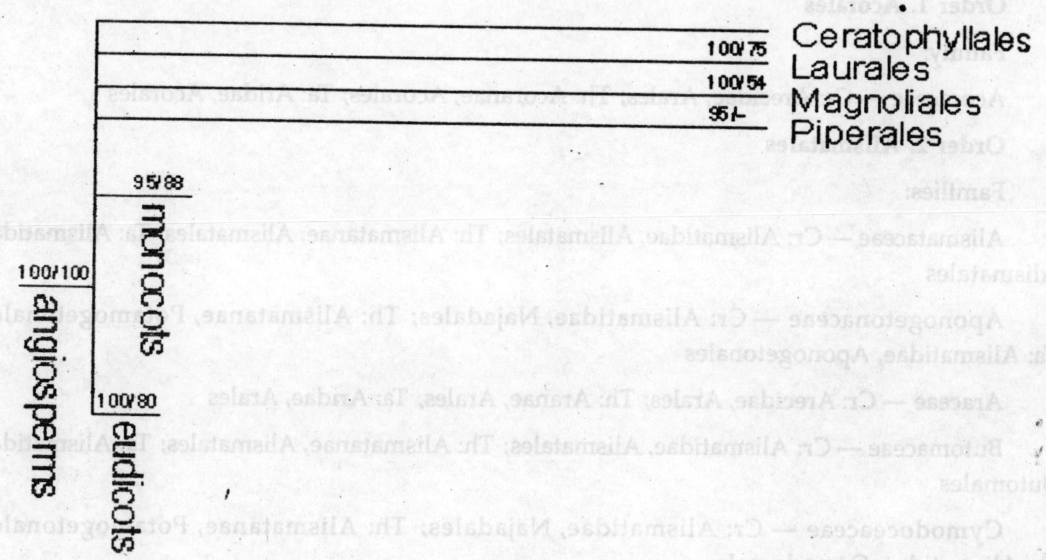

Fig. 7.23 : Phylogenetic interrelationships of the orders of flowering plants, compiled from recent cladistic analyses. Jackknife support is given on the branches (a dash for values <50%), first jackknife values from analysis of 545 sequences of the rbcL, atpB, and 18S rDNA genes (D. E. Soltis, M. W. Chase, P. S. Soltis, D. Albach, M. E. Mort, V. Savolainen, M. Zanis, and J. S. Farris, unpublished, in preparation) and second jackknife values from analysis of 2538 rbcL sequences (Källersjö et al., 1998).

III. APG Group : Monocots

Monocots and Eudicots are not formally ranked and named because it is not yet clear at which level they should be recognized. The same problems occur with Commelinoids (a phylogenetically derived subgroup of Monocots) and with Rosids and Asterids (Subgroups of Eudicots), although these are commonly known as Subclasses Commelinidae, Rosidae, and Asteridae, respectively (Fig. 7.24).

1. APG Group: Non-Commelinoid Monocots

1. Not Assigned to Order

Families:

Corsiaceae — Cr: Liliidae, Orchidales; Th: Lilianae, Burmanniales; Ta: Liliidae, Burmanniales

Japonoliriaceae

Nartheciaceae

Petrosaviaceae — Cr: Alismatidae, Triuridales; Th: Triuridanae, Triuridales; Ta: Triurididae, Petrosaviales

Triuridaceae — Cr: Alismatidae, Triuridales; Th: Triuridanae, Triuridales; Ta: Triurididae, Triuridales

Order 1. Acorales

Family:

Acoraceae — Cr: Arecidae, Arales; Th: Acoranae, Acorales; Ta: Aridae, Acorales

Order 2. Alismatales

Families:

Alismataceae — Cr: Alismatidae, Alismatales; Th: Alismatanae, Alismatales; Ta: Alismatidae, Alismatales

Aponogetonaceae — Cr: Alismatidae, Najadales; Th: Alismatanae, Potamogetonales; Ta: Alismatidae, Aponogetonales

Araceae — Cr: Arecidae, Arales; Th: Aranae, Arales; Ta: Aridae, Arales

Butomaceae — Cr: Alismatidae, Alismatales; Th: Alismatanae, Alismatales; Ta: Alismatidae, Butomales

Cymodoceaceae — Cr: Alismatidae, Najadales; Th: Alismatanae, Potamogetonales; Ta: Alismatidae, Cymodoceales

Hydrocharitaceae — Cr: Alismatidae, Hydrocharitales; Th: Alismatanae, Alismatales; Ta: Alismatidae, Hydrocharitales

Juncaginaceae — Cr: Alismatidae, Najadales; Th: Alismatanae, Potamogetonales; Ta: Alismatidae, Juncaginales

Limnocharitaceae — Cr: Alismatidae, Alismatales; Th: Alismatanae, Alismatales; Ta: Alismatidae, Alismatales

Posidoniaceae — Cr: Alismatidae, Najadales; Th: Alismatanae, Potamogetonales; Ta: Alismatidae, Posidoniales

Potamogetonaceae — Cr: Alismatidae, Najadales; Th: Alismatanae, Potamogetonales; Ta: Alismatidae, Potamogetonales

Ruppiaceae — Cr: Alismatidae, Najadales; Th: Alismatanae, Potamogetonales; Ta: Alismatidae, Potamogetonales

Scheuchzeriaceae — Cr: Alismatidae, Najadales; Th: Alismatanae, Potamogetonales; Ta: Alismatidae, Jucaginales

Tofieldiaceae

Zosteraceae — Cr: Alismatidae, Najadales; Th: Alismatanae, Potamogetonales; Ta: Alismatidae, Zosterales

Order 3. Asparagales

Families:

Agapanthaceae — Cr = Liliaceae; Th: Lilianae, Orchidales; Ta = Alliaceae

Agavaceae — Cr: Liliidae, Liliales; Th: Lilianae, Orchidales; Ta: Liliidae, Amaryllidales

Alliaceae — Cr = Liliaceae; Th: Lilianae, Orchidales; Ta: Liliidae, Amaryllidales

Amaryllidaceae — Cr = Liliaceae; Th: Lilianae, Orchidales; Ta: Liliidae, Amaryllidales

Anemarrhenaceae

Anthericaceae — Cr = Liliaceae; Th: Lilianae, Orchidales; Ta: Liliidae, Asparagales

Aphyllanthaceae — Cr = Liliaceae; Th: Lilianae, Orchidales; Ta: Liliidae, Asparagales

Asparagaceae — Cr = Liliaceae; Th: Lilianae, Orchidales; Ta: Liliidae, Asparagales

Asphodelaceae — Cr = Liliaceae; Th: Lilianae, Orchidales; Ta: Liliidae, Asparagales

Asteliaceae — Cr = Liliaceae; Th: Lilianae, Orchidales; Ta: Liliidae, Asparagales

Behniaceae

Blandfordiaceae — Cr = Liliaceae; Th: Lilianae, Orchidales; Ta: Liliidae, Asparagales

Boryaceae

Convallariaceae — Cr = Liliaceae; Th: Lilianae, Orchidales; Ta: Liliidae, Asparagales

Doryanthaceae — Cr = Agavaceae; Th: Lilianae, Orchidales; Ta: Liliidae, Asparagales

Hemerocallidaceae — Cr = Liliaceae; Th: Lilianae, Orchidales; Ta: Liliidae, Amaryllidales

Herreriaceae — Cr = Liliaceae; Th: Lilianae, Orchidales; Ta: Liliidae, Asparagales

Hesperocallidaceae — Cr = Liliaceae; Th: Lilianae, Orchidales; Ta: Liliidae, Amaryllidales

Hyacinthaceae — Cr = Liliaceae; Th: Lilianae, Orchidales; Ta: Liliidae, Amaryllidales

Hypoxidaceae — Cr = Liliaceae; Th: Lilianae, Orchidales; Ta: Liliidae, Hypoxidales

Iridaceae — Cr: Liliidae, Liliales; Th: Lilianae, Orchidales; Ta: Liliidae, Iridales

Ixioliriaceae

Lanariaceae

Laxmanniaceae

Orchidaceae — Cr: Liliidae, Orchidales; Th: Lilianae, Orchidales; Ta: Liliidae, Orchidales

Tecophilaeaceae — Cr = Liliaceae; Th: Lilianae, Orchidales; Ta: Liliidae, Tecophilaeales

Themidaceae

Xanthorrhoeaceae — Cr: Liliidae, Liliales; Th: Lilianae, Orchidales; Ta: Liliidae, Xanthorrhoeales

Xeronemataceae

Order 4. Dioscoreales

Families:

Burmanniaceae — Cr: Liliidae, Orchidales; Th: Lilianae, Burmanniales; Ta: Liliidae, Burmanniales

Dioscoreaceae — Cr: Liliidae, Liliales; Th: Lilianae, Taccales; Ta: Liliidae, Dioscoreales

Taccaceae — Cr: Liliidae, Liliales; Th: Lilianae, Taccales; Ta: Liliidae, Taccales

Thismiaceae — Cr = Burmanniaceae; Th: Lilianae, Burmanniales; Ta: Liliidae, Burmanniales

Trichopodaceae — Cr = Dioscoreaceae; Th: Lilianae, Taccales; Ta: Liliidae, Dioscoreales

Order 5. Liliales

Families:

Alstroemeriaceae — Cr = Liliaceae; Th: Lilianae, Liliales; Ta: Liliidae, Alstroemeriales

Campynemataceae

Colchicaceae — Cr = Liliaceae; Th: Lilianae, Liliales; Ta: Liliidae, Colchicales

Liliaceae — Cr: Liliidae, Liliales; Th: Lilianae, Liliales; Ta: Liliidae, Liliales

Luzuriagaceae — Cr = Smilacaceae; Th: Lilianae, Liliales; Ta: Liliidae, Smilacales

Melanthiaceae — Cr = Liliaceae; Th: Lilianae, Liliales; Ta: Liliidae, Melanthiales

Philesiaceae — Cr = Smilacaceae; Th: Lilianae, Liliales; Ta: Liliidae, Smilacales

Ripogonaceae

Smilacaceae — Cr: Liliidae, Liliales; Th: Lilianae, Liliales; Ta: Liliidae, Smilacales

Order 6. Pandanales

Families:

Cyclanthaceae — Cr: Arecidae, Cyclanthales; Th: Pandananae, Cyclanthales; Ta: Aridae, Cyclanthales

Pandanaceae — Cr: Arecidae, Pandanales; Th: Pandananae, Pandanales; Ta: Aridae, Pandanales

Stemonaceae — Cr: Liliidae, Liliales; Th: Lilianae, Taccales; Ta: Liliidae, Stemonales

Velloziaceae — Cr: Liliidae, Liliales; Th: Pandananae, Velloziales; Ta: Commelinidae, Velloziales

2. APG Group : Commelinoid Monocots

1. Not Assigned to Order

Families:

Abolbodaceae - Cr = Xyridaceae; Ta = Xyridaceae

Bromeliaceae — Cr: Zingiberidae, Bromeliales; Th: Commelinanae, Bromeliales; Ta: Commelinidae, Bromeliales

Dasypogonaceae — Cr = Xanthorrhoeaceae; Th: Arecanae, Dasypogonales; Ta: Liliidae, Xanthorrhoeales

Hanguanaceae — Cr: Liliidae, Liliales; Th: Arecanae, Hanguanales; Ta: Liliidae, Hanguanales

Mayacaceae — Cr: Commelinidae, Commelinidales; Ta: Commelinidae, Mayacales

Rapateaceae — Cr: Commelinidae, Commelinidales; Th: Commelinanae, Bromeliales; Ta: Commelinidae, Rapateales

Order 1. Arecales

Family:

Arecaceae - Cr: Arecidae, Arecales; Th: Arecanae, Arecales; Ta: Arecidae, Arecales

Order 2. Commelinales

Families:

Commelinaceae — Cr: Commelinidae, Commelinidales; Th: Commelinanae, Commelinales; Ta: Commelinidae, Commelinales

Haemodoraceae — Cr: Liliidae, Liliales; Th: Commelinanae, Commelinales; Ta: Commelinidae, Haemodorales

Philydraceae — Cr: Liliidae, Liliales; Th: Commelinanae, Commelinales; Ta: Commelinidae, Philydrales

Pontederiaceae — Cr: Liliidae, Liliales; Th: Commelinanae, Commelinales; Ta: Commelinidae, Pontederiales

Order 3. Poales

Families:

Anarthriaceae — Cr = Restionaceae; Th: Commelinanae, Restionales; Ta: Commelinidae, Restionales

Centrolepidaceae — Cr: Commelinidae, Restionales; Th: Commelinanae, Restionales; Ta: Commelinidae, Centrolepidales

Cyperaceae — Cr: Commelinidae, Cyperales; Th: Commelinanae, Juncales; Ta: Commelinidae, Cyperales

Ecdeiocoleaceae — Cr = Restionaceae; Th: Commelinanae, Restionales; Ta: Commelinidae, Restionales

Eriocaulaceae — Cr: Commelinidae, Eriocaulales; Th: Commelinanae, Xyridales; Ta: Commelinidae, Eriocaulales

Flagellariaceae — Cr: Commelinidae, Restionales; Th: Commelinanae, Restionales; Ta: Commelinidae, Flagellariales

Hydatellaceae — Cr: Commelinidae, Hydatellales; Th: Hydatellanae, Hydatellales; Ta: Commelinidae, Hydatellales

Joinvilleaceae — Cr: Commelinidae, Restionales; Th: Commelinanae, Restionales; Ta: Commelinidae, Restionales

Juncaceae — Cr: Commelinidae, Juncales; Th: Commelinanae, Juncales; Ta: Commelinidae, Juncales

Poaceae — Cr: Commelinidae, Cyperales; Th: Commelinanae, Restionales; Ta: Commelinidae, Poales

Prioniaceae

Restionaceae — Cr: Commelinidae, Restionales; Th: Commelinanae, Restionales; Ta: Commelinidae, Restionales

Sparganiaceae — Cr: Commelinidae, Typhales; Ta: Aridae, Typhales

Thurniaceae — Cr: Commelinidae, Juncales; Th: Commelinanae, Juncales; Ta: Commelinidae, Juncales

Typhaceae — Cr: Commelinidae, Typhales; Th: Commelinanae, Typhales; Ta: Aridae, Typhales

Xyridaceae — Cr: Commelinidae, Commelinidales; Th: Commelinanae, Xyridales; Ta: Commelinidae, Xyridales

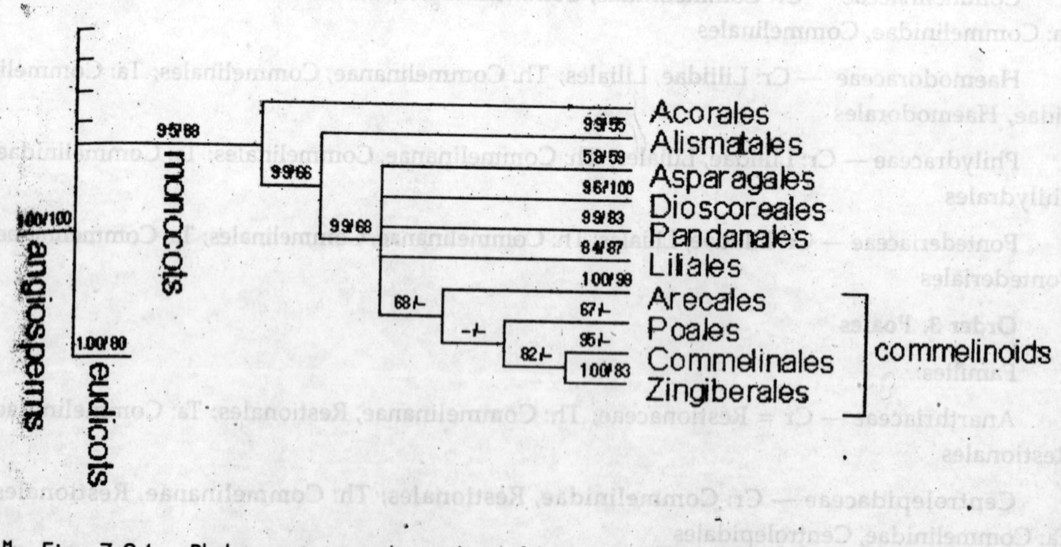

Fig. 7.24 : Phylogenetic interrelationships of the orders of flowering plants, compiled from recent cladistic analyses. Jackknife support is given on the branches (a dash for values <50%), first jackknife values from analysis of 545 sequences of the rbcL, atpB, and 18S rDNA genes (D. E. Soltis, M. W. Chase, P. S. Soltis, D. Albach, M. E. Mort, V. Savolainen, M. Zanis, and J. S. Farris, unpublished, in preparation) and second jackknife values from analysis of 2538 rbcL sequences (Källersjö et al., 1998).

Order 4. Zingiberales

Families:

Cannaceae — Cr: Zingiberidae, Zingiberales; Th: Commelinanae, Cannales; Ta: Commelinidae, Cannales

Costaceae — Cr: Zingiberidae, Zingiberales; Th: Commelinanae, Cannales; Ta: Commelinidae, Zingiberales

Heliconiaceae — Cr: Zingiberidae, Zingiberales; Th: Commelinanae, Cannales; Ta: Commelinidae, Musales

Lowiaceae — Cr: Zingiberidae, Zingiberales; Th: Commelinanae, Cannales; Ta: Commelinidae, Lowiales

Marantaceae — Cr: Zingiberidae, Zingiberales; Th: Commelinanae, Cannales; Ta: Commelinidae, Cannales

Musaceae — Cr: Zingiberidae, Zingiberales; Th: Commelinanae, Cannales; Ta: Commelinidae, Musales

Strelitziaceae — Cr: Zingiberidae, Zingiberales; Th: Commelinanae, Cannales; Ta: Commelinidae, Musales

Zingiberaceae — Cr: Zingiberidae, Zingiberales; Th: Commelinanae, Cannales; Ta: Commelinidae, Zingiberales

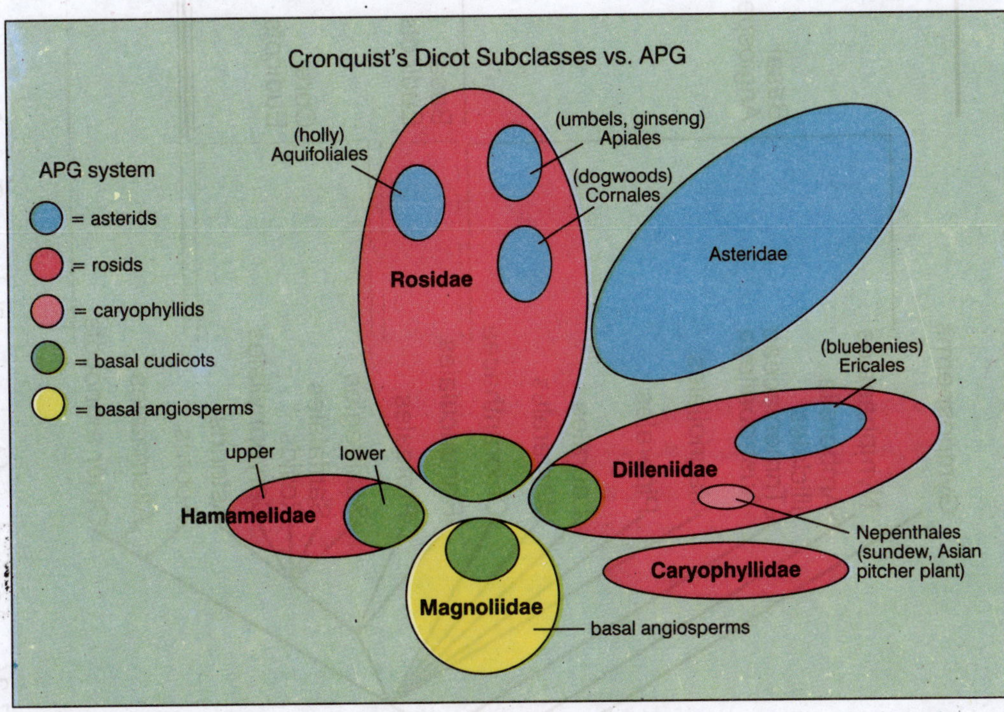

Fig. 7.25 : Figure showing the compative placement of Dicot subclasses according to Cronquist's and The Angiosperm Phylogeny Group (APG) system

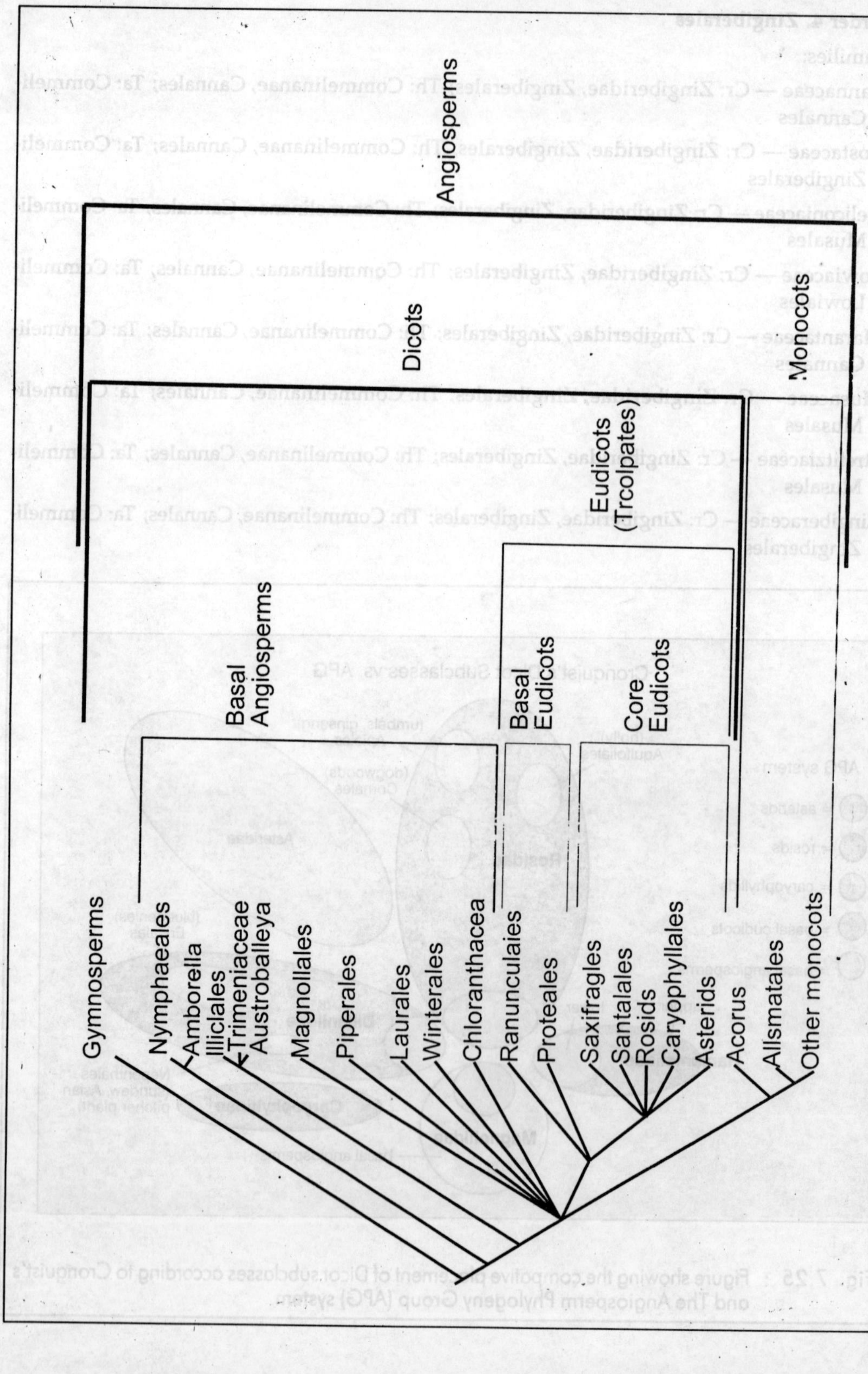

Fig. 7.26 : An overview of Angiosperm Phylogeny Group System (APG)

SUGGESTED READING

- Bentham, G. and J.D. H ooker, 1862-1883, *Genera Plantarum*, 3 Vols., London.
- Bessey, C.E., 1915, Phylogenetic taxonomy of flowering plants. *Ann. Mo. Bot. Gard.*, 2:109-164.
- Candolle, A.P. de., 1813, *Theorie elementaire de la botanique*, Paris.
- Candolle, A.P. de., 1824-1873, *Prodromus systematics naturalis regni Vegetabilis*, 17 Vols., Paris.
- Cronquist, A., 1981, *An Integrated system of Classification of Flowering Plants*, Columbia University Press, New York.
- Dahlgren, R., 1980, A revised system of classification of angiosperms, *Bot. J. Linn. Soc.*, 80:91-124.
- Dahlgren, R., S. Rusendal-Jenses and B.J. Nielsen,1981, A revised classification of the angiosperms with comments on the co-relation between chemical and other characters. In: *Phytochemistry and Angiosperm Phylogeny* (D.A. Young and D.S. Seigler, eds.), Praeger, New York.
- Dahlgren, R., S. Rusendal-Jenses and B.J. Nielsen,1983, General aspects of angiosperm evolution and microsystematics. *Nord. J. Bot.*, 3:119-149.
- Eichler, A.W., 1875-1878, *Bluthendiagramme construirt und erlautert*, 2 Vols., Leipzig.
- Engler, A. and K. Prantl, 1887-1915, *Die Naturlichen Pflanzenfamilien*, 23 Vols., Leipzig.
- Hilu, K. W., T. Borsch, K. Müller, D. E. Soltis, P. S. Soltis, V. Savolainen, M. W. Chase, M. P. Powell, L. A. Alice, R. Evans, H. Sauquet, C. Neinhuis, T. A. B. Slotta, J. G. Rohwer, C. S. Campbell and L. W. Chatrou, 2003, Angiosperm phylogeny based on matK sequence information. *American Journal of Botany*, 90(12):1758-1776.
- Hutchinson, J., 1964-1967, *The Genera of Flowering Plants*, 2 Vols., Claredon, Oxford.
- Hutchinson, J., 1969, *Evolution and Phylogeny of Flowering Plants*, Academic Press, Oxford.
- Hutchinson, J., 1973, *The Families of Flowering Plants*, 3rd ed., Claredon, Oxford.
- Jussieu, A.L. de., 1789, *Genera Plantarum Secundum Ordines naturals disposita*, Paris.
- Linnaeus, C., 1735, *Systema naturae*, Lugduni Batavorum.
- Linnaeus, C., 1737, *Genera Plantarum*, Lugduni Batavorum.
- Linnaeus, C., 1753, *Species Plantarum*, 2 Vols., Stockholm.
- Magallón, S., P. R. Crane and P. S. Herendeen, 1999, Phylogenetic pattern, diversity and diversification of eudicots. *Annals of the Missouri Botanical Garden*, 86:297-372.
- Nandi, O. I., M. W. Chase and P. K. Endress, 1998, A combined cladistic analysis of angiosperms using rbcL and non-molecular data sets. *Annals of the Missouri Botanical Garden*, 85:137-212.
- Qiu, Y. L., J. H. Lee, F. Bernasconi-Quadroni, D. E. Soltis, P. S. Soltis, M. Zanis, E. A. Zimmer, Z. D. Chen, V. Savolainen and M. W. Chase, 2000, Phylogeny of basal angiosperms: Analyses of five genes from three genomes. *International Journal of Plant Sciences*, 161:S3-S27.
- Savolainen, V., M. W. Chase, S. B. Hoot, C. M. Morton, D. E. Soltis, C. Bayer, M. F. Fay, A. Y. De Bruijn, S. Sullivan, and Y. L. Qiu, 2000a, Phylogenetics of flowering plants based on combined analysis of plastid atpB and rbcL gene sequences. *Systematic Biology*, 49:306-362.
- Savolainen, V., M. F. Fay, D. C. Albach, A. Backlund, M. van der Bank, K. M. Cameron, S. A. Johnson, M. D. Lledó, J.-C. Pintaud, M. Powell, M. C. Sheahan, D. E. Soltis, P. S. Soltis, P. Weston, W. M. Whitten, K. J. Wurdack and M. W. Chase, 2000b, Phylogeny of the eudicots: a nearly complete familial analysis based on rbcl gene sequences. *Kew Bulletin*, 55:257-309.

Contd.

- Soltis, D. E., A. E. Senters, M. J. Zanis, S. Kim, J. D. Thompson, P. S. Soltis, L. P. Ronse De Craene, P. K. Endress and J. S. Farris, 2003, Gunnerales are sister to other core eudicots: implications for the evolution of pentamery. *American Journal of Botany*, 90(3):461-470.
- Soltis, D. E., P. S. Soltis, M. W. Chase, M. E. Mort, D. C. Albach, M. Zanis, V. Savolainen, W. H. Hahn, S. B. Hoot, M. F. Fay, M. Axtell, S. M. Swensen, L. M. Prince, W. J. Kress, K. C. Nixon and J. S. Farris, 2000, Angiosperm phylogeny inferred from 18S rDNA, rbcL, and atpB sequences. *Botanical Journal of the Linnean Society*, 133:381-461.
- S. B. Hoot, S. Magallón and P. R. Crane, 1999, Phylogeny of basal eudicots based on three molecular data sets: *atp*B, *rbc*L, and 18S nuclear ribosomal DNA sequences. *Ann. Missouri Bot. Gard.*, 86(1):1-32.
- Takhtajan, A., 1980, Outline of the classification of flowering plants (Magnoliophyta). *Bot. Rev.*, 46:225-359.
- Thorne, R.F., 1976, A phylogenetic classification of the angiospermae. *Evol. Biol.*, 9:35-106.
- Thorne, R.F., 1983, Proposed new realignments in the angiosperms. *Nord. J. Bot.*, 3:85-117.

8 TAXONOMY — A SYNTHETIC SUBJECT

AT A GLANCE

The principal aim of plant taxonomy has been to provide a system of classification based on evolutionary relationships. Most of the natural systems of plant classification have been based on exomorphic characters, placing greater emphasis on the characters of the flower as reproductive characters have been considered to be more important and conservative than the vegetative characters. However, since evolution has taken divergent courses in different groups of plants, with various characters showing difference in their extent and speed of evolutionary changes, the goal of attaining a truly natural system of classification remains unfulfilled. Hence, it is very essential to have a comprehensive knowledge of the concerned taxa from various aspects and various branches of botany like anatomy, embryology, cytology, phytochemistry, ultrastructure, etc. taking the advantage of the newly developed techniques of molecular biology and newer aspects of taxonomy like chemotaxonomy, cytotaxonomy, numerical taxonomy, etc., not only for the appropriate placement of taxa in the taxonomic systems but also to solve the discrepancies prevailing in the classification systems based on classical taxonomy. Thus, the 'synthetic' approach has become the modern basis of plant classification, which evaluates all types of characters related to plant form, structure and function and which has helped reveal the mechanisms of evolution at various levels of taxonomic hierarchy. Many of the recent systems of classification such as those of Cronquist, Takhtajan and Dahlgren have utilized the synthetic approach and these newer evidences in deducing plant relationships and classification of Angiosperms. In this chapter, the taxonomical evidences used for the classification of plants have been discussed.

8.1 Morphology in Relation to Taxonomy

Morphological characters of the plants have provided the foundation and framework for taxonomy and they have been used extensively in the preparation of classification systems, diagnostic keys, etc. Although, in recent years the synthetic approach has fast developed, yet the traditional method of plant taxonomy based upon comparative external morphological characters is still indispensable to systematics and this traditional method continues to predominate and has a bigger say over other sources of taxonomic characters in plant classification, mainly due to the following reasons :

(a) The morphological characters are easily observable. Since they have innumerable variants, they help in delimitation and identification.

(b) To analyze morphological characters, one need not have sophisticated laboratory arrangements. A hand lens or dissecting microscope, or possibly a light microscope is enough to study these characters.

(c) There is a well-knit terminology to describe variations as morphological characters have been in use for plant classification for several centuries.

(d) The time and effort needed to obtain information from other sources such as phytochemistry and molecular biology ensure the advantage of morphological characters over others in plant taxonomy.

8.1.1 External Morphology

External morphological characters provide majority of the evidences used in the construction of classification system. It is often argued that exomorphic characters are less important than the endomorphic or other characters. But in case of plants this assumption may not always be true, as some exomorphic characters are stable in certain groups and show considerable variation in others. It has been concluded by Bailey (1951), that endomorphic characters such as anatomy are neither more nor less reliable than exomorphic features.

Further, as evident from most of the classificatory systems, floral characters, in general, have been preferred over vegetative characters. There may be a wide variation in several superficial vegetative attributes, especially at higher levels. As for example, structure, size and shape of leaves may be extremely variable within a genus or even within a species. Vegetative characters are often environmentally controlled, particularly plant pubescence. For example, a particular member of a species may develop into an extremely hairy glabrous form, while a different member of the same species growing in a different place, in a different environment may not show such form. On the other hand, reproductive features are concerned to be more constant and conservative against fluctuating environmental pressures than their vegetative counterparts particularly because, in flowering plants, reproductive characters are subjected to a much lesser degree of evolutionary pressure than the vegetative parts, as the sexual reproductive parts are produced only for a very brief period in most species. However, reproductive characters are liable to strong selection pressures originating from various aspects of reproductive biology, e.g. the pollinator relationship (Clifford & Lavarack, 1974). Thus, equal emphasis need to be given to both vegetative and reproductive characters as both are subject to different types of selection pressure, to which they react in different ways.

Some of the vegetative and reproductive characters, which are employed in plant taxonomy and in deducing phylogeny, include growth habit, phenological characters, underground organs, stem, leaves, petiole and stipules. These vegetative features can be profitably used in identification of trees and shrubs, as they are less variable in such groups. Floral characters such as types of inflorescence and flower and associated structures, play a major role. Types of fruit and seed also provide good diagnostic features useful at various levels of classification. In recent years, various biometric and graphic techniques are being used to interpret morphological features. Thus, many of the morphological features previously overlooked are being carefully studied now and their importance in taxonomic practices being assessed.

(a) **Vegetative features :**

(i) **Growth habit (herbs, shrubs, and trees)** — It may be variable or constant within a genus or a family, e.g. all Brassicaceae members are herbaceous while members of Asteraceae are both herbaceous as well as woody.

(ii) **Growth pattern** — has been of particular interest in defining taxonomic groups and in interpreting interrelationships above generic level in Zingiberaceae. A very good example is the genus *Zingiber*. There has been a lot of controversy regarding its placement. Some authors have placed it under Alpineae, some under Hedychieae, while many have assigned it to Zingibereae. However, study of growth patterns of leaves, which are arranged distichously to the rhizome, and based on the plane of distichy which may differ from parallel to transverse, it was suggested that *Zingiber* is more closely related to Hedychieae and Zingibereae (where the leaves are parallel to the rhizome), than to Alpineae (where the leaves are transverse to the rhizome).

PLATE 7

pg245ch8pl1

The leaves of *Desmodium vespitospermum* (Leguminosae) — resembling a butterfly

pg245ch8pl2

The beautiful leaf margin of *Acalypha* sp.

PLATE 8

The dense prickles on a branch of *Rubus ellipticus* (Rosaceae)

The prickles of *Zanthoxylum* sp. (Rutaceae)

PLATE 9

↳ The exquisite arrangement of spines of *Calamus* sp.

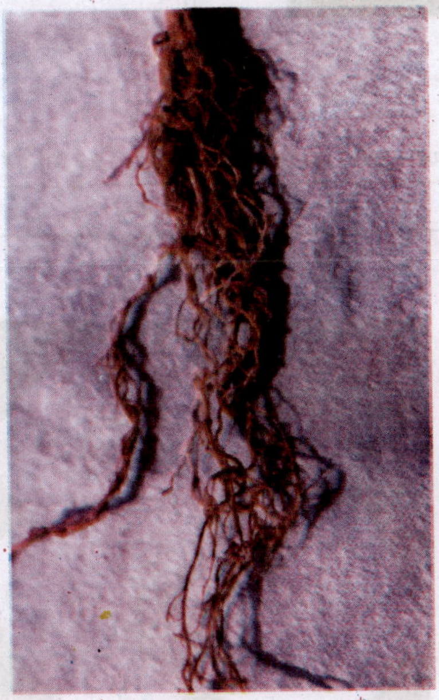

↳ The roots of *Withania somnifera* (Solanaceae)

PLATE 10

pg248ch8pl1

The unique flower of *Couroupita guianensis*

The minute flowers of *Cuscuta reflexa*

pg248ch8pl3

Ottelia alismoides — in full bloom

PLATE 11

pg249ch8pl1

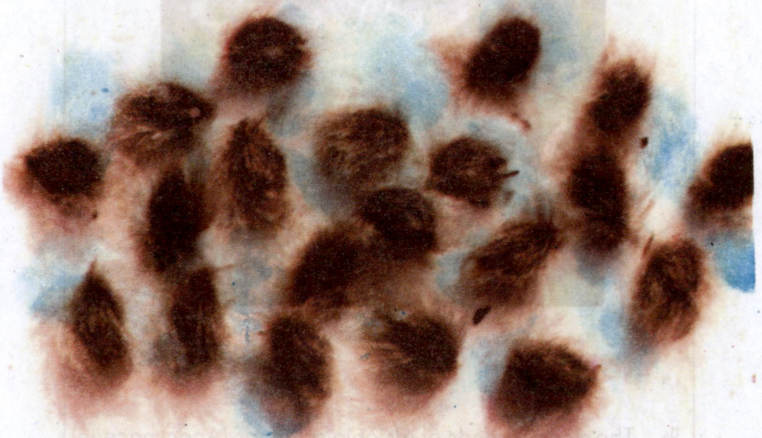

K

↬ Seeds of *Ipomoea fistulosa* (Convolvulaceae) — showing the surface hairs

pg249ch8pl2

↬ The surface structure of the seeds of *Piper longum* (Piperaceae)

PLATE 12

The slender seeds of *Alstonia scholaris* (Apocynaceae) with their fine wooly haris, used for dispersion

The beautiful seeds of *Abrus precatorius* (Fabaceae)

(iii) **Leaf characters** — This includes leaf structure, size and shape, arrangement, type, form, duration and venation. For example, leaflet size, shape and arrangement on the rachis has been successfully employed for distinguishing three species of *Dalbergia* (Papilionaceae), viz. *D. latifolia, D. sissoo* and *D. sympathetica*. However leaf structure, size and shape is often variable in herbaceous species and thus cannot be used reliably for generic delimitation. For example species of *Crotalaria* exhibits continuous variation in this respect. Similarly, stipule character, which is fairly constant within a species except for Rosaceae, has also been used as a taxonomic character. Leaf venation, which also serves as an important taxonomical character, is of particular interest to palaeobotanists who carefully study leaf venation for describing fossils.

(iv) **Underground parts** — such as structure and morphology of roots, tubers, etc. are of some taxonomic significance in plants. For example, in *Dioscorea* (Dioscoreaceae), the structure and morphology of root-tubers has been used as a taxonomical character in the delineation of a large number of species (Burkill & Prain, 1936). Similarly, root morphology has been used as a character to distinguish two species of *Chlorophytum* (Liliaceae), viz. *C. tuberosum* and *C. borivilianum* (Chouard, 1931).

(v) **Seedling characters** — Juvenile characters, especially seedling characters, such as type of germination, cotyledonary characters, leaf dimorphism, etc. have been of systematic value and taxonomic significance in certain families like Brassicaceae (Gomez-Campo & Maria, 1974), Convolvulaceae (Sampathkumar, 1982), etc.

(vi) **Plant pubescence** — It is a character, which has been long neglected in morpho-taxonomic studies. Although the degree of pubescence within the species is largely governed by ecology, the type of trichomes, stylar hairs, etc. have been successfully used as taxonomic characters and in tracing phylogeny in certain groups of plants. For example, Inamdar (1967) confirmed the systematic position of *Nyctanthus* in the family Oleaceae on the basis of the structure and ontogeny of its trichomes. Similarly, Raji (1971) used the stylar hair patterns of the tribe Viceae (Papilionaceae) to deduce phylogenetic pathways of its constituent genera.

(b) **Reproductive features :**

(i) **Inflorescence and flower** — Floral characters such as the type and position of inflorescence, flowers and associated structures such as structure of perianth, floral symmetry, the number, size, shape and union of floral leaves in each whorl, types of androecium and stamens, gynoecium and carpels, ovules, and also the characters of bracts, bracteoles and pedicels are very important from the taxonomic point of view. For example, type of calyx has been used to distinguish species in *Anthyllis*. Similarly, nectaries and floral discs have proved to be of great diagnostic value in Brassicaceae. Other important floral characters are those associated with pollination i.e., different plant groups pollinated by different agents (wind and insects) have different floral adaptations. Those plants, which are wind pollinated i.e., anemophilous, bear flowers which are small, inconspicuous, unisexual and mono- or achlamydous and are usually aggregated into inconspicuous spikes or catkins with a large output of pollen grains. On the other hand, those plants, which are pollinated by insects, i.e. entomophilous, have flowers, which are large, conspicuous, brightly colored and are often nectariferous or scented. Adaptations for different kinds of pollen vector have also resulted in specialization in the calyx and/or corolla, in flower positioning, etc.

(ii) **Fruit characters** — Characters of fruits have been used in the construction of diagnostic keys especially at and above the generic level, as fruit morphology appears to be fairly constant within the genus. Their adaptations for different types of dispersal or of opening has been of great help to taxonomists in analyzing taxonomic problems. For example, fruit characters have been used to distinguish different families of Rhoeadales.

(iii) **Seed characters** — Seeds have been successfully used particularly at the specific level. Seed characters such as number, color, shape and sculpturing have proved helpful. For example, *Anthericum* and *Chlorophytum* (Uliaceae) have been distinguished on the basis of number and shape of the seeds. Similarly, length and color of hairy outgrowths on the testa of the seeds have helped in the distinction of genera and/or species in the Acanthaceae, Asclepiadaceae, Convolvulaceae and Malvaceae.

8.1.2 Microscopic Morphology (epidermal characters) or Ultrastructural Systematics

The development and use of Scanning Electron Microscopy in recent years, has opened a new chapter in the field of plant classification and systematics. The ability of this technique to reveal details of the plant surface, has allowed morphologists and taxonomists consider epidermal features such as hairs, trichomes, cell-types, nectaries, etc., besides the gross morphological characters, which have so far been neglected in the construction of groups or identification of taxa. These microscopic morphological features have proved to be very useful in plant taxonomy when gross features have proved inadequate. Following are some of the epidermal features that are now being used frequently in taxonomic studies:

(i) **Trichomes** — Trichomes are epidermal outgrowths, which exhibit great variation in structure and function and are thus of much taxonomic value. They may be glandular or non-glandular, simple or branched with the branches dendritic (arising like those of a tree) or star-like, terete, flattened, basally swollen, whip like or T-shaped, or of various other forms in different groups of plants (Fig. 8.1). Hence, plant-hair types have been successfully used in the classification of genera and species and in the recognition of hybrids within certain groups. A very good example of the use of trichomes in plant classification is that of *Parthenium* (Asteraceae). In a series of observations (1944, 1945, 1946, 1949) on the trichomes of two species of *Parthenium* and their natural hybrids, Rollins (1941) found that *P. argentatum* had T-shaped trichomes while in *P. incanum* they were whip-like with a long thread. Intermediate types of trichomes were observed in the hybrids between these species. Another good example is the generic key for the Indian members of Asteraceae proposed by Ramayya (1969) on the basis of his critical studies of trichomes in the family. Besides the distribution and anatomy of the actual trichomes, the degree of modification of the adjacent epidermal cells and the scars left by lost trichomes are important aspects of variation.

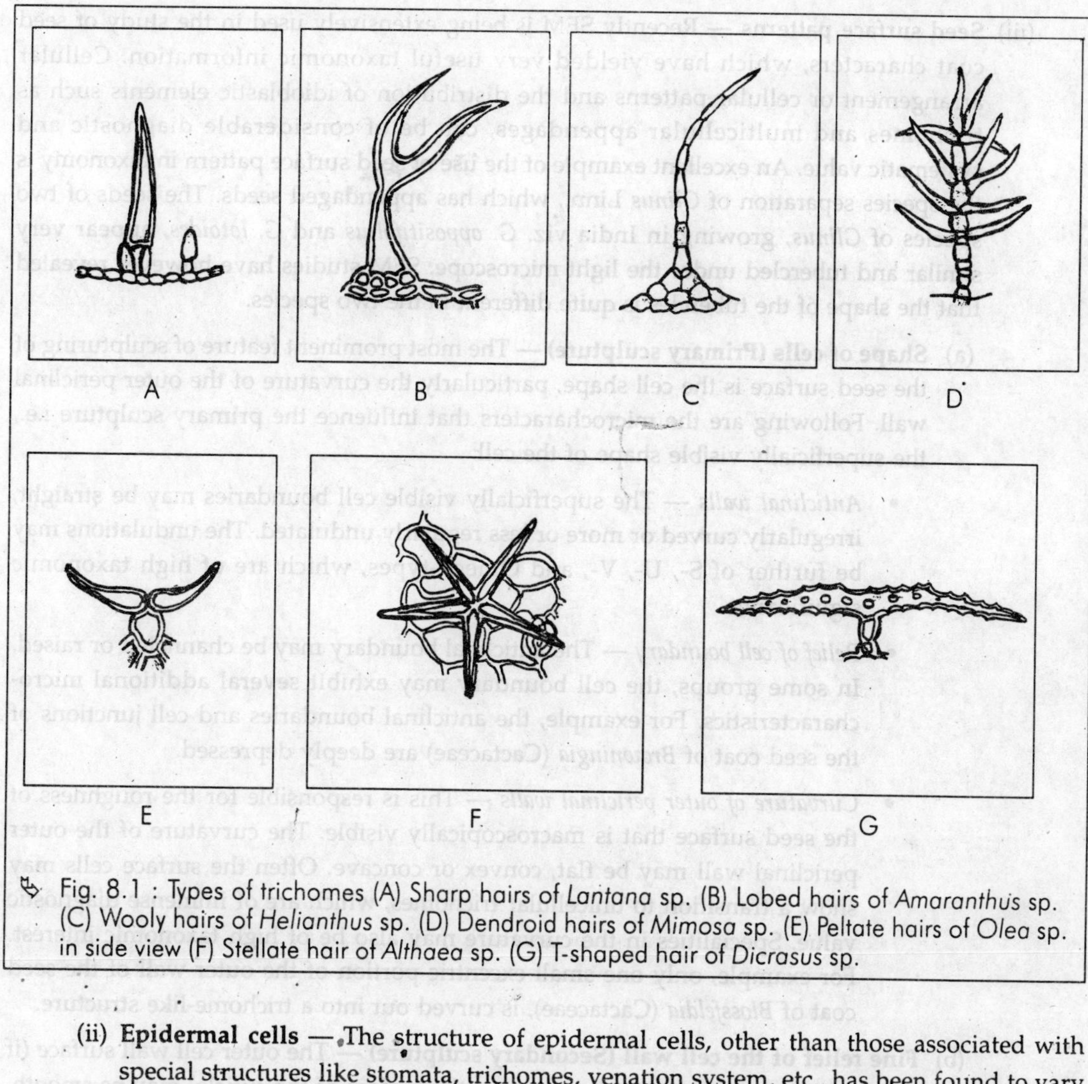

Fig. 8.1 : Types of trichomes (A) Sharp hairs of *Lantana* sp. (B) Lobed hairs of *Amaranthus* sp. (C) Wooly hairs of *Helianthus* sp. (D) Dendroid hairs of *Mimosa* sp. (E) Peltate hairs of *Olea* sp. in side view (F) Stellate hair of *Althaea* sp. (G) T-shaped hair of *Dicrasus* sp.

(ii) **Epidermal cells** — The structure of epidermal cells, other than those associated with special structures like stomata, trichomes, venation system, etc., has been found to vary in certain groups of plants and has been used as a taxonomic character in recent years. This variation in structure, include, variation in size, shape, orientation, undulation of anticlinal wall, curvature of periclinal wall, etc. The expression of the distribution of cell types in a systematized way is termed as *dermogram*, and the dermograms may be indicated for each species as a dermotype (Prat, 1932). Dermotype studies have been employed with success in taxonomy e.g., in the taxonomy of species of *Puccinellia* (Poaceae) [Sorensen, 1953; Tutin, 1955], *Agrostis* (Poaceae) and related genera (Bjorkman, 1960), etc. Recently, Ramayya & Rajgopal (1968, 1971) have even proposed a generic key for the Indian members of Alzoaceae and Portulacaceae based on the dermotype studies of these families. Often the epidermal cells have been found to be modified along leaf margins and also may be calcified, silicified, suberized, gelatinized or contain various types of crystals, which may be of significant taxonomic importance. For example, silicified and suberized epidermal cells are of taxonomic value in the family Poaceae.

(iii) **Seed surface patterns** — Recently SEM is being extensively used in the study of seed coat characters, which have yielded very useful taxonomic information. Cellular arrangement or cellular patterns and the distribution of idioblastic elements such as trichomes and multicellular appendages, can be of considerable diagnostic and systematic value. An excellent example of the use of seed surface pattern in taxonomy is the species separation of *Glinus* Linn., which has appendaged seeds. The seeds of two species of *Glinus*, growing in India viz. *G. oppositifolius* and *G. lotoides*, appear very similar and tubercled under the light microscope. SEM studies have however revealed that the shape of the tubercles is quite different in the two species.

(a) **Shape of cells (Primary sculpture)** — The most prominent feature of sculpturing of the seed surface is the cell shape, particularly the curvature of the outer periclinal wall. Following are the microcharacters that influence the primary sculpture i.e., the superficially visible shape of the cell:

- *Anticlinal walls* — The superficially visible cell boundaries may be straight, irregularly curved or more or less regularly undulated. The undulations may be further of S-, U-, V-, and Omega-types, which are of high taxonomic significance.
- *Relief of cell boundary* — The anticlinal boundary may be channeled or raised. In some groups, the cell boundary may exhibit several additional micro-characteristics. For example, the anticlinal boundaries and cell junctions of the seed coat of *Browningia* (Cactaceae) are deeply depressed.
- *Curvature of outer periclinal walls* — This is responsible for the roughness of the seed surface that is macroscopically visible. The curvature of the outer periclinal wall may be flat, convex or concave. Often the surface cells may show a transition to unicellular trichomes, which are of immense diagnostic value. Specialities in the curvature may also be of high taxonomic interest. For example, only one small excentric portion of the outer wall of the seed coat of *Blossfeldia* (Cactaceae), is curved out into a trichome-like structure.

(b) **Fine relief of the cell wall (Secondary sculpture)** — The outer cell wall surface (if the testa consists of intact cells, it means the surface of the cuticle) may be smooth, or exhibit micro-ornamentation, which may be striate, reticulate or micropapillate (verrucose), resulting from very different portions of the cell wall. The major structural categories are as follows :

- *Cuticular sculptures* — This may occur as regular or irregular striations. In angiosperms, the cuticular folds show high micromorphological diversity, which may serve as good diagnostic characters, but their systematic significance is limited.
- *Secondary wall thickenings* — This may occur as helical to reticulate patterns on the inner side of the outer periclinal and on the anticlinal walls. Usually they occur in the form of reticulations or striations and are of high taxonomic significance. For example, dense reticulate pattern is seen on the seed coat of *Chondrorhyncha* (Orchidaceae).

PLATE 13

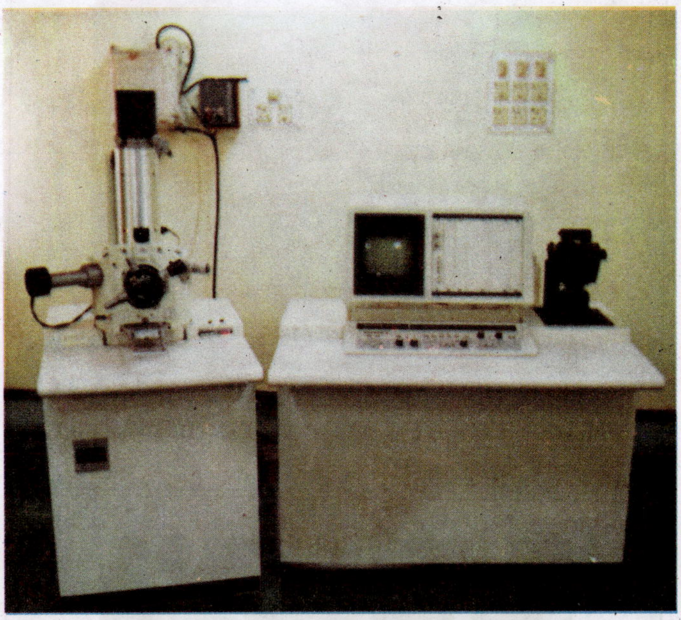

↬ The Scanning Electron Microscope at the Instrumentation Centre of The University of Burdwan

PLATE 14

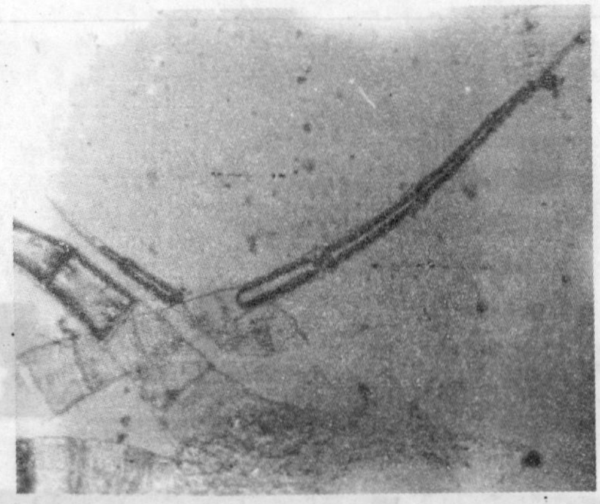

↬ Trichomes of *Mucuna prurita* (Fabaceae) under the Light Microscope (100 X)

↬ Scanning electron microscopic photograph of trichomes of leaf surface of *Diospyros* sp.

(c) **Epicuticular secretions (Tertiary sculpture)** – Seed coat exudes of different chemical nature, particularly mucilaginous adhesive substances are very common and their nature and origin may be of taxonomic and systematic significance and used in the delimitation of genera, families, and even orders. The micromorphology of epicuticular waxes and related solid lipophilic substances can be easily viewed with the help of SEM. For example, only certain epidendroid and vandoid tribes and subtribes under Orchidadeae, possess epicuticular waxes on their seed coats (e.g. Bulbophyllinae).

(iv) **Pollen wall ornamentation** – SEM has had an enormous impact on palynology and has proved very useful to palynologists primarily concerned with taxonomy. Like seed surface patterns, the different types of surface ornamentation of the pollen wall i.e. exine have also yielded very useful taxonomic information (Echlin, 1968; Straka, 1971). Variation in exine structure and sculpture, together with variation in the number, position and complexity of apertures in the pollen grain wall gives morphological diversity of the pollen grains.

(v) **Stomata** – Systematic studies on the development and morphology of stomata and associated epidermal cells are proving to be increasingly important from the taxonomic point of view, as they may be expected to give clues to the various evolutionary trends and help assign taxa of uncertain affinity to their true position. Though certain types of stomata are characteristic of some families and subfamilies, some authors are of the view that the taxonomic value of types of stomata is lowered when exploited at lower taxonomic levels, as one may find different types of stomata on the same leaf (Sachdeva & Malik, 1986). The distribution of stomata, frequency and structure differs in different taxonomic groups. There are different types of classification of stomata based on different criteria.

Following are some of the types of stomata based on the arrangement of subsidiary cells in the mature stomatal complex of vascular plants (Dichler, 1974) :

1. **Anomocytic (Ranunculaceous type)** : Stomata and guard cells surrounded by a limited number of cells that are indistinguishable in shape, size or form from those of normal epidermal cells.

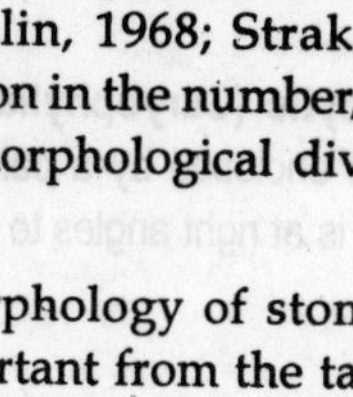

2. **Cyclocytic** : Stomata and guard cells surrounded by a ring of four or more small subsidiary cells.

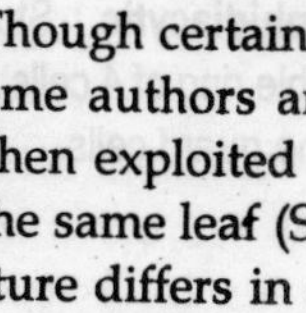

3. **Amphicyclocytic** : Stomata and guard cells surrounded by a double ring of small cells.

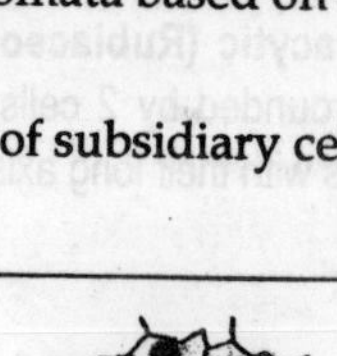

4. **Actinocytic** : Stomata and guard cells surrounded by a single ring of enlarged and radially elongated subsidiary cells.

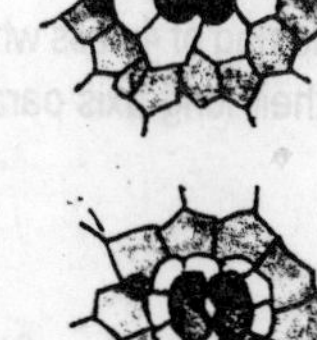

5. **Anisocytic (Cruciferous type)** : Stomata and guard cells surrounded by 3 cells of which 2 are larger and 1 is smaller.

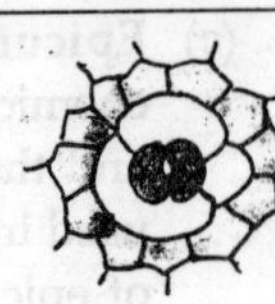

6. **Amphianisocytic** : Stomata and guard cells enclosed by 2 rings of cells with the inner ring consisting of 3 cells of which 2 are larger and 1 is smaller and the outer ring comprising of 2-3 or even 4 cells.

7. **Diacytic (Caryophyllaceous type)** : Stomata and guard cells enclosed by a pair of subsidiary cells whose common wall is at right angles to the guard cells.

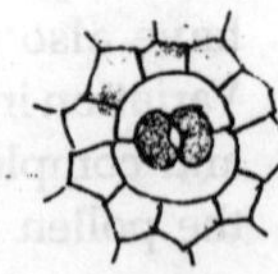

8. **Amphidiacytic** : Stomata and guard cells surrounded by a double ring of 4 cells which are at right angles to the long axis of the guard cells.

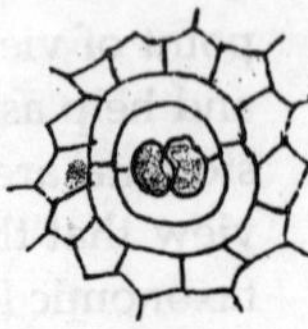

9. **Paracytic (Rubiaceous type)** : Stomata and guard cells surrounded by 2 cells which completely enclose the guard cells with their long axis parallel to long axis of the guard cells.

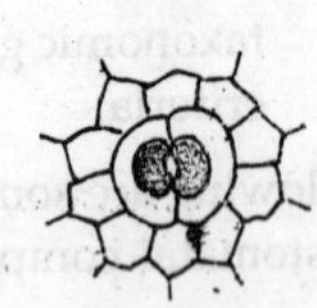

10. **Amphiparacytic** : Stomata and guard cells surrounded by double ring of 4 cells which completely enclose the guard cells with their long axis parallel to long axis of the guard cells.

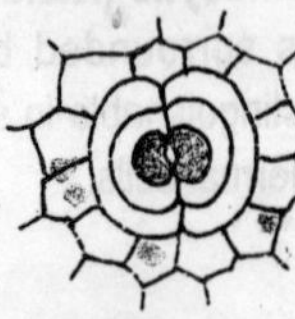

11. **Brachyparacytic** : Stomata and guard cells flanked by 2 cells which do not completely enclose the guard cells, may or may not elongate, with their long axis parallel to long axis of the guard cells.

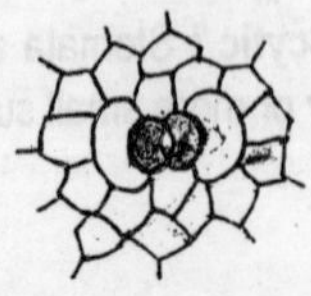

12. **Amphibrachyparacytic** : Stomata and guard cells flanked by 4 cells which do not completely enclose the guard cells, may or may not elongate, with their long axis parallel to long axis of the guard cells.

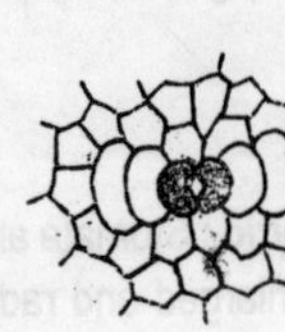

13. **Hemiparacytic :** One of the cells adjacent to one guard cell enclosing it and parallel to its long axis, while the other guard cell surrounded by 3 or more normal epidermal cells.

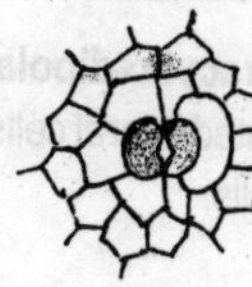

14. **Paratetracytic :** Stomata and guard cells flanked by 2 elongated lateral cells parallel to the guard cells and 2 narrow polar cells.

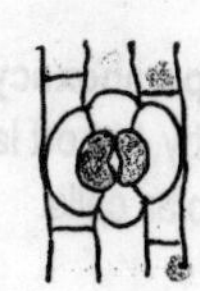

15. **Amphiparatetracytic :** Stomata and guard cells flanked by 2 elongated lateral cells parallel to the guard cells and 2 narrow polar cells, all of which are surrounded by a ring of small cells.

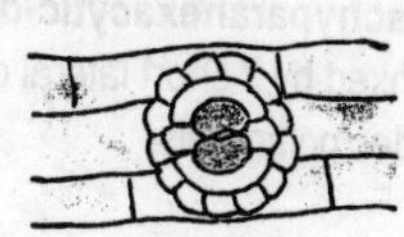

16. **Brachyparatetracytic :** Stomata and guard cells flanked by 2 short lateral cells parallel to the guard cells and 2 wider polar cells.

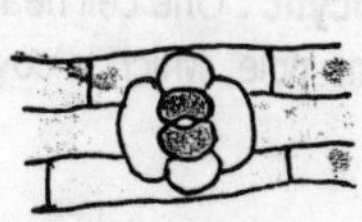

17. **Amphibrachyparatetracytic :** Stomata and guard cells flanked by 2 short lateral cells parallel to the guard cells and 2 wider polar cells, all of which are surrounded by a ring of small cells.

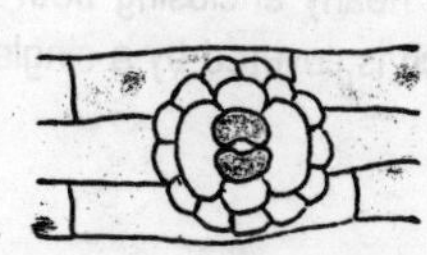

18. **Staurocytic :** Stomata and guard cells surrounded by 4 cells more or less equal in size, with anticlinal walls of the subsidiary cells extending at right angles from the poles and middle of the guard cells.

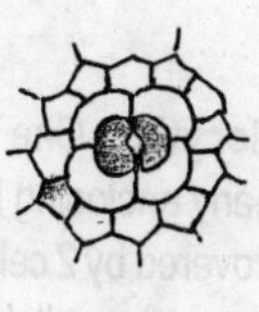

19. **Anomotetracytic :** Stomata and guard cells surrounded by 4 cells in an irregular and variable pattern.

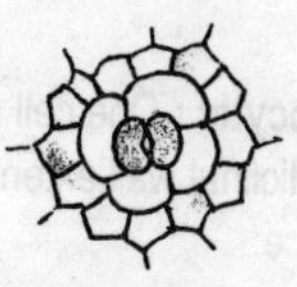

20. **Parahexacytic-monopolar :** Stomata and guard cells flanked by 4 elongated lateral cells parallel to the guard cells and 2 narrow polar cells.

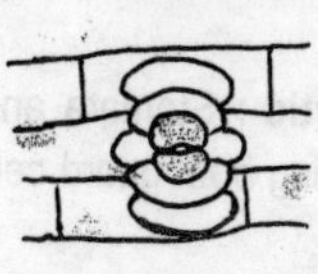

21. **Parahexacytic-dipolar :** Stomata and guard cells flanked by 2 elongated lateral cells parallel to the guard cells and 4 narrow polar cells.

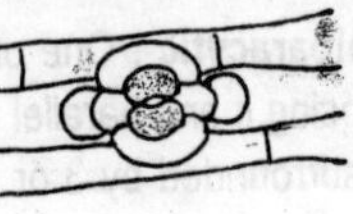

22. **Brachyparahexacytic-monopolar :** Stomata and guard cells flanked by 4 short lateral cells parallel to the guard cells and 2 wider polar cells.

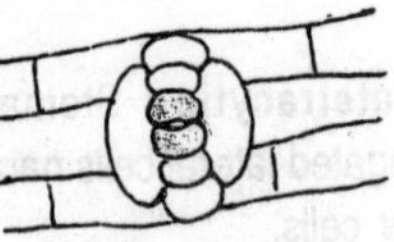

23. **Brachyparahexacytic-dipolar :** Stomata and guard cells flanked by 2 short lateral cells parallel to the guard cells and 4 wider polar cells.

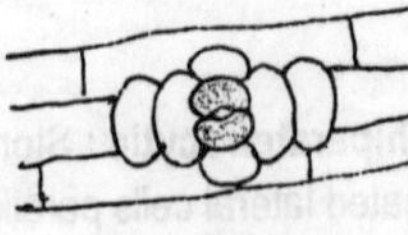

24. **Polocytic :** One cell nearly enclosing both guard cells except at one pole, which is covered by a single epidermal cell.

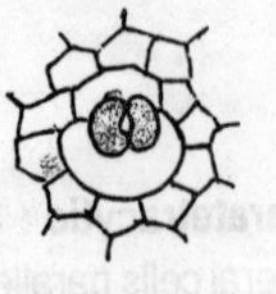

25. **Copolocytic :** One cell (subtended by a crescent-shaped cell) nearly enclosing both guard cells except at one pole, which is covered by a single epidermal cell.

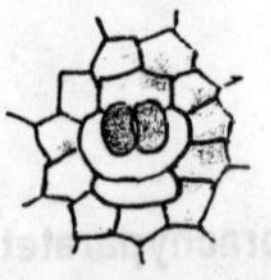

26. **Axillocytic :** One cell nearly enclosing both guard cells except for one free pole.

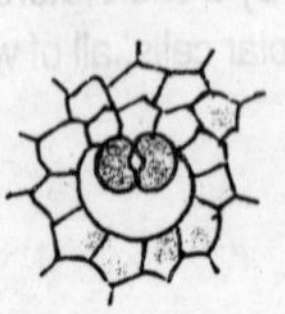

27. **Coaxillocytic :** One cell (subtended by a crescent-shaped cell) nearly enclosing both guard cells except for one free pole that is covered by 2 cells with a common anticlinal wall extending from the pole parallel to the long axis of the guard cells.

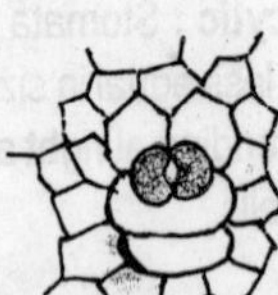

28. **Desmocytic :** One cell nearly enclosing both guard cells with one anticlinal wall extending from one of the poles cutting the cell once.

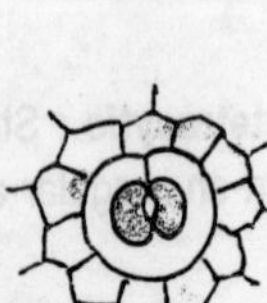

29. **Pericytic :** Stomata and guard cells surrounded by 1 cell enclosing both guard cells.

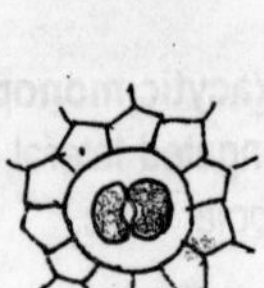

30. **Copericytic :** Stomata and guard cells surrounded by 1 cell (subtended by a crescent-shaped cell) enclosing both guard cells.

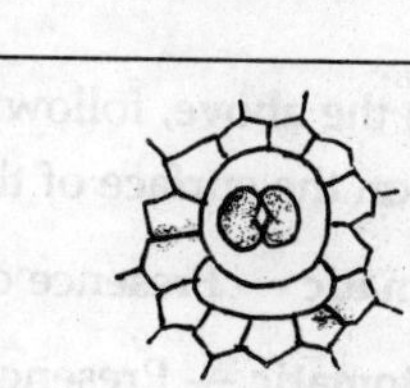

31. **Amphipericytic :** Stomata and guard cells surrounded by 1 cell enclosing both guard cells, which in turn is enclosed by a second single cell.

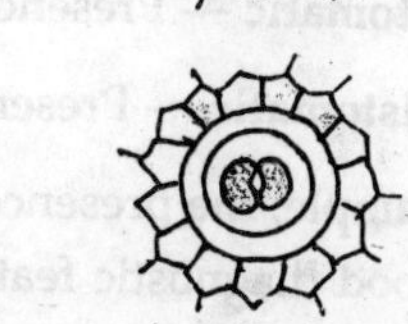

Examples: In the family Acanthaceae the stomata are diacytic, whereas in the closely related Scrophulariaceae the stomata are anomocytic. In the family Combretaceae, the stomata are paracytic in the subfamily Strephonematoideae and anomocytic in the subfamily Combretoideae except the tribe Laguncularieae where they are cyclocytic.

Further, depending on the development of stomata, Pant (1965) has classified stomata into following three types :

1. **Perigenous :** Stomata in which the subsidiary cells do not have a common origin with the guard cells, but are formed by cells lying around the meristemoid that divides to form the guard cells.

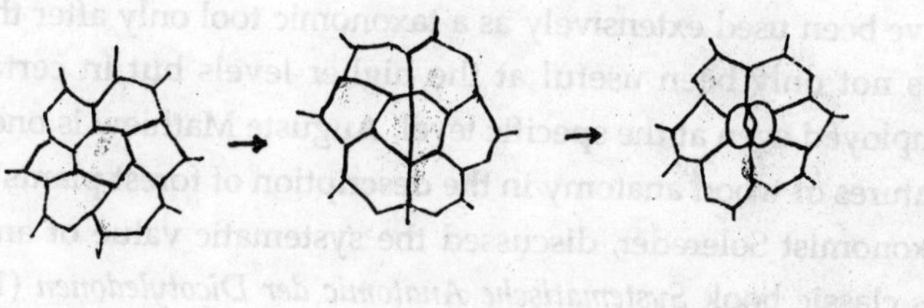

2. **Mesogenous :** Stomata in which the subsidiary cells are produced from the same meristemoid (initial) as the guard cells.

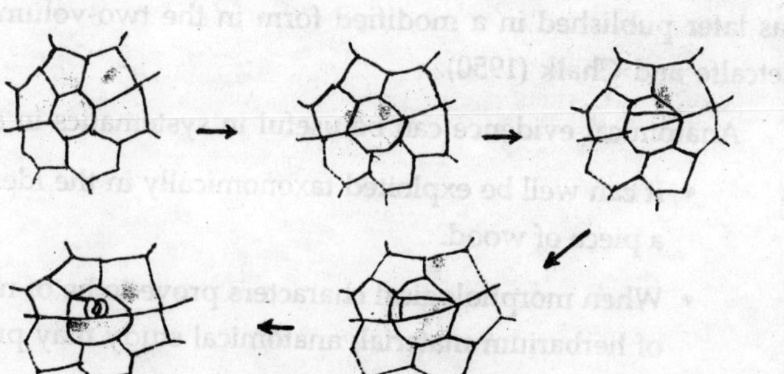

3. **Mesoperigenous :** Stomata in which at least one of the subsidiary cells has a common origin with the guard cells, but others do not.

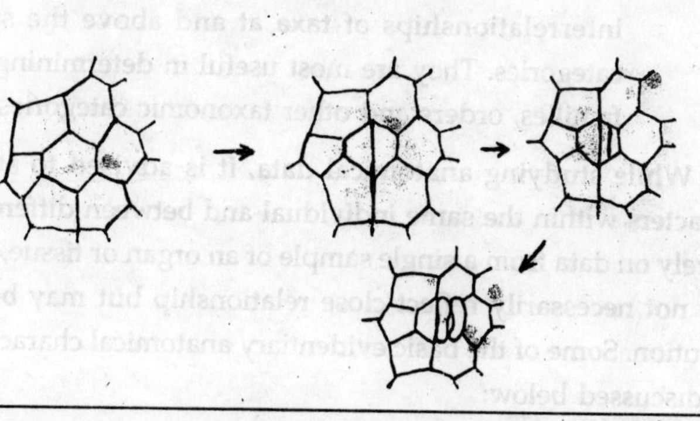

Besides the above, following types are also used in taxonomy, depending on the presence of the stomata on the surface of the leaf :

Epistomatic — Presence of stomata on the upper side of the leaf.

Hypostomatic — Presence of stomata on the lower side of the leaf.

Amphistomatic — Presence of stomata on both sides of the leaf.

For example, the presence or absence of stomata on the upper surface of a leaf has been used as a very good diagnostic feature in the separation of *Ainus subcordata* C.A. Mey and *A. orientalis* Decne. (Betulaceae), which are otherwise very difficult to distinguish.

8.2 Anatomy in Relation to Taxonomy

Anatomical characters of vegetative and floral parts of flowering plants have been successfully employed to solve taxonomic problems and for the elucidation of phylogenetic relationships. It was Bureau (1864), who for the first time used anatomical characters in plant classification for the delimitation of taxa of various levels, within the family Bignoniaceae. However, anatomical data have been used extensively as a taxonomic tool only after the nineteenth century. Anatomical data has not only been useful at the higher levels but in certain instances, have been successfully employed even at the specific level. Auguste Mathiew is one of the pioneer taxonomists, who used features of wood anatomy in the description of forest plants in *Florae forestiere* (1858). Later, another taxonomist Solereder, discussed the systematic value of anatomical structures in dicotyledons in his classic book *Systematische Anatomic der Dicotyledonen* (1899), the english translation of which was later published in a modified form in the two-volume book *Anatomy of the Dicotyledons* by Metcalfe and Chalk (1950).

Anatomical evidence can be useful in systematics in several ways:

- It can well be exploited taxonomically in the identification of fragmentary material, say a piece of wood.
- When morphological characters prove to be of no help in the preliminary identification of herbarium material, anatomical study may prove helpful.
- Anatomical data has proved to be very useful in discerning evolutionary trends and interrelationships of taxa at and above the species level and at higher taxonomic categories. They are most useful in determining relationship between different genera, families, orders and other taxonomic categories.

While studying anatomical data, it is advised to study the ranges of variability of these characters within the same individual and between different individuals of the same species and not rely on data from a single sample of an organ or tissue, as similarities in structural organization may not necessarily reflect close relationship but may be the result of parallel and convergent evolution. Some of the basic evidentiary anatomical characters of well-established taxonomic value are discussed below:

8.2.1 Vegetative Anatomy

(a) **Leaf anatomy** — Leaf anatomy provides various characters of taxonomic importance as has been rightly stated by Carlquist (1961), that "the leaf is perhaps anatomically most varied organ of angiosperms and its anatomical variations often concur closely with generic and specific and occasionally familial lines'. Leaf anatomy has been used widely in several taxonomically different groups such as Euphorbiaceae, Cyperaceae and Gramineae of Angiosperms and Coniferae of Gymnosperms. It has been one of the most reliable characters in grass systematics. For example, the leaf anatomy of several species of Cyperaceae, was studied by Koyama (1967) and Govindrajalu (1966, 1968, 1974, 1975 and 1976) and they formulated keys to identify various species of *Cyperus, Fuirena,* etc. Brown (1958) surveyed, 72 genera of grasses and on the basis of their tissue arrangement, six main types were recognized. However, they could not, be segregated into the two traditional subfamilies, Pooideae and Panicoideae. Similarly, Vidakovie (1957) have used several characters of leaf anatomy in differentiating species in *Pinus.* Taxonomic implication of leaf anatomy of several genera of Musaceae, Zingiberaceae, Xanthorrhoeaceae and Ericaceae have also been established by several workers.

Some of the important characters of taxonomic significance in leaf anatomy include the following :

- Nature and thickness of epidermis — The size and shape of epidermal cells is of great value in the taxonomy of several taxa (Fig. 8.2). Cuticular characters of the epidermis and stomata have also proved to be of great value. For example, Conde (1975) studied 5 species of the genus *Opuntia* with respect to cuticular thickness, epidermal papillosity, stomatal size and frequency, hypodermal thickness, vessel number, etc. and found that each species was distinct in respect of the degree of papillosity of epidermal cells, hypodermal thickness and vessel width. The information from trichome anatomy has also proved useful in certain taxa, e.g. trichomes have furnished as diagnostic characters in certain species of *Veronica* (Faust & Jones, 1973).
- Structure and types of mesophyll, storage parenchyma, mid vein structure, bundle sheath, secretory apparatus, etc. — For example, Anderson & Crech (1975) suggested precise groupings of *Solidago* and other species of Asteraceae based on their study of leaf anatomy, including qualitative and quantitative differences in mesophyll, storage parenchyma, secretory apparatus, bundle-sheath extension and midvein structure.
- Pattern of sclerenchyma — Patterns of the distribution of sclerenchyma in *Carex* and *Festuca* have been used in distinguishing species (Metcalfe, 1968).
- Silica bodies — Silica bodies in the epidermal cells of members of certain families like Zingiberaceae, Musaceae and Palmae among Monocotyledons (Tomlinson, 1956, 1959a,b) and Rosaceae in the Dicotyledons have been used as diagnostic character in systematics at generic as well as specific levels.
- Chloroplast structure can also prove to be of taxonomic significance.

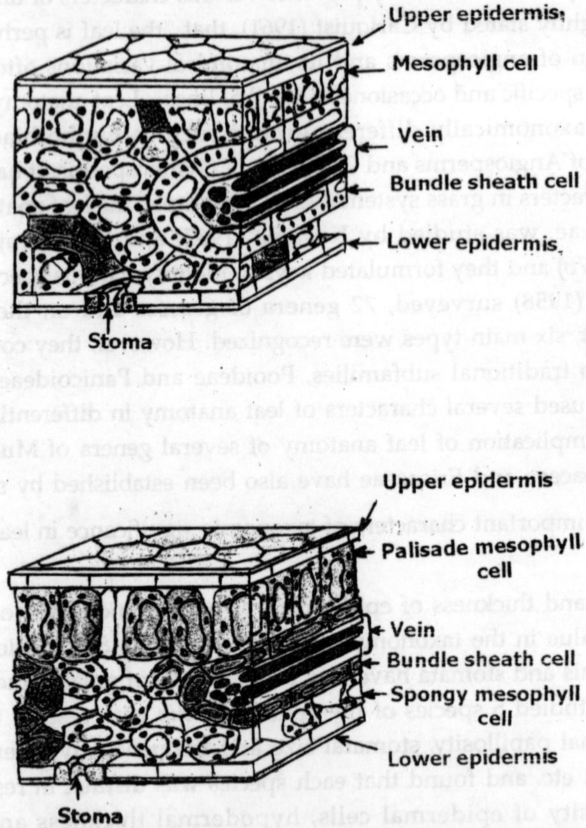

Fig. 8.2 : Cross-section of different types of leaves showing anatomical characters of taxonomic significance

(b) **Stem anatomy** — Stem anatomy has also been long relied on as a taxonomic tool (Fig. 8.3). The two-volume work by Metcalfe & Chalk (1950) is an excellent example of an illustrated encyclopedia of this and of other aspects of plant anatomy, which reveals the taxonomical significance of anatomical characters in plant classification and can be used at various levels from Dicotyledon-Monocotyledon distinction, to the separation of various species of the same genus. Stem anatomy has particularly proved to be of diagnostic value in the herbaceous members. For example, anatomy of stems has been successfully employed in the delimitation of species of *Dioscorea* which otherwise are not easily separable on exomorphic grounds. Carlquist (1957) has used anatomical features of the genus *Fitchia* (Asteraceae) in the classification of various species. Further, it is also possible to identify parents of several hybrids on anatomical grounds.

The anatomical features, which can be taken into account as diagnostically useful characters of the stem include :

- Degree of elevation of stem ridges.
- The distribution and abundance of collenchyma.

- Pattern of collenchyma thickenings.
- Transformation of ground tissue cells of cortex into transfusion cells — e.g. *Casuarina* (Metcalfe & Chalk, 1950).
- Distribution of fibres — e.g. *Genista* (Pellegrin, 1908).
- Variation in the structure of the cells of stem endodermis — e.g. in families like Piperaceae, Asteraceae and Lamiaceae.
- Features of the stem pith — e.g. species of *Dubantia* and *Fitchia* have been distinguished on the basis of anatomical differences in pith (Carlquist, 1961).
- Shape and size of sclerenchyma girders — e.g. used to distinguish species of the subgenus *Genuini* of *Juncus* (Stace, 1970).
- Arrangement and type of vascular bundles — e.g. two species of *Dioscorea*, viz. *D. cayenensis* and *D. rotundata* have been distinguished on the basis of arrangement of vascular bundles in the stem, which otherwise are difficult to distinguish on exomorphic grounds (Ayensu, 1970).

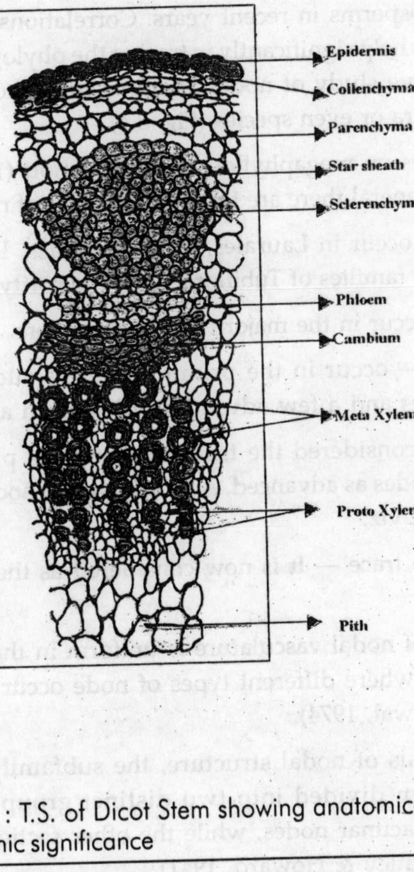

↳ Fig. 8.3 : T.S. of Dicot Stem showing anatomical features of Taxonomic significance

(c) **Petiole anatomy** — Metcalfe and Chalk (1950) and Howard (1963) have suggested that the petiole anatomy might also be of taxonomic significance (Fig. 8.4). According to Howard (1963) families, genera and even species in some cases may he identified by petiole characters. Some of the important diagnostic characters of petiole anatomy include:

- Position of petiole on stem.
- Presence or absence of stipules.
- Petiole outline.
- Number of layers of parenchyma in the cortex.
- Vascularization of petioles.
- Distribution of perivascular fibres.
- Number of traces.

Example: Petiole anatomy of 64 species of *Baphia* of Leguminosae have been studied by Soladoye (1982), and some species of *Phlomis* and *Eremostachys* of Labiatae by Azizian and Cutler (1982), which provide clear support of its use in the taxonomy of these genera.

(d) **Nodal anatomy** — Nodal anatomy has also gained much imporiance in taxonomy and phylogeny of angiosperms in recent years. Correlations of nodal anatomy with some other features might help significantly in tracing the phylogeny of angiosperms (Dickison, 1975). A comparative study of nodal anatomy may show important relationships or distinctness of genera or even species (Fig. 8.4).

Based on his studies on megaphyllous plants, Sinnott (1914) provided a classification of nodal types. In general there are three major types of nodes:

- Unilacunar — occur in Laurales, Caryophyllales, Ericales, Ebenales, Primulales, Myrtales, some families of Tubiflorae and a majority of families of Asteridae.
- Trilacunar — occur in the majority of Dicotyledons.
- Multilacunar — occur in the primitive orders such as Magnoliales, Piperales, Trochodendrales and a few advanced orders such as Umbellales and Asterales.

  Sinnott (1914) considered the trilacunar node as primitive, and unilacunar and multilacunar nodes as advanced. A fourth type of node was discovered by Marsden & Bailey (1955), viz.

- Unilacunar two trace — It is now considered as the basic type of node found in angiosperms.

Usually, the mode of nodal vasculature is uniform in the family, but exceptions have also been reported, where different types of node occur even in the same individual plant (Sehgal & Paliwal, 1974).

Example: On the basis of nodal structure, the subfamily Icacinoideae of the family Icacinaceae has been divided into two distinct groups i.e., one section, which is characterized by trilacunar nodes, while the other section, which is characterized by unilacunar nodes (Bailey & Howard, 1941).

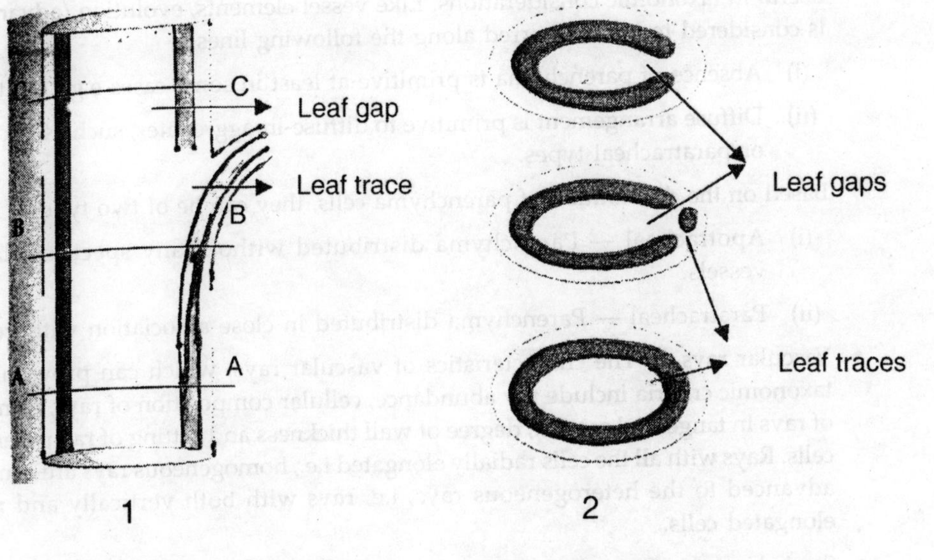

Fig. 8.4 : Diagrammatic representation of leaf trace and leaf gaps ; (1) L.S. through a node showing leaf trace and leaf gap, (2) T.S. through the node illustrated in 1 at levels A-A, B-B, and C-C respectively.

(e) **Wood anatomy** — Wood anatomy has been used at almost all taxonomic levels. Because of their conservative nature, anatomical features of the secondary wood have been very useful in taxonomy and phylogeny. Along with other lines of evidence, it has been successfully used in deciding the systematic position of primitive vesselless families such as Amborellaceae, Tetracentraceae, Trochodendraceae and Winteraceae, all included under the Magnoliales of angiosperms. Similarly, due to the presence of specialized wood, it has been agreed by all phylogenists that the Englerian group of primitive angiosperms, namely, Amentiferae (including families like the Salicaceae, Betulaceae, Fagaceae, Juglandaceae, etc.) cannot be considered primitive (Tippo, 1938). Some of the important features of wood anatomy of taxonomical significance are as follows:

- Vessel elements — They are considered to have been derived from the tracheids and their evolution (advancement) is considered to have occurred along the following lines:
  - (i) Decrease in the length of vessel element.
  - (ii) Transition from vessels with angular outline to nearly circular outline.
  - (iii) Loss of borders and decrease of bars on perforation plates.
  - (iv) Alteration from oblique to nearly transverse angle of end wall.
  - (v) Pitting of lateral walls of vessels showing evolutionary series from scalariform to transitional to opposite to alternate.

Apart from the structure, the abundance of vessels, distribution of vessels, and sculpturing on vessel walls have also proved to be of taxonomic significance (Fig. 8.5). Thus, solitary vessels are considered to be primitive to aggregate groupings, such as pore clusters, pore multiples and pore chains. Similarly, diffuse-porous woods are considered primitive to ring-porous woods.

- Axial parenchyma – The distribution and characteristics of the cells of axial parenchyma and the length and thickness and lignification of their walls, can be useful in taxonomic considerations. Like vessel elements, evolution (advancement) is considered to have occurred along the following lines :
  (i) Absence of parenchyma is primitive at least in some cases e.g., Winteraceae.
  (ii) Diffuse arrangement is primitive to diffuse-in-aggregates, such as apotracheal or paratracheal types.

  Based on the distribution of parenchyma cells, they can be of two types:
  (i) Apotracheal — Parenchyma distributed without any specific relation to vessels.
  (ii) Paratracheal — Parenchyma distributed in close association with vessels.
- Vascular rays — The characteristics of vascular rays, which can prove useful as taxonomic criteria include ray abundance, cellular composition of rays, dimensions of rays in tangential section, degree of wall thickness and pitting of ray parenchyma cells. Rays with all the cells radially elongated i.e., homogeneous rays are considered advanced to the heterogeneous rays, i.e. rays with both vertically and radially elongated cells.
- Storied wood — Storied structure of wood refers to the planes of divisions of cambial initials. Woods with non-stratified cells are considered primitive to storied structures.
- Presence or absence of latex vessels, resins, gums, crystals, etc. in the wood are also the characters of taxonomic importance.

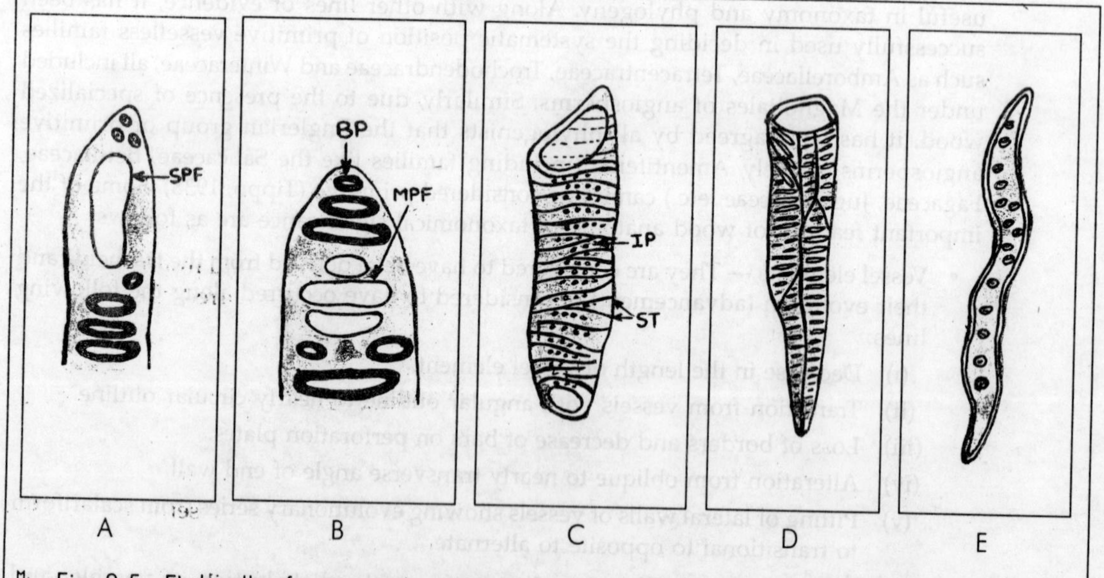

Fig. 8.5 : End-walls of some tracheae (vessel members) – (A) With simple perforation plate (SPF); (B) with multiple (scalariform) perforation plates (MPF); (C) Complete tracheae with simple perforation plates, intervessel pitting (IP) and spiral thickening (ST); Tracheids - (D) with scalariform (E) with bordered pits (BP).

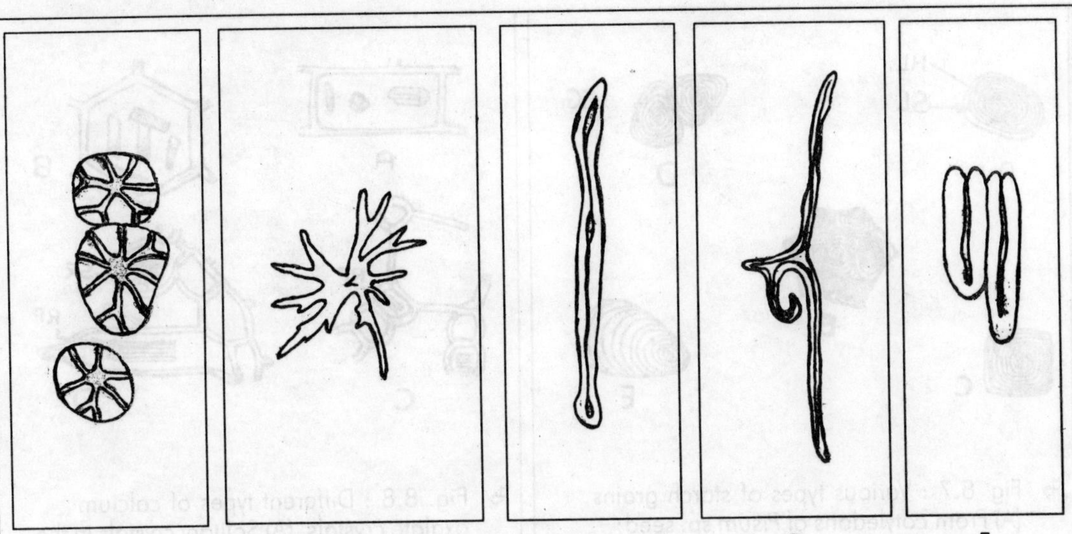

Fig. 8.6 : Different forms of sclereids. (A) Brachysclereids from the pulp of *Pyrus* fruit; (B) Astrosclereids from the leaf of *Trochodendron*; (C) Macrosclereid from the pulp of apple (*Malus*); (D) Trichosclereid from the leaf of *Olea*; (E) Osteosclereid from testa of *Pisum* seed.

(f) **Sclereids** — Sclereids, i.e. cells with very thick lignified walls, which are widely distributed in the plant body, have been used as diagnostic tools in several taxa like Connaraceae, Nymphaeaceae, Oleaceae, Theaceae, Umoniaceae, and a few genera of Araceae, Acanthaceae, Ericaceae and Melastomaceae (Fig. 8.6). In dicots, they are more common in woody forms than in herbaceous ones, but they are extremely rare in monocots, except in certain genera of Araceae, Agavaceae, Arecaceae and a few other families. As they exhibit various shapes, sizes and characteristics of their walls, they have been of some taxonomic significance. Two main types of sclereids have been recognized, viz. isomorphic and polymorphic types. The sclereid forms may be characteristic of a particular species and thus of taxonomic value.

(g) **Cellular contents** — Many types of microscopic characters of cell contents i.e., chemical deposits, can serve as important diagnostic tools, and at times prove extremely helpful in delineating species, genera and families. Crystals and crystalliferous cells have been found to be of systematic importance in several families of angiosperms such as Euphorbiaceae (Biesboer & Mahlberg, 1981), Leguminosae (Leelavathi et al., 1984), Verbenaceae (Mathew & Shaw, 1984), etc. Some of the important chemical deposits of systematic significance are as follows:

- Albuminoids — e.g. *Laportea* (Guerin, 1923).
- Starch grains — The immense diversity in the types of starch grains (Fig. 8.7) may turn out to be a good taxonomic character for the angiosperms in general (Czaja, 1978), e.g. *Solanum tuberosum*.
- Protein bodies — Deposition of solid protein depositions have systematic use, e.g. some Cactaceae (Metcalfe & Chalk, 1950).
- Large silica bodies — Silica bodies in the epidermal cells of various families like Arecaceae, Musaceae and Zingiberaceae (Tomlinson, 1956, 1959 a,b) and Rosaceae can be used at generic as well as specific levels.

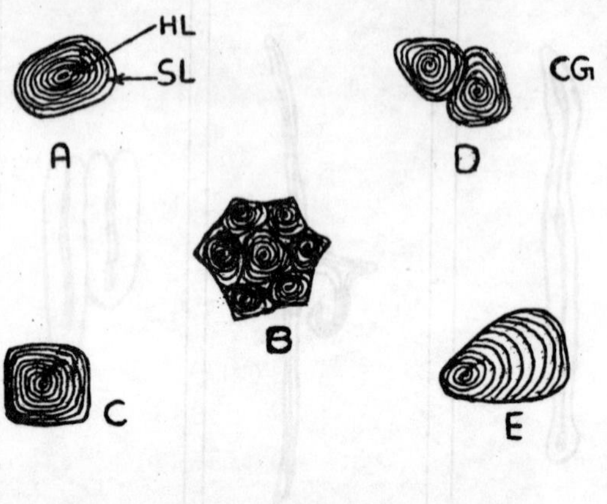

Fig. 8.7 : Various types of starch grains (A) From cotyledons of *Pisum* sp. seed; (B) From endosperm of *Oryza* sp. grain; (C) From endosperm of *Zea* sp. grain; (D) From cotyledons of *Cicer* seed; (E) From tuber of *Solanum tuberosum*. HL- Hilum, SL- stratified layers, CG- Compound grain

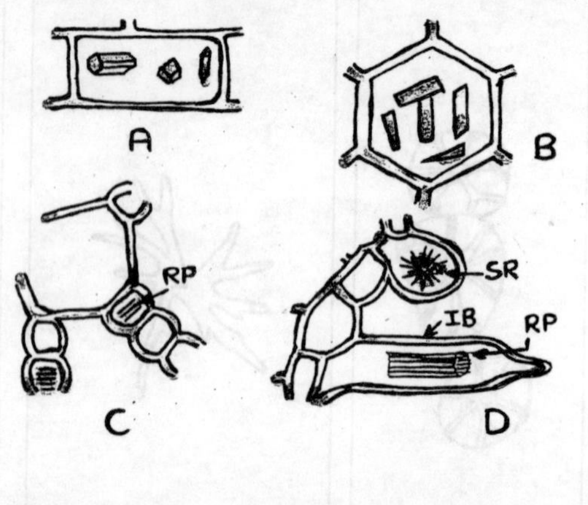

Fig. 8.8 : Different types of calcium oxalate crystals. (A) Solitary crystals in the tunic of *Allium cepa*; (B) Solitary crystals in a cell of *Leonurus* sp. (C) Raphides (RP) in the cells of the petiole of *Colocasia* sp. (D) Raphides and sphaeraphides (SR) in the idioblast (IB) cells of the leaf of *Pistia* sp.

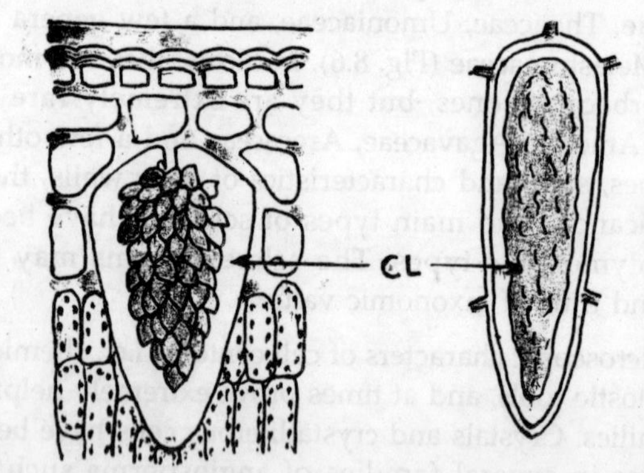

Fig. 8.9 : Different types of calcium carbonate crystals (Cystolith - CL). (A) In the leaf of *Ficus elastica* (B) In a leaf-cell of *Ruellia* sp.

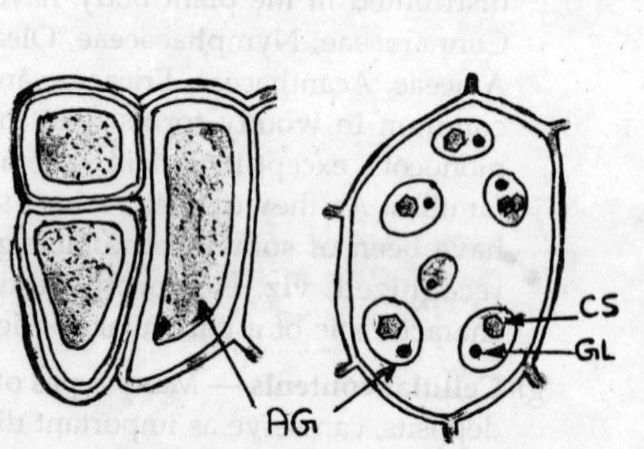

Fig. 8.10 : Structures of aleurone grains (AG). (A) In *Zea mays* grain; (B) In endosperm cells of *Ricinus* sp. seed. CS – crystalloid, GL - globoid

- **Calcium oxalate crystals — They are widely distributed in plants and are of different types, like prismatic, styloid and idioblasts (Fig. 8.8). Their distribution is very specific for a particular taxon and hence of taxonomic importance, e.g. *Eichhornia*, *Allium* (Chartschenko, 1932, Rothert, 1900).**
- **Cystoliths (calcium carbonate crystals) (Fig. 8.9) — e.g. Cannabinaceae, Moraceae and Urticaceae (Carlquist, 1961).**

- Tanniniferous cells — e.g. The presence and absence of tanniniferous cells in the root cortex of related families of Rapateaceae and Xyridaceae can be used as systematic criterion.

Apart from these, presence or absence of Laticifers, which are cells or a series of fused cells containing latex, and their structure, has also been of some taxonomic value. They are common features of many succulent plants and other plants of arid regions, and vary widely in their structure and the latex in their composition. For example, certain species in Aroideae lack laticifers or any related structures, while others have longitudinal rows of elongated, cylindrical, sac-like cells (De Bary, 1884).

8.2.2 Floral anatomy

As the reproductive organs show a high degree of conservation, they have been widely used in the classifications (Fig. 8.11A). At the same time, it is also quite likely that the vascular supply to these floral organs is also conservative and thus more reliable in taxonomic and phylogenetic interpretations (Fig. 8.11B). The distribution and course of vascular bundles within the receptacle and floral parts have proved to be of systematic significance, particularly in ranking taxa of higher order such as genera and families. Even specific characters may be quite clear in some cases. The significant role played by the floral anatomy in the solution of morphological problems has been greatly emphasized by Puri (1951). Unlike other branches of anatomy, the application of floral anatomy to taxonomy is limited due to technical and interpretative difficulties. However with the development of rapid clearing techniques, this branch of investigation received a great impetus. The floral anatomical characters of families and genera are generally well marked and have been useful in solving some fundamental questions, like the nature of flower, carpel, inferior ovary and also several problems related with homologies, phylogeny and taxonomy.

Following are some of the examples of the contribution of floral anatomy in resolving the taxonomic position of some disputed taxa:

- Confirmation of the origin of the families of Annonaceae, Calycanthaceae and Menispermaceae from Ranunculaceae.
- Separation of *Paeonia* from Ranunculaceae and its inclusion under a separate family Paeoniaceae.
- Derivation of Polemoniaceae and Caryophyllaceae from a caryophyllaceous stock.
- Cyperaceae and Poaceae were formerly treated together in one single order. Later Hutchinson (1973) separated them and placed them in Cyperales and Poales respectively, which has been confirmed by floral anatomical studies of both the families.
- Inclusion of Solanaceae and Scrophulariaceae under one single order, Scrophulariales due to uniformity in floral vasculature.
- Confirmation of the close relationship between *Cyrtandromoea* and members of Scrophulariaceae based on the presence of several lateral traces in carpels, a bilocular ovary, and absence of a disc in both.
- Support for the removal of *Lilaea* from Scheuchzeriaceae and be placed under an independent family Lilaeaceae, because both differ in their vascular supply of flower and number of ovules.
- Confirmation of the transfer of *Hydrocotyle asiatica* L. to the genus *Centella* in the form of *Centella asiatica* L.

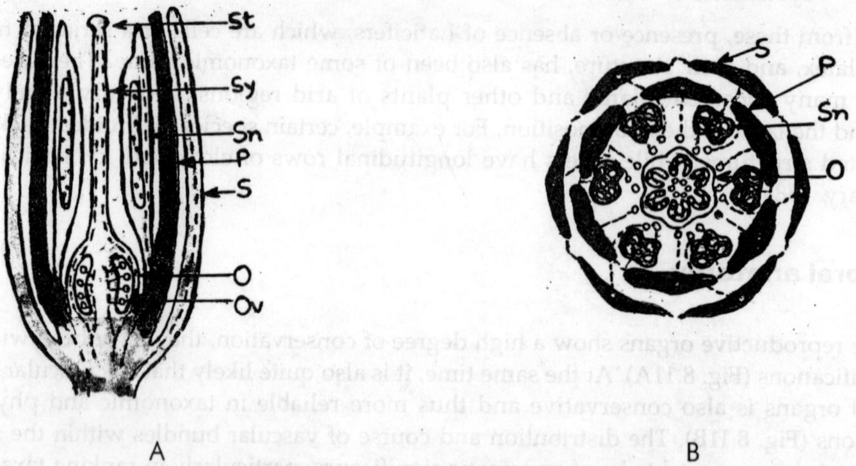

Fig. 8.11 : (A) Diagram of a flower in longitudinal section; (B) Same in cross-section ; broken lines are the course of vascular strands and their interconnections. S - Sepal, P - Petal, Sn– Stamen, St – Stigma, Sy - Style, O – Ovary, Ov - Ovules.

8.2.3 Ultrastructural Systematics

Electron microscopy has brought revolution in all biological fields, and so also in the field of taxonomy. Heywood and Dakhshini (1971) and Meywood (1967, 1968, 1969) have demonstrated and reviewed the benefits of scanning electron microscopy to plant systematics and have suggested that this ultrastructural device represents one of the most powerful taxonomic tools now available for systematic research. Like the role of SEM in plant micromorphology, Transmission electron microscopy (TEM), aided by ultramicrotome techniques, have proved a powerful tool in studying various anatomical aspects of taxonomic significance. However, till date only a few ultrastructural characters have been exploited and applied in plant classification. Some of these characters are as follows:

(a) **Dilated Cisternae (DC) —** These structures are of common occurrence in the Cruciferae (Iverson & Flood, 1970) and of the order Capparales. Dilated Cistemae (DC) in the endoplasmic reticulum was first reported by Bonnet & Newcomb (1965) in the root cells of *Raphanus sativus* (radish). Later they were also reported in the phloem parenchyma of foliar veins in *Brassica chinensis* (Favali & Gerola, 1968) and in *Capparis cynophallophora* (Jorgensen et al.,1977). The significance of the endoplasmic reticulum in sieve elements has been focused by Spanner & Moattari (1978), in the light of its evolutionary origin. Depending upon their location in the cells, the internal structure of Dilated Cistemae are of following types :

- They usually contain filamentous structures when present in the root cells.
- They contain protein tubules when present in other parts.

(b) **Sieve-tube plastids** — The potential uses of ultrastructural features of sieve-element plastids, which are of systematic value, were noted almost a decade ago. Since then more than 1500 species from 380 families have been investigated by Behnke (1972, 1977, 1981) with regard to this character. The plastid elements are of following types, depending upon their accumulation of starch and protein:

- S-type —They are the plastid elements, which accumulate starch. About 65% of the flowering plants have such plastids in the sieve tube elements (Behnke, 1977). They are present in the subclasses Dilleniidae, Hamamelidae, Rànunculidae, a great majority of the orders in the Rosidae, half the members of the Magnoliidae, a few orders in Caryophyllidae, Rhamnaceae (Behnke, 1974 b)
- P-type — They are the plastid elements, which accumulate protein. They are further differentiated on the basis of their number and shape of the crystalloids as well as the nature of the filaments surrounding them. They are present only in Pinaceae among Gymnosperms (Behnke, 1974 a) and among the Angiosperms, in all the 21 families of Monocotyledons and a few groups among Dicotyledons, such as Vitaceae and Leeaceae (Behnke, 1974 b) and half the members of the Magnoliidae.
- S_0-type — They are the plastid elements, which have neither starch, nor protein accumulations, e.g. Crassulaceae, Rafflesiaceae and some species of Moraceae, Ulmaceae and Urticaceae.

According to Behnke, classification and delimitation of higher taxa in flowering plants can be aided by utilizing the different types of sieve-element plastids. P-type is considered to be ancestral and the S-type is considered to have been derived from the P-type by loss of protein. The heterogenity of the Magnoliidae as to its plastid types is attributed to its basal position in the evolution of flowering plants (Behnke, 1975). The presence of P-type plastids in all groups of Monocotyledons, and the preponderance of S-type in the basal groups of Dicotyledons, have been interpreted as an additional evidence for the independent origin of Monocotyledons and Dicotyledons (Meeuse, 1970). A very good example of the use of plastids has been made in elucidating the inter-relationships and circumscribing the order Caryophyllales (Centrospermae) [Behnke, 1976].

8.3 Cytology in Relation to Taxonomy

Cytology has made an outstanding contribution to taxonomy during the last few decades, in the elucidation of many systematic problems. Cytology, which includes the study of the cell, or more appropriately karyology, the study of chromosome, has made significant contributions in taxonomy. Karyotype is defined as the phenotypic appearance of the somatic chromosomes (Levitsky, 1924), while the diagrammatic representation of the karyotype is termed as idiogram. Cytotaxonomy is the utilization of the cytological characters in the elucidation of taxonomic problems. The chromosome characteristics used in most cytotaxonomic studies include :

(a) Chromosome number

(b) Chromosome size

(c) Chromosome morphology

(d) Chromosome behavior at meiosis

(e) Banding patterns

Since these characteristics are directly derived from nucleus, the seat of the hereditary material, they collectively yield information, which can be used in understanding evolutionary relationships amongst taxa at various levels.

8.3.1 Chromosome number

The number of chromosomes in a species is usually constant and this makes it an important taxonomic character. However there are exceptions where chromosome number varies. These changes usually occur in chromosomes during the process of division, and these changes may affect the gene sequence, their number or even there may be loss of chromosomes themselves. The perpetuation of this slow process, results in the evolution of new chromosomal races. The chromosome number shows a wide range in vascular plants. The lowest chromosome number in flowering plants is recorded in *Haplopappus gracilis* (Asteraceae) [2n = 4] and the highest in *Poa litorosa* (Poaceae) [2n = 265]. *Ophioglossum* species (Pteridophyte) has the highest number of chromosomes in the plant kingdom (2n = 1240). This great diversity of chromosome numbers and their relative constancy within populations and species provide an important character for taxonomic groupings of large number of plants.

The chromosome number relationship with taxonomic groups can be broadly classified into the following three classes:

- **Constant number :** In certain groups of vascular plants, the chromosome number is constant throughout the whole group, e.g., *Quercus* and other members of the Fagaceae have the same basic number, n = 12. In such cases chromosome number is not of any help in distinguishing various taxa within the group.
- **Euploidy :** When chromosome numbers in various members of a taxon are in the proportion of exact multiples, the series is described as euploidy. For example in Malvaceae, the somatic numbers in various species range from 10,15,20, 25, ...40; from 12, 18, 24 to 30; from 14, 28, 42, 56 to 84 and so on.
  - ➢ ***Basic number*** — In an euploid series, the various members may be unified by a basic number (x) which is the gametic number of a diploid species. As in the above example of Malvaceae, the basic number x = 5. The other species in the series are described as triploids (3x), tetraploids (4x), hexaploids (6x) ... polyploids (nx). The basic number is usually constant for a genus or higher taxa and has proved useful in supraspecific studies.
  - ➢ ***Primary and secondary basic numbers*** — In many cases, more than one basic number can be present in a group. For example, in the living species of *Chlorophytum* (Liliaceae), the chromosome numbers vary from 14 to 28, 42, 56, 84, etc. and also from 16 to 32, which means that *Chlorophytum* has two basic numbers x = 7 and x = 8. In such cases, the inferred base numbers ranging between 2 and 13 may be referred to as the primary basic numbers in the absence of living diploid members, while the remaining are termed secondary basic numbers. In case of *Chlorophytum* however, Naik (1976a) from his detailed analysis of the meiotic behavior of chromosomes in one of the species, *C. laxum*, has shown that the two base numbers 7 and 8 should be considered as secondary, most probably derived from the primary basic number x = 4.
  - ➢ ***Polyploid pairs*** — Closely related species in certain groups of plants may be cytologically distinct i.e., one may be diploid while the other a tetraploid. Such related pairs are termed polyploid pairs. For example, *Cardamine hirsuta* (2n = 16) and *C. flexuosa* (2n = 32) of the Cruciferae is a polyploid pair.

➢ ***Dibasic polyploidy*** — It is now a well known fact that hybridization has a dominant role in evolution and such hybridization may involve crossing of any two genetically unlike individuals, which have different base numbers of chromosomes. Such hybrids undergo polyploidy since each chromosome is doubled as the pairing at meiosis is restored, and this type of polyploidy is termed as dibasic polyploidy. An excellent example of this type of polyploidy is Karpechenko's (1927, 1928) artificially synthesized *Raphanobrassica,* which is a hybrid between *Brassica oleracea* (2n = 18) and *Raphanus sativus* (2n = 20) and has 2n = 38 chromosomes.

- **Aneuploidy :** If the chromosome numbers in the different members within a group bear no simple numerical relationship to each other, then the series is termed as an aneuploid series or simply aneuploidy. Various species of *Carex* (Cyperaceae) for example show a wide range of chromosome numbers from n = 6 to 112 with multiples of 5, 6, 7, and 8 thus exhibiting aneuploidy. Aneuploidy may result due to either of the following:

(i) **Change in the basic number** — An increase or decrease in the number of chromosomes may take place, whereby the same genetic material becomes distributed in a different number, leading to a change in the basic number. This phenomenon is important from the taxonomic and evolutionary point of view, as change in the basic number results in new variations and recombinations, leading to the evolution of new varieties and ultimately to new species. The changes in the basic number of chromosomes can be brought about by a process called **polysomaly.** This is very common and results in increased sets of genes wherein there is a duplication of one or a pair of chromosomes. This means that any one pair may undergo **polyploidy**. Due to this some of them can afford to lose one or two chromosomes and get stabilised in nature with this new chromosome number. This naturally brings about a different basic number. For example, most of the species of *Dahlia* (Asteraceae), have x=8, but *D. merckii* is regarded as a polyploid, with n= 18, resulting by addition of two pairs of chromosomes. Polyploidy has also been reported in *Datura* (Blakeslee and Belling, 1924), *Nicotiana* (Goodspeed and Avery, 1939), etc. Polysomics can be of various types:

➢ ***Trisomics*** — Plants containing one extra chromosome are known as trisomics i.e., 2n + 1. In trisomics the extra chromosome produces a certain amount of unbalance and thus limiting their taxonomic significance.

➢ ***Tetrasomics*** — Plants containing two extra chromosomes are known as tetrasomics i.e., 2n + 2, etc.

➢ ***Monosomics*** — Plants with one chromosome less are known as monosomics i.e., 2n – 1. Normally diploid monosomics are inviable.

➢ ***Nullisomics*** — Plants with two chromosomes less are known as nullisomics i.e., 2n – 2, etc.

Polysomics are usually unstable and as they are not isolated genetically from their relatives under natural conditions, they would lose their identity through crossing with normal plants followed by selection for more viable, genetically balanced normal disomic types (Naik, 1984).

(ii) **Basic number remaining unaltered** — The basic number may remain unaltered, but the genetic material present may be changed due to the addition or loss of chromosomes. This phenomenon is less important from the evolutionary point of view as the genetic make-up of a taxon becomes unbalanced due to loss or addition, resulting in an unstable condition, which cannot be perpetuated and thus cannot give rise to well differentiated novelties. This type of aneuploid alteration of the basic number has been reported in *Crepis* (Asteraceae) and its relatives (Babcock, 1942, 1947; Babcock & Cameron, 1934; Babcock et al., 1937). This loss of chromosome/s, fragmentation or misdivision of centromere can take place as irregularities during the cell division even in the diploid species.

(iii) **B-chromosomes** — They are one or more *accessory* or supernumerary chromosomes, in addition to normal chromosomes and have been detected in a large number of plants (Darlington, 1937; Fernandez, 1946; Hakansson, 1945). In plants, when compared to the other members of the chromosome complement, they are generally of a much smaller size and are of unknown origin. They reduce fertility or increase the vigor of plants. It has been found that they perpetuate in certain natural populations and may have some evolutionary significance.

8.3.2 Chromosome size

The individual chromosomes of some taxa show marked differences in shape and size at mitotic metaphase. The size of chromosome varies greatly in different families and also amongst members of the same family. The monocotyledons usually have larger chromosomes than the dicotyledons. In general, woody plants have smaller chromosomes than their herbaceous relatives. Nearly all the members of the families of Araceae, Commelinaceae, Cyperaceae, Dioscoreaceae and Zingiberaceae possess small chromosomes. Iridaceous members reveal small to medium sized chromosomes, while taxa of Amaryllidaceae exhibit large sized chromosomes. Members of the family Liliaceae are characterized by the presence of all types of chromosomes.

Chromosome length is usually used to characterize the size of the karyotype. However sometimes chromosome volume, expressed as a function of DNA content, is used for this purpose (Rees, 1963). The chromosome length in most plants varies from 0.5-30 μ (Warmke, 1941). Chromosome size is not related to the phylogeny of angiosperms in general but is characteristic of only certain groups and families (Stebbin, 1938). Phylogenetic reduction in chromosome size, was first of all described by Delaunay (1926) in the genus *Muscari* of Liliaceae. In this genus, species with greater morphological specialization possessed smaller absolute chromosome size as compared to the relatively primitive species. Reduction in chromosome size has also been reported in *Crepis* (Babcock & Cameron, 1934) and in *Dianthus* (Rohweder, 1934). Similarly phylogenetic increase in chromosome size has been reported in the family Poaceae (Avdulov, 1931), in *Galium* (Fagerlind, 1937) and in *Godetia* (Hakansson, 1941,1943). The two genera *Trillium* and *Paris* possess the largest chromosomes and are definitely specialized in their vegetative characteristics, and are considered to have descended from Uvularieae (Hutchinson, 1934), the members of which have also large chromosomes but nevertheless smaller in contrast to those found in *Trillium* and *Paris.* The evolutionary significance of chromosome size was further elaborated by Stebbins (1950). The co-adaptive nature of genetic system in plants was first pointed out by Darlington (1939). He suggested that chiasma frequency often depended on the length of the chromosomes, the larger chromosomes means high chiasma frequency resulting in high recombination index while small chromosomes mean low chiasma frequency resulting in low recombination index. Therefore, there is a trend to linking of small chromosome size with multiplicity in their number as in tree species (Mehra, 1972).

8.3.3 Chromosome morphology

Apart from the number and size of the chromosomes of many genera and families of flowering plants, conspicuous differences in appearances of the karyotype of the chromosomes have also been found in species having the same chromosome number. The chromosomes are best discernible at mitotic metaphase. The karyotype of the chromosomes can be characterized on the following basis:

- Relative length of the arms of chromosomes
- Position of the centromere
- Presence of satellites

Accordingly the chromosomes can be characterized as following types:

➢ *Symmetrical* — A karyotype consisting of chromosomes all essentially similar to each other in size and with median or sub-median centromeres and with two equal arms are termed as a symmetrical chromosomes. Depending on the position of the centromere they may be of further two types (Fig. 8.12) :

(i) **Metacentric or V-shaped** — Chromosomes with median centromere.

(ii) **Sub-metacentric or L-shaped** — Chromosomes with sub-median centromere.

↳ Fig. 8.12 : Symmetrical chromosomes

Karyotypes of this nature are the most common ones and Levitzky (1931) considered them as generalized types. For example, the chromosomes in the karyotype of the primitive genus *Helleborus* (Ranunculaceae), differ little from each other in size and most of them are V-shaped with median or submedian centromeres.

➢ *Asymmetrical* — This type of karyotype possess many chromosomes with sub-terminal or terminal centromeres, or great differences in size between the largest and the smallest chromosomes, or both (Stebbins, 1950). Depending on the position of the centromere they may be further of two types (Fig. 8.13) :

(i) **Acrocentric or J-shaped** — Chromosomes with sub-terminal centromere.

(ii) **Telocentric or I-shaped** — Chromosomes with terminal centromere.

Karyotypes of this nature are considered as specialized types. For example, in the advanced genera *Aconitum* and *Delphinium* (Ranunculaceae), the flowers have the largest number of J-shaped chromosomes.

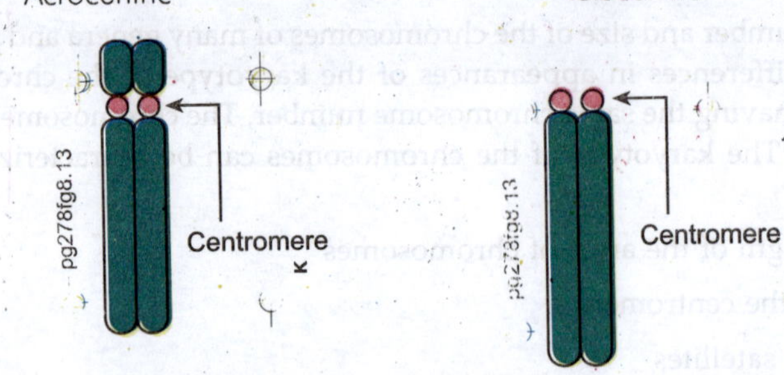

Fig. 8.13 : Asymmetrical chromosomes

➢ ***Secondary constrictions and satellites*** — The karyotypes can also be differentiated on the basis of secondary constrictions, which are small bead-like appendages. Occasionally they occur at the terminal ends of one or more pairs of chromosomes in many species and are known as satellites (Fig. 8.14). These structures are widely distributed in the plant kingdom, which shows that they are a valuable, if not essential, part of the chromosomal complement. However, very little is known about the evolutionary changes in the satellites and nucleoli.

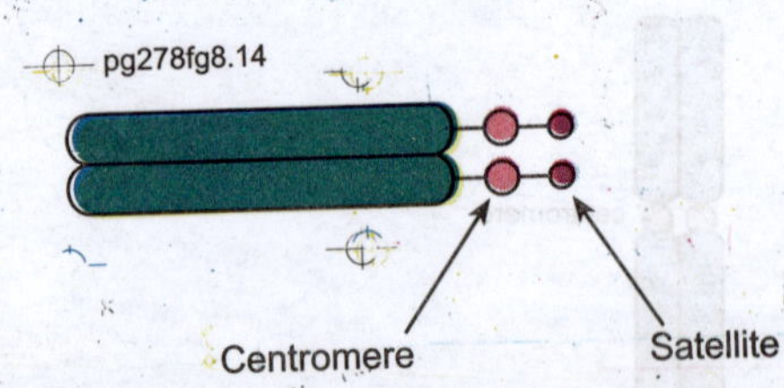

Fig. 8.14 : Acrocentric (Subterminal) satellited chromosome

Generally, the asymmetrical karyotypes are most common in plants, which are usually specialized morphologically, while symmetrical ones are found in more or less generalized plants, but also occur in morphologically specialized ones too. The reason for this increasing asymmetry in karyotype evolution is not known. However, Stebbins (1950) suggested that unequal translocations and inversions involving the centromere seem to be responsible for this. Generally symmetrical karyotypes are taken to reflect primitive status of taxa, while various specialized types (asymmetrical karyotypes), are considered to have been derived, from the symmetrical karyotypes. However, Jones (1978) has challenged this view by postulating that symmetrical karyotypes originally evolved from asymmetrical ones by end-to-end fusion of telocentric chromosomes. This reverse trend of specialization, i.e. from telocentric chromosomes to metacentric ones, has been demonstrated in certain members of the Commelinaceae (Jones, 1970).

8.3.4 Chromosome behavior at meiosis

A study of chromosome behavior at meiosis can provide some valuable information about the relationship of populations and species. Pairing behavior at meiosis is mostly determined by chromosome number and chromosome homology. The kind and degree of pairing can:

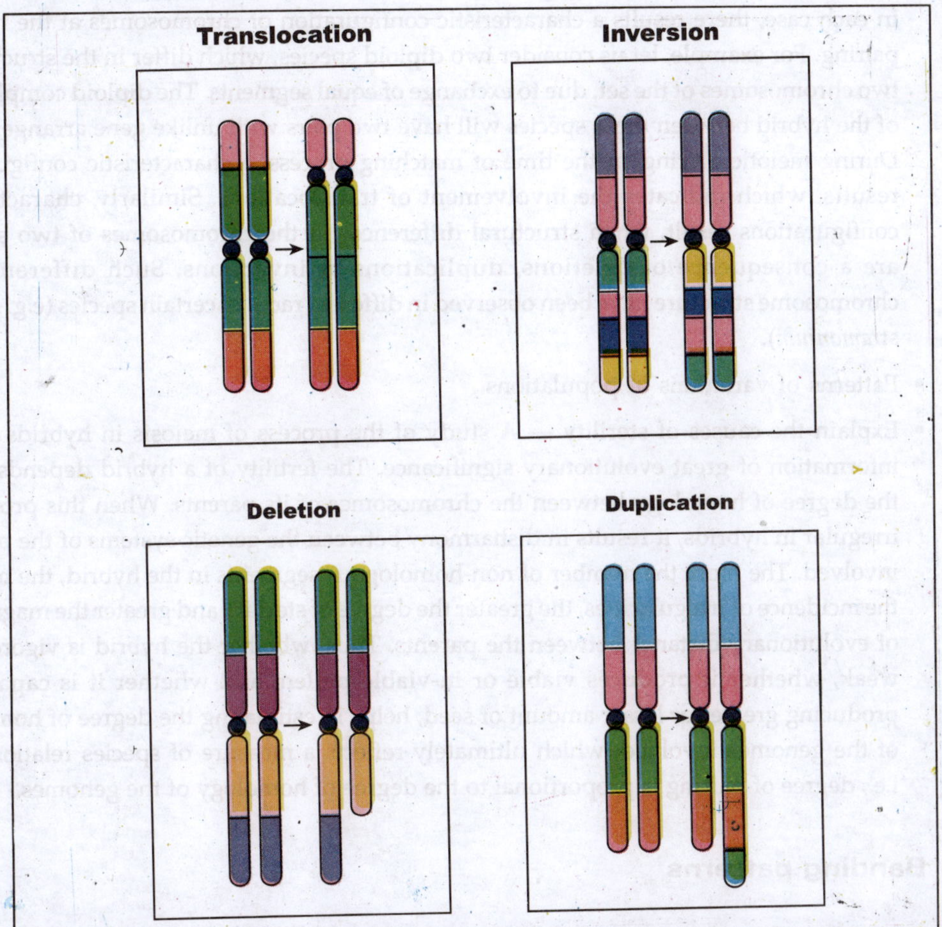

Fig. 8.15 : Consequences leading to structural differences in the chromosomes

- Show whether hybridization has occurred - The degree of chromosome homology, in hybrids is an indication of the degree of relationship of the parental species.
- Indicate structural differences in the parental chromosomes - Meiotic pairing behavior in hybrids can also point to structural differences in the chromosomes of its parents. Such differences may be the result of (Fig. 8.15) :

➢ **Translocations** — a chromosomal segment is removed from one place and reinserted somewhere else in the genome, either in the same or in some other chromosome.

➢ **Inversion of segments** — a segment of a chromosome becomes reinserted in the same chromosome but the opposite way around. It can be further of two types:

(i) Paracentric — inversions involving only one chromosomal arm.

(ii) Pericentric — inversion incorporating the centromere i.e., involving both chromosomal arms.

➢ **Deletions** — an interstitial or terminal chromosomal segment is lost.

➢ **Duplications** — when a segment of the chromosome is represented two or more times in a chromosome of a homologous pair.

In each case, there results a characteristic configuration of chromosomes at the time of pairing. For example, let us consider two diploid species, which differ in the structure of two chromosomes of the set, due to exchange of equal segments. The diploid complement of the hybrid between these species will have two pairs with unlike gene arrangements. During meiotic pairing, at the time of matching process, a characteristic configuration results, which indicates the involvement of translocations. Similarly, characteristic configurations result, when structural differences in the chromosomes of two species are a consequence of deletions, duplications or inversions. Such differences in chromosome structure have been observed in different races of certain species (e.g. *Datura stramonium*).

- Patterns of variations of populations.
- Explain the causes of sterility — A study of the process of meiosis in hybrids yields information of great evolutionary significance. The fertility of a hybrid depends upon the degree of homology between the chromosomes of its parents. When this process is irregular in hybrids, it results in disharmony between the genetic systems of the parents involved. The more the number of non-homologous segments in the hybrid, the more is the incidence of irregularities, the greater the degree of sterility and greater the magnitude of evolutionary distance between the parents. Thus, whether the hybrid is vigorous or weak, whether it produces viable or in-viable pollen, and whether it is capable of producing greater or lesser amount of seed, helps in estimating the degree of homology of the genomes involved, which ultimately reflects a measure of species relationships i.e., degree of pairing is proportional to the degree of homology of the genomes.

8.3.5 Banding patterns

In recent years new staining techniques have developed, using Giemsa and fluorochrome dyes, which stain chromosomes in a consistent banding pattern instead of with uniform intensity as in the case of previous usual techniques of staining with basic fuchsin (Fuelgen reagent). DNA is inevitably non-homogeneous along the arms of a chromosome. During mitosis and meiosis, local variations in the degree of condensation of the chromatin to form heterochromatin may be apparent in metaphase preparations. The new techniques have thus, allowed to distinguish morphologically between chromosomes with enhanced efficiency. These techniques, for example, have been utilized by Ehrendorfer (1976) and Ehrendorfer et al. (1977) in studying the systematic position of the genus *Anacyclus*, which confirms with the results from other fields of studies, particularly morphological and phytochemical approaches.

8.3.6 Systematic value of cytotaxonomy

Cytological evidence is of immense importance in interpreting a classification and establishing relationships. The information collected from chromosome characteristics has been found to be valuable at all levels of the taxonomic hierarchy (Jackson, 1971) and the usefulness of cytological data varies from group to group. A few examples are undermentioned :

- **Cytological variation at family level** : Cytological data have been fruitfully employed to unravel affinities and inter-relationships at the family level. Following are a few examples:

  (i) Cytological data has provided logical basis for improved arrangement of tribes and genera in some families such as the Ranunculaceae, Brassicaceae and Poaceae. For example in Ranunculaceae, all the genera with n = 7, 8 and 9, as well as genera with long chromosomes and short chromosomes have been placed under Anemoneae and Helleboreae. Based on karyological similarity i.e., with same base number n = 7 and small-sized chromosomes, *Thalictrum* and *Anemonella* of Anemoneae, and *Isophyrum* and *Aquilegia* of Helleboreae have been grouped in tribe Thalictreae. Similarly two other genera of the Helleboreae, *Coptis* and *Zanthorhiza* with very small chromosomes and the base number 9 have been placed under an additional tribe, Coptideae.

  (ii) The number and size of the chromosomes are the basis of the major sub-divisions of the family Poaceae as recognized currently.

  (iii) Some earlier workers have placed the genera *Agave* and *Yucca* under separate families, Amaryllidaceae and Liliaceae respectively. Cytological studies reveal that both have 5 long and 25 short chromosomes. The karyotype similarity, thus, justify their inclusion in the family Agavaceae as has been done by Hutchinson.

  (iv) Delimitation of the tribes in Asteraceae has been done on the basis of chromosome numbers.

- **Cytological variation at generic level** : There are several examples where chromosome numbers support generic status. A few are cited below:

  (i) The genus *Cistus* (Cistaceae) was formerly included in *Helianthemum*. However, *Cistus* has base chromosome number 8 and *Helianthemum* has 9. This supports the recognition of *Cistus* as a separate genus.

  (ii) Several taxonomists treated the genera *Physaria* and *Lesquerella* of Brassicaceae as a single genus. However cytological details suggest that they should be treated as two different genera.

  (iii) Based on cytological studies, a new classification of the genus *Narcissus* of Amaryllidaceae has been proposed by Fernandes (1951).

  (iv) The two genera *Physaria* and *Lesquerella*, under the family Brassicaceae, were recognized by many workers, as a single genus. Cytological evidence, however, suggests that these two genera should remain separated.

- **Cytological variation at specific and infra-specific level** : There are several examples of cytological variation occurring at specific and infra-specific level. Following are a few examples :

  (i) All the species under the genus *Tephrosia* (Fabaceae) are with 2n = 22 except for *T. constricta*, which has 2n = 16. Thus, this species has been separated as a distinct genus, *Sphinctospermum* and the chromosome number supports this separation.

(ii) Cheenaveeraiah (1962), on the basis of karyotypic studies suggested that the section *Sitopsis* of the species *Aegilops*, should be shifted to *Triticum* from *Aegilops*, or it should be given the rank of a new genus.

(iii) Based on cytological studies, Brandt (1961) recognized two races in *Veronica prostrata* of the family Scrophulariaceae, and suggested that they should be treated as subspecies *prostrata* (n = 8 and with small pale blue flowers) and subspecies *scheereri* (n = 16 and with larger dark blue flowers).

(iv) Previously *Monotropa hypopitys* of the family Monotropaceae, was treated as a single species with two varieties, var. *hirsuta* and var. *glabra*. However, cytological examination revealed the former to be a hexaploid, with 2n = 48, and the latter a diploid, with 2n = 16. Thus the hexaploid was retained as the species *M. hypopitys* and the var. *glabra* was raised to the specific rank as *M. hypophegea*.

8.4 Embryology in Relation to Taxonomy

Embryology is the study of micro- and megasporogenesis, gametophyte development, fertilization, and development of endosperm, embryo, and seed coats. Although the possibility of utilizing embryological characters in taxonomy was indicated by some earlier botanists like Hofmeister and Strasburger, the role of embryology in solving taxonomic problems was first brought into prominence by a German embryologist, Schnarf in 1931. Embryological characters, being less prone to adaptive stress, are relatively stable and have acquired great significance in plant taxonomy, especially when external morphological characters are inconclusive and misleading as a result of convergence (Kapil & Bhatnagar, 1980). Thus, embryological evidences have been used in solving taxonomical problems at almost all levels, and have helped resolve the doubtful systematic positions of several taxa. However, according to Jones and Luchsinger (1987), the embryological characters have proved to be of significant help in determining relationships within families, genera and species, but less useful at the rank of order, subclass, or class. Maheshwari (1950) and John (1963), on the basis of their extensive work on embryology, have provided lists of families, tribes, etc. where embryology has either supported earlier classifications or has proposed a new systematic position for the taxon concerned. In the later years, similar excellent work on the role of embryology have been done by Johansen (1950), Maheshwari (1950,1964), Cave (1953, 1959, 1962), Kapil (1962), Murgai (1962), Rau (1962), Arekal (1963), Johri (1963,1967), Davis (1966), Sporne (1969), Bhandari (1971), Rembert (1971), Philipson (1974), Dickison (1974), Falser (1975), Bhojwani and Bhatnagar (1979), Kapil and Bhatnagar (1980), etc. These plant taxonomists have utilized embryologicial characters in various ways:

- In the analysis of evolutionary trends.
- In the circumscription and delimitation of taxa.
- In making decisions on systematic position, etc.

BASAL : Araliaceae, Caprifoliaceae, Commelinaceae, Cyperaceae, Magnoliaceae, Nymphaceae, Oleaceae - *Fraxinus*, Papaveraceae, Ranunculaceae, Rubiaceae, Umbelliferae

BASOLATERAL : Gramineae

BENT : Aceraceae, Anacardiaceae, Cactaceae, Leguminosae, Loasaceae, Papaveraceae, Polygonaceae, Scrophulariaceae, Zygophyllaceae

COILED : Potamogetonaceae, Solanaceae

ENCIRCLING : Aizoaceae, Amaranthaceae, Cactaceae, Caryophyllaceae, Chenopodiaceae, Cleomaceae, Malvaceae, Menispermaceae, Moraceae, Nyctaginaceae, Phytolacaceae, Portulacaceae

FOLDED : Aceraceae, Anacardiaceae, Convolvulaceae, Cruciferae, Geraniaceae, Leguminosae, Ulmaceae

LINEAR : Amaryllidaceae, Caprifoliaceae, Asteraceae, Cupressaceae, Empetraceae, Ericaceae, Pinaceae, Haloragidaceae, Linaceae, Loasaceae, Loganaceae, Oleaceae - *Fraxinus*, Oxalidaceae, Platanaceae, Polemoniaceae, Rubiaceae, Sparganiaceae, Verbenaceae

SAGITTATE : Vitaceae

SPATHULATE - WITH ENDOSPERM : Asclepiadaceae, Berberidaceae, Boraginaceae, Cornaceae, Ericaceae, Euphorbiaceae, Hamamelidaceae, Linaceae, Oleaceae, Oxalidaceae, Passifloraceae, Polemoniaceae, Polygalaceae, Polygonaceae, Rubiaceae, Urticaceae, Violaceae

SPATHULATE - WITHOUT ENDOSPERM : Betulaceae, Elaeagnaceae, Labiateae, Leguminosae, Pedaliaceae, Plantaginaceae, Rhamnaceae, Rosaceae, Sapotaceae, Sterculiaceae

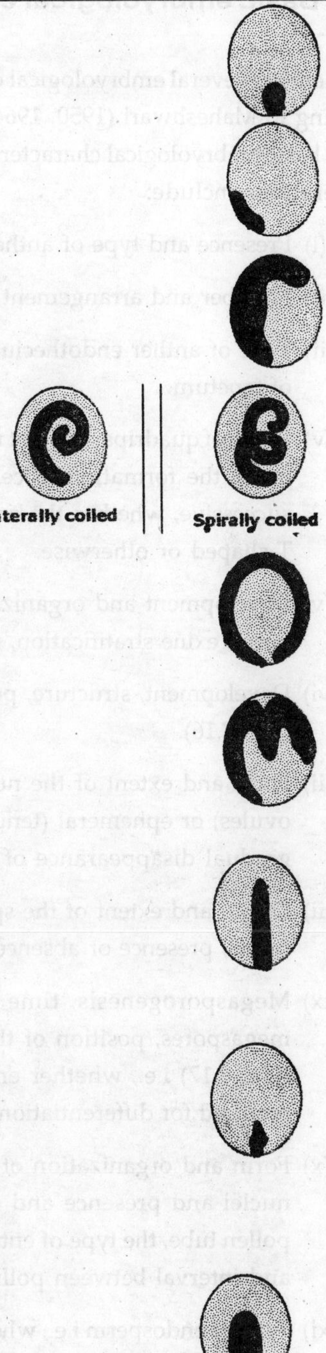

Fig. 8.16 : Types of Embryo (After Martin & Barkley, 1961)

8.4.1 Basic embryological characters of taxonomic significance

There are several embryological characters, which have proved to be of taxonomic significance. According to Maheshwari (1950, 1964), Bhojwani and Bhatnagar (1979) and Radford (1986), some of these basic embryological characters, which have proved to be of special importance in taxonomic considerations include:

(i) Presence and type of anther tapetum, whether glandular or amoeboid.

(ii) Number and arrangement of anther loculi.

(iii) Type of anther endothecium i.e., structure and thickenings of endothecium and nature of tapetum.

(iv) Mode of quadripartition of microspore mother cell, i.e., whether it takes place by furrowing or by the formation of cell plates, whether the mode of division is simultaneous or successive, whether the arrangement of microspores is tetrahedral, isobilateral, linear, T-shaped or otherwise.

(v) Development and organization of the pollen grains i.e., number and position of germ pores, exine stratification, number of cells at the time of anthesis, etc.

(vi) Development, structure, position, vasculature of integuments, and orientation of ovule (Fig. 8.16).

(vii) Form and extent of the nucellus i.e., whether it is broad and massive (crassinucellate ovules) or ephemeral (tenuinucellate), presence or absence of hypostase, persistence or gradual disappearance of nucellus.

(viii) Origin and extent of the sporogenous tissue in ovule i.e., whether one-celled or many-celled, presence or absence of wall layers, type of division of nucellar epidermal cells.

(ix) Megasporogenesis, time of wall formation in megasporogenesis, arrangement of megaspores, position of the functioning megaspore, and development of embryo sac (Fig. 8.17) i.e., whether embryo sac is mono-, bi- or tetrasporic, number of divisions required for differentiation of mature egg.

(x) Form and organization of the mature embryo sac i.e., number and distribution of its nuclei and presence and absence of endosperm haustoria. Fertilization, pathway of pollen tube, the type of entry of the pollen tube into the ovule i.e., micropylar or chalazal and interval between pollination and fertilization.

(xi) Type of endosperm i.e., whether nuclear, cellular or helobial type, nature of food reserves, persistence or gradual disappearance in the mature seed.

(xii) Seed-coat.

(xiii) Certain abnormalities of development such as parthenogenesis, apogamy, adventive embryony and polyembryony.

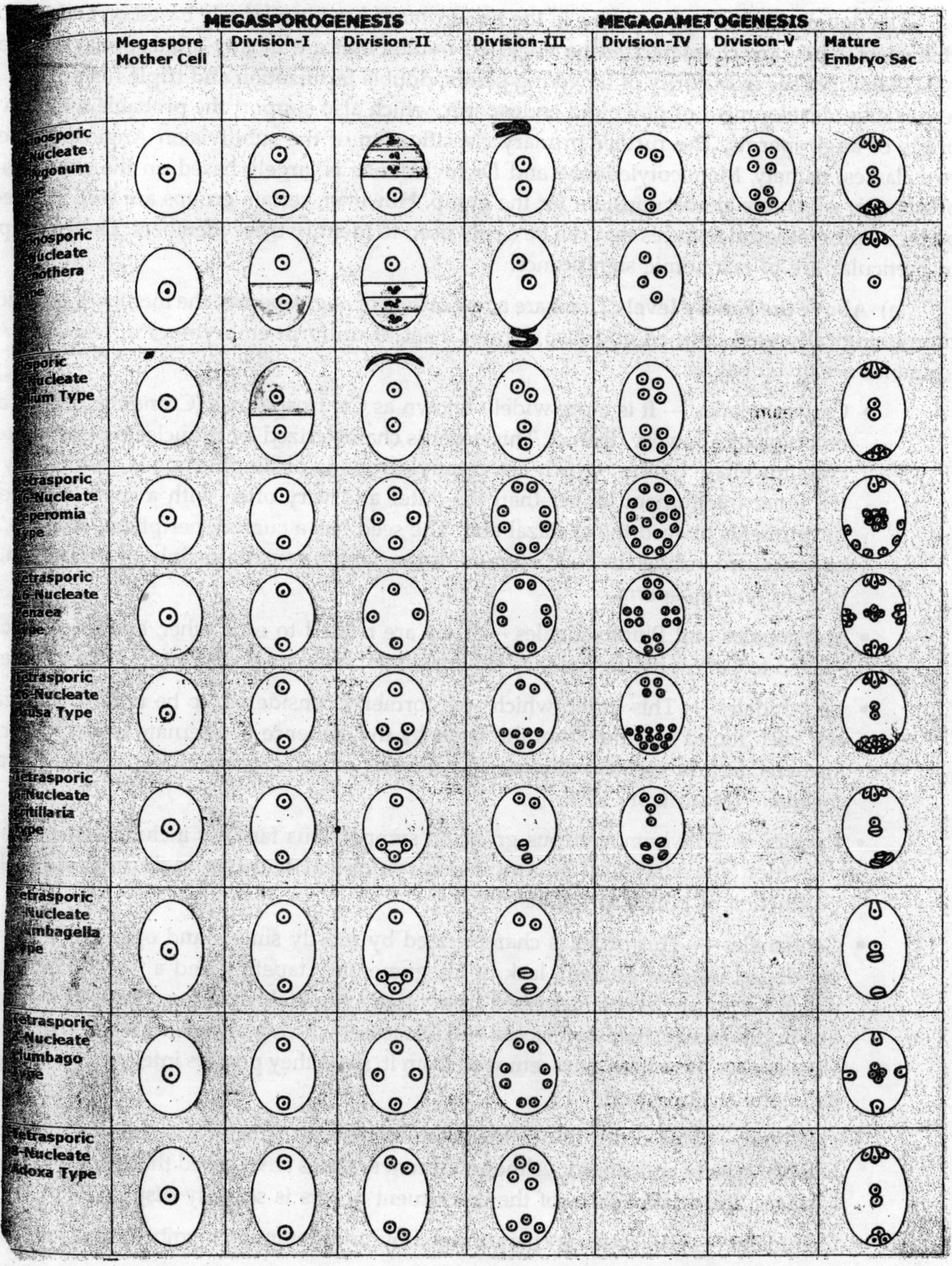

Fig. 8.17 : Types of Embryo Sac in Angiosperms (After Maheshwari, 1950)

8.4.2 Application of embryological data in taxonomy

The delimitation of angiosperms in the first instance has been largely based on embryological characters. They are characterized by enclosed ovules that are mostly anatropous, polygonum type of embryo sac in majority of flowering plants, double fertilization and triple fusion, and post-fertilization development of polyploid endosperm, which also support the probable monophyletic origin of angiosperms. The further primary classification of the subdivision Angiospermae into two classes, namely, Monocotyledonae and Dictoyledonae is largely based on the characteristics of embryo, which are almost constant for the group. However, certain groups are well marked out by their embryological features that can be easily used to identify these members. These exceptions in particular are of taxonomic significance.

(a) **Above the Family level :** There are a few orders of angiosperms, the members of which are morphologically well defined and also exhibit a set of uniform embryological features. A few examples are given below:

- Caryophyllales — It is more widely known as Centrospermae (Cronquist, 1968a) and is embryologically very distinct. This order is characterized by trinucleate pollen, bitegmic crassinucellate ovules, which are campylotropous or amphitropous, commonly with the inner integument longer than the outer and very often with a space between the integuments toward the chalazal end. The seed has a curved, peripheral embryo, more or less surrounding the food storage tissue, which consists mainly of perisperm, with little or no endosperm.
- Polygonales and Plumbaginales — They are related to each other, and are also clearly related to Caryophyllales as they exhibit similar embryological features to a great extent.
- Lecythidales — This order, which was formerly considered, to be related to Myrtales, has been found to differ from the latter by the presence of alternate leaves, centrifugal stamens, lack of internal phloem and a series of embryological features in its various members (Mauritzon, 1939).
- Ericales — This homogeneous group of sympetalous families includes members, most of which share several common embryological features including, notably, the production of pollen in tetrads.
- Gentianales — This order is characterized by mostly simple and opposite leaves, well developed internal phloem, lack of integumentary tapetum and a nuclear endosperm, which helps it to be distinguished from other sympetalous orders. Some of the families like Buddlejaceae, Menyanthaceae and Oleaceae, which had been formerly placed under Gentianales, have been now removed from it, since they possess integumentary tapetum and cellular endosperm.
- Helobiae — This order under the monocotyledons (Engler), is characterized by the helobial type of endosperm. Although recent systems have raised the order to the level of subclass, the relationships of the constituent orders is scarcely disputed.
- Orchidales — This order is characterized by undifferentiated embryo and very little or no endosperm.

(b) **At and below the Family level :** Embryological evidences have strongly supported the proper placement of certain disputed families and genera, once regarded to have a doubtful systematic position. A few examples are given below :

- Onagraceae — This family is characterized by the presence of peculiar monosporic four-nucleate Onagrad-type of embryo-sac in all its genera except *Trapa*. Such an embryo sac is derived from the micropylar megaspore of the tetrad. The lower three chalazal megaspores do not disintegrate. In this type of embryo sac, antipodals are absent. Different systematists have placed the genus *Trapa* in several ways:
  - ➢ Some have placed it under Onagraceae
  - ➢ Some have placed it as an isolated member of Haloragaceae, and
  - ➢ Some have placed it as the only genus of the monotypic family Trapaceae.

  Embryological evidence however strongly supports the last view. *Trapa* has a well-developed suspensor haustorium, in addition to the eight-nucleate embryo sac, which are absent in any other members of Onagraceae. Further, in *Trapa,* the ovary is semi-inferior, bilocular with one ovule in each locule and the fruit is a large one-seeded drupe, while in other members of Onagraceae, the ovary is inferior and tetralocular with numerous ovules on axile placentation and fruit is generally a loculicidal capsule.
- Empetraceae — This family, was considered, to be closely related to Euphorbiaceae and Celastraceae by Don (1827). However, Bentham and Hooker (1880) treated them as an anomalous family under Monochlamydeae. Later, several taxonomists thought that they are related to Ericaceae on the basis of, several morphological features. This latter view, has been supported by Samuelsson (1913) on the basis of embryological data.
- Cyperaceae — Earlier systematists considered the families of Cyperaceae and Poaceae to be closely associated. However morphological and anatomical features and the structure of the spikelets reveal that the two families are quite distinct. The Cyperaceae-Poaceae alliance is also contradicted by the embryological findings (Shah, 1967) due to the following:
  - ➢ The microsporogenesis is of simultaneous type in Cyperaceae and the tetrad members do not separate (three degenerate and only one is functional), whereas in the Poaceae, microsporogenesis is of successive type and the tetrad members separate.
  - ➢ The parietal cell is present in the archesporium of the Cyperaceae, while it is absent in the Poaceae.
  - ➢ The antipodals are ephemeral in Cyperaceae, while they form a complex or become coenocytic in Poaceae.
  - ➢ The embryogeny is of Onagrad type with Juncus variation in Cyperaceae and it is variable in Poaceae.
  - ➢ The testa and pericarp are distinctly free in the Cyperaceae and the former comprises both the integuments, whereas in the Poaceae, testa and pericarp are fused and either both the integuments are obliterated or only the inner integument forms the testa.

  Embryological evidence on the other hand, reveals a close relationship of Cyperaceae with Juncaceae, although the members of Juncaceae produce pollen grains in tetrads (which is actually a monad also containing the three non-functional nuclei of the other three microspores of the tetrad). They are similar in the following aspects:
  - ➢ The pollen of the Cyperaceae is considered to be a reduced form of that of the Juncaceae.

> ➢ Both the families resemble in the simultaneous cytokinesis, tenuinucellar ovules with obturator, Polygonum type of embryo sac with ephemeral antipodals, Onagrad type of embryogeny with Juncus variation and periclinal division of the cells of quadrat.

- Cactaceae — Different taxonomists have treated the family Cactaceae in different ways due to the presence of a large number of variable characters in the different members of this family. Wettstein (1935) placed them under Centrospermae but others assigned them either an ordinal rank or placed them near Passiflorales. Embryological evidence however strongly support the opinion of Wettstein (Mauritzon, 1934; Newmann, 1935), which is also further supported by the data from phytochemistry (Mabry, 1966).
- Garryaceae — There has been a long dispute regarding the systematic position of Garryaceae. Engler and Gilg (1924) and Engler and Diels (1936) have referred them to Amentiferae, while Bentham and Hooker (1880) and Wangerin (1910) included them high up in the polypetalous families near Cornaceae. However, the latter placement is more natural, which has been confirmed by morphological and embryological studies of Hallock (1930). Comparative morphological and phylogenetic studies of the family (Mosely and Beeks, 1955) also support the view of their inclusion in the Cornales.
- Liliaceae — There has also been a rearrangement of the tribes, and subfamilies within the Liliaceae based on embryological studies. A large number of genera also have been assigned their positions on morphological as well as embryological grounds (Cave, 1951). The Liliaceae and Amaryllidaceae are considered as polyphyletic families, and certain groups of both these families are related to each other. Based on these resemblances, Hutchinson (1959) has united three tribes of the subfamily Dracaenoideae (family Liliaceae), Yucceae. Nolineae and Dracaeneae (with the exception of two genera *Astelia* and *Milligania)* together with the subfamily Agavoideae (family Amaryllidaceae) and *Phormium* of the tribe Hemerocallideae into the Agavaceae. Accordingly he recognized six tribes in the Agavaceae viz. Yucceae, Dracaeneae, Phormieae, Nolineae, Agaveae and Polyantheae. On the other hand on the basis of embryological data, Wunderlich (1950) has concluded that Hutchinson's Agavaceae is not an uniform family and the question whether these fairly heterogeneous groups should be united in one family remains open. She has however suggested that Hutchinson's Agavaceae should be separated into four groups viz. Yucca-Agave, Nolineae, Dracaena-Sanseviera, and Cordyline, and the genera *Phormium* and *Doryanthes* should be excluded.
- Lemnaceae — The Lemnaceae are considered, to have been derived either from the Araceae or from the Helobiales. However, embryological studies of S.C. Maheshwari (1954, 1956, 1958) clearly indicate that the Lemnaceae have evolved from the Araceae, and they are not related to the Helobiales. The resemblances between Lemnaceae and Araceae include:
  - ➢ The presence of bitegmic ovules with a micropylar cap of nucellus.
  - ➢ Cellular endosperm with a chalazal haustorium.
  - ➢ Irregular sequence of the divisions in the development of embryo.
  - ➢ Short and stocky suspensor and integumentary operculum in the seed.

On the other hand, the Helobiales are characterized by ovules that do not have any micropylar cap, helobial or nuclear endosperm, very regular sequence of divisions in the development of embryo and the hypertrophy of a prominent basal cell in the embryo. The members of Lemnaceae, are also now considered to be related to Hydrophyllaceae and Boraginaceae, and Cronquist (1968) has placed them in Polemoniales somewhere near the point of origin of Lamiales.

- *Paeonia* — Many taxonomists have treated this genus as a member of the monogeneric tribe Paeonieae of the family Ranunculaceae. But *Paeonia* differs from Ranunculaceae not only in its chromosome number, vascular anatomy, floral anatomy, but also in embryological details in the following ways:
  - ➢ *Paeonia* shows an unique embryogeny with its young zygote first becoming coenocytic and then the cell formation begins from its periphery. The embryo initials develop from some of the peripheral cells (Yakovlev & Yoffe, 1959). The coenocytic phase although common in the gymnosperms, is unknown elsewhere within the angiosperms.
  - ➢ The seeds in *Paeonia* are arillate, and fruit, a follicle while in Ranunculaceae the seeds are non-arillate and the fruit is an achene.

  The embryological details thus suggests that *Paeonia* belongs to an independent family Paeoniaceae.

- *Butomus* — *Butomus* of the family Butomaceae is characterized by the presence of the Polygonum type of embryo sac, while all the other genera under this family have the Allium type of embryo sac. This suggests that only *Butomus* should be retained in Butomaceae (Pichon, 1946). The other genera of this family should be transferred either to Alismataceae or to Limnocharitaceae. Cronquist (1968, 1981) and Takhtajan (1969, 1980) also retained only *Butomus* under Butomaceae and transferred the remaining genera to Limnocharitaceae.
- *Exocarpus* — The genus *Exocarpus* was removed from Santalaceae of angiosperms and was treated as a member of the family Exocarpaceae near Taxaceae in gymnosperms, because of the presence of a naked ovule and pollen chamber. But due to the presence of a typical angiospermic flower, Polygonum type of embryo sac, cellular endosperm and some other embryological characters, Ram (1956) confirmed that *Exocarpus* belongs to the family Santalaceae of angiosperms, and not with gymnosperms.
- *Peganum* — This genus has been variously treated as a member of Rutaceae (Bentham & Hooker, 1862-1867) or Zygophyllaceae (Engler & Prantl, 1897). But its embryological details resemble those of Linaceae, and differs from those of Rutaceae and Zygophyllaceae. The recent embryological findings, therefore, suggest its inclusion in an independent family, Peganaceae near Linaceae (Soueges, 1953).
- *Parnassia* — There has also been a dispute regarding the systematic position of this genus. It has generally been treated as a member of the family Saxifragaceae. But the total dissimilarity of its embryological details with the other genera of Saxifragaceae suggests its removal from Saxifragaceae and inclusion in a separate family Parnassiaceae.

PLATE 15

The lotus flower (*Nelumbo nucifera*) showing the inner stamens and gynoecium

The spongy torus of the flower of *Nelumbo*

The aggregate fruit of indehiscent nutlets of *Nelumbo*

PLATE 16

Individual nutlets of *Nelumbo*

Section through the seeds of *Nelumbo* showing the straight embryos and starchy exalbuminous endosperm

8.5 Palynology in Relation to Taxonomy

The gametophytes, which hold the key to reproduction and perpetuation of the species, are of special significance in plant evolution and taxonomy. Of the two types of gametophytes, the male gametophyte i.e., the pollen grains, are structurally and functionally adapted, for their efficient transfer to the receptive stigma, and these characters are of value to systematists in deducing the phylogeny and taxonomy of the group. Pollen morphological characters were fully appreciated as early as 1830-1840, by authors such as Lindley, von Mohl and Fritzsche. But due to lack of sufficient data, these characters were generally overlooked by taxonomists. The use of the Electron Microscope in recent times has greatly accelerated the pace of progress in palynological research, and pollen characters are being extensively used in taxonomic and evolutionary studies (Blackmore, 1984). The first comprehensive book on palynology was written by Wodehouse (1935) and subsequently a treatise on pollen analysis was contributed by Erdtman (1943). The first manual written by Erdtman, dealing with angiosperms entitled *Pollen Morphology and Plant Taxonomy,* appeared in 1952, which provided a summary of pollen morphological characteristics of all angiosperm families. Following this there has been a great increase in the comparative analytical studies of pollen morphology and wall structure, which have proved to be of considerable systematic and phylogenetic value.

Palynological characters have been used in several ways in taxonomy, including :

- The repositioning of several disputed taxa (Nair, 1980)
- Interpretation of problems relating to the origin and evolution of different groups (Nair, 1980).
- In providing classification of angiosperms (Cronquist, 1981).

Pollen morphology has proved to be of great value in the classification of Acanthaceae (Raj, 1961), Asclepiadaceae (Radford. *et al.*, 1974), Caryophyllidae (Cronquist, 1981), Asteraceae (Slix. I960), Euphorbiaceae (Kohler, 1965), Gentianaceae (Nilsson, 1967), Saxifragaceae (Nair, 1970), etc. Erdtman (1963) used the pollen characters in discussing and solving the taxonomic problems of 105 families.

8.5.1 Characters of pollen used in taxonomy

The main characters of taxonomic value in pollen grains are :

- Variation in shape, size and symmetry.
- Pollen association.
- Pollen nuclear number.
- Number and position of furrows.
- Type, number, position and structure of apertures.
- Architecture of its wall, exine stratification, sculpture and structure.

(a) **Shape, size and symmetry of pollen** : Size and shape of pollen are important diagnostic characters. The shape may be circular, triangular, etc, in polar view, whereas it may be peroblate, spheroidal, subspheroidal, suboblate, oblate, oblate spheroidal, prolate, prolate spheroidal, subprolate, perprolate, etc. in equatorial view (Fig. 8.19). There is a considerable variation in the size of pollen in different plants. In modern flowering plants, the size of the pollen grains varies from about 5×2.4 μm to 200 μm or slightly more in certain members of Cucurbitaceae. Several size classes, have been recognized depending on the length of pollen (Walker & Doyle, 1975), which are as follows:

➢ Minute grains	<10 μm
➢ Small grains	10-24 μm
➢ Medium-sized grains	25-49 μm
➢ Large grains	50-99 μm
➢ Very large grains	100-199 μm
➢ Gigantic grains	>200 μm

Regarding symmetry, they may be symmetric or asymmetric (Fig. 8.18). Symmetric pollen as a rule may be either radially or bilaterally symmetrical. The bilateral pollen further may be plano-convex, concavo-convex or biconvex in lateral view.

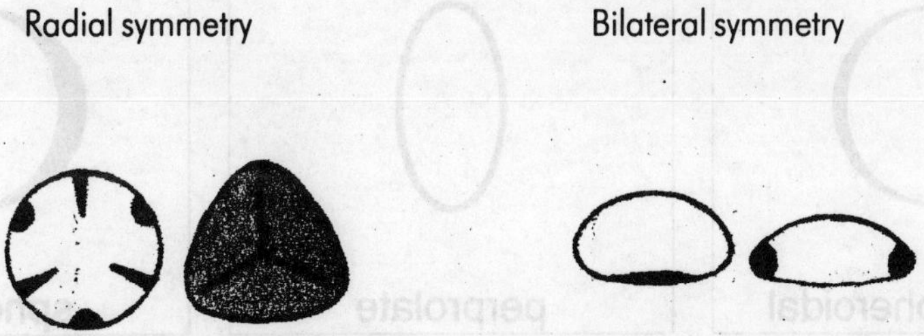

Fig. 8.18 : Symmetry of pollen grains

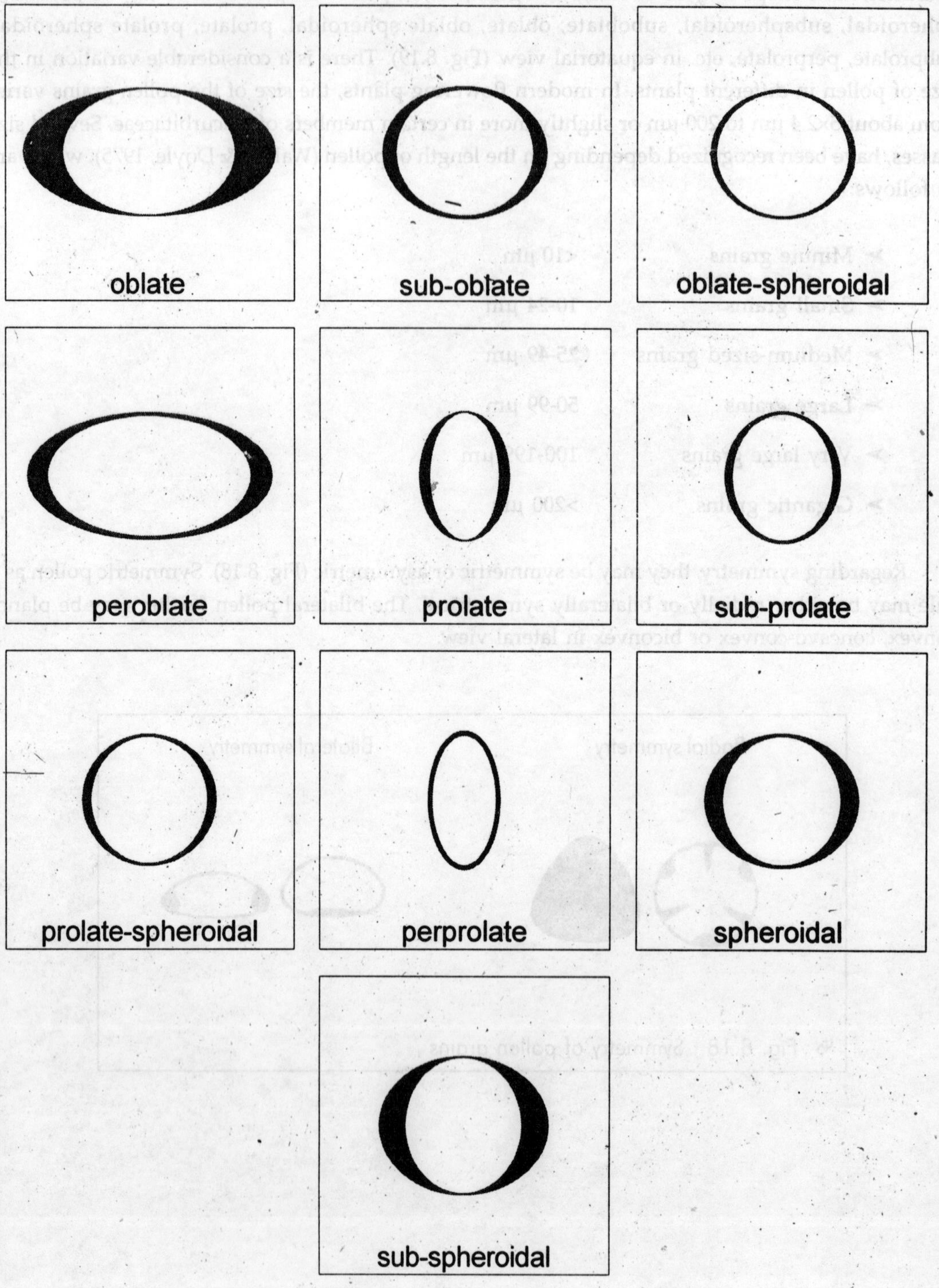

Fig. 8.19 : Shapes of pollen grains

(b) **Pollen association** : Pollen associations, such as their arrangement in tetrads, presence of pollinia, etc. are characteristic of several angiospermic families (Kuprianova, 1974). They are seen in diverse families of the dicotyledons, such as Annonaceae, Droseraceae, Mimosaceae, Monimiaceae and Winteraceae. However, in the monocotyledons, they are restricted to such advanced families as the Cyperaceae and Juncaceae. Extreme cohesion of pollen in the form of pollinia is seen in the Asclepiadaceae and Orchidaceae. These associations can be of the following types (Fig. 8.20) :

- **Monads** — Single pollen grains which when mature, do not remain united with other pollen grains e.g. Poaceae.
- **Dyads** — Pollen grains when united in pairs e.g. *Scheuchzeria palustris.*
- **Tetrads** — When four pollen grains are united together. They may be further categorized into the following types:
  - ➢ **Tetrahedral** — One the pollen grains lie centrally over the other three e.g., *Drimys.*
  - ➢ **Tetragonal** (isobilateral) — All pollen grains lie in one plane in a bluntly quadrangular manner e.g., in monocotyledons.
  - ➢ **Rhomboidal** — e.g., *Anona muricata.*
  - ➢ **Decussate** (cross) — The pollen grains lie pair-wise at right angles to each other e.g., *Magnolia grandiflora.*
  - ➢ **T-shaped** — In this case, the first division of the PMC is transverse and a dyad cell is formed of which either the upper or the lower dyad cell instead of undergoing a transverse division, divides vertically or longitudinally, yielding either a straight, or inverted T-shaped configuration e.g., *Aristotolochia polyanthus.*
  - ➢ **Linear** — In this case, the first division is always transverse, giving rise to two dyad cells. The second division is also transverse, which results in a linear tetrad e.g., *Mimosa pudica.*
- **Pseudomonad** (cryptotetrad) — It is a degraded tetrad in which monad-like tetrads are formed without dividing walls between the four components. One nucleus is normally developed, whereas the remaining three are obliterated e.g., Cyperaceae.
- **Massula** (ae) — They are also called polyads. In this case each of the tetrad cells divides once or twice resulting in a massulae of 8-64 cells, which remain together after maturity and are held together in small units e.g., Mimosaceae.
- **Pollinium** (ia) — This includes a unit of all grains produced by one theca (anther loculus). The pollinium apparatus is the functional unit of a 'corpusculum' with its two attached arms and pollinia e.g., *Cryptostegia, Calotropis.*

The taxonomic usefulness of pollen association in the multipalynous family Orchidaceae has been pointed out by Vij and Kashyap (1978) who studied 50 species under Orchidaceae and observed that single pollen grains were found in the relatively primitive groups (*Cephalanthera ensifolia, Cypripedium cordigerum, Paphiopedilum villosum*), tetrads in the tribe Neottieae and Epidendrieae and perfect massulae in the Orchideae and some Neottieae, where the shape of the massulae can be of taxonomic value for lower-level classifications.

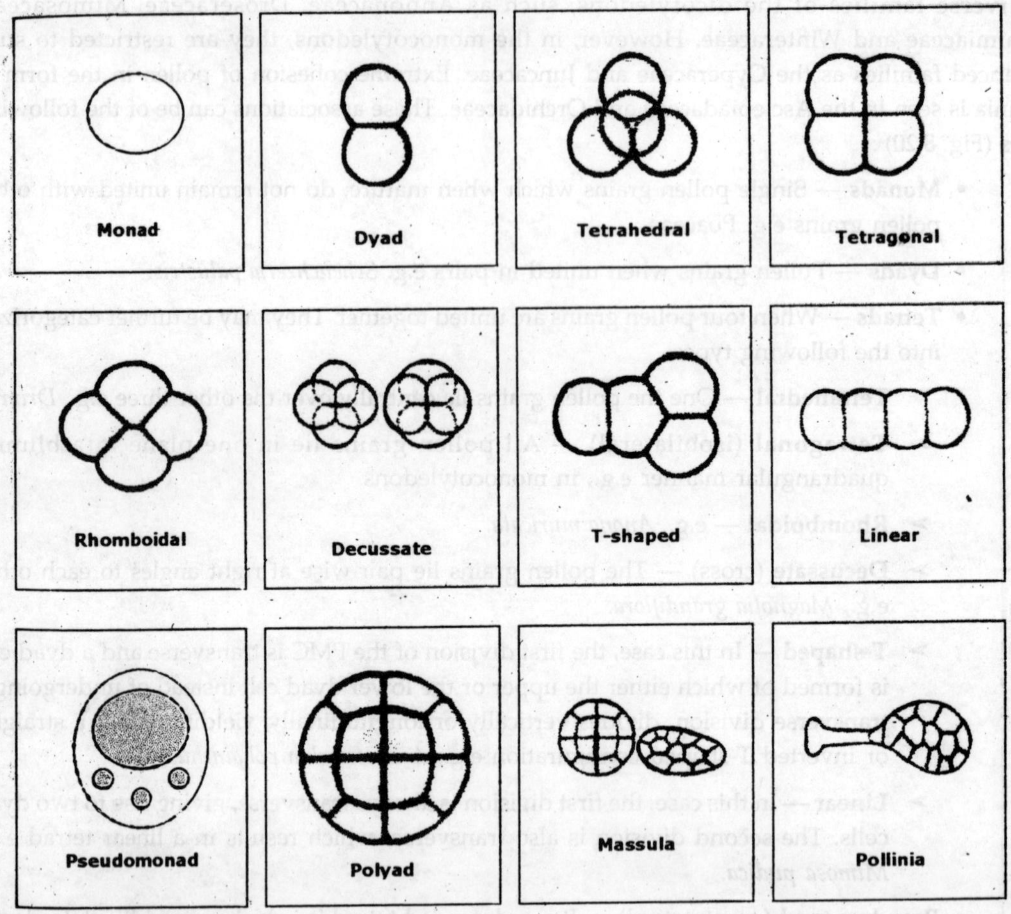

Fig. 8.20 : Pollen units

(c) Pollen nuclear number : The number of nuclei in the pollen grains at the time of their release is a valuable taxonomic and phylogenetic character. The number of nuclei in a pollen grain may be regarded as a function of the stage at which the pollen grains are released. In the angiosperms, the pollen show a definite pattern at which stage a pollen is released (Brewbaker, 1967; Dahlgren, 1975b; Schurhoff, 1924, 1926), and they may be released either:

- At the **two-nucleate stage,** before the division of the generative nucleus, or
- At the **three-nucleate stage,** after the division of the generative nucleus.

Further it has been found that :

- In angiosperms as a whole and within individual taxa, the two-nucleate pollen grains are more primitive.

- The three-nucleate pollen grains have arisen from the two-nucleate ones.
- The change is irreversible.

The stage at which the pollen are released varies in different groups of angiosperms:

- Dicotyledons – The pollen nuclear number do not show a definite pattern.
  - ➢ Certain orders have consistently trinucleate pollen grains, e.g. Asterales, Caryophyllales, Plumbaginales, Polygonales (Dahlgren, 1975b).
  - ➢ Some orders are heterogenous in having both two-and three-nucleate pollen and both the types may even occur in one and the same genus, as has been observed in *Burmannia, Ipomoea* and *Lobelia* (Brewbaker, 1967).
- Monocotyledons – In the Monocotyledons, the highly reduced wind-pollinated groups, aquatics and achlorophyllous saprophytes show the three-nucleate pollen grains.

(d) **Apertural forms** : The form, number, distribution and position of apertures are important palynological criteria in assessing relationships and phylogeny of plants. Almost all palynological discussions on plant relationships and phylogeny are based on the form, number, distribution and position of the apertures (Nair, 1971). Apertures are generally thin-walled areas in the outer pollen wall or exine, through which the pollen tube usually emerges at the time of germination. There are two basic forms of apertures:

- **Simple Apertures** – Apertures are either long or short and rounded and may be of following types (Fig. 8.21) :
  - ➢ **Sulcus** – It is an elongated latitudinal ectoaperture situated at the distal or proximal pole of a pollen grain. A sulcus has the same shape as a colpus, but differs in orientation [e.g. *Nymphaea* (Nymphaeaceae), *Sparganium* (Typhaceae)]. Sulcate grains are either:
    - ❑ **Anasulcate** (distal)
    - ❑ **Catasulcate** (proximal)
    - ❑ **Zonosulcate** (extend right around the grain)
  - ➢ **Ulcus** – A more or less rounded aperture situated at the distal or proximal pole of a pollen grain is termed ulcus and may be derived from a sulcus or a sulcoid aperture [e.g. *Sparganium* (Sparganiaceae), *Typha* (Typhaceae)].
  - ➢ **Colpus** – It is an aperture, which is long-furrow-like i.e. twice as long as broad (with a length and breadth ratio greater than 2) and the grain is referred to as colpate. Two basic types of pollen occur in the angiosperms:
    - ❑ **Monocolpate** – It is a characteristic of the monocotyledons and some dicotyledonous families such as woody Ranales, and have a single furrow on one side of the pollen remote from the point of contact in the tetrad.
    - ❑ **Tricolpate** – It is a characteristic of most of the dicotyledons, and are provided with three meridionally placed furrows.

    Two other types of pollen also occur, which are considered to have been derived from both the monocolpate and tricolpate types as a result of specialization. They are :

    - ❑ **Acolpate** – They are pollen without a furrow.
    - ❑ **Pantocolpate** - They are pollen with many furrows.

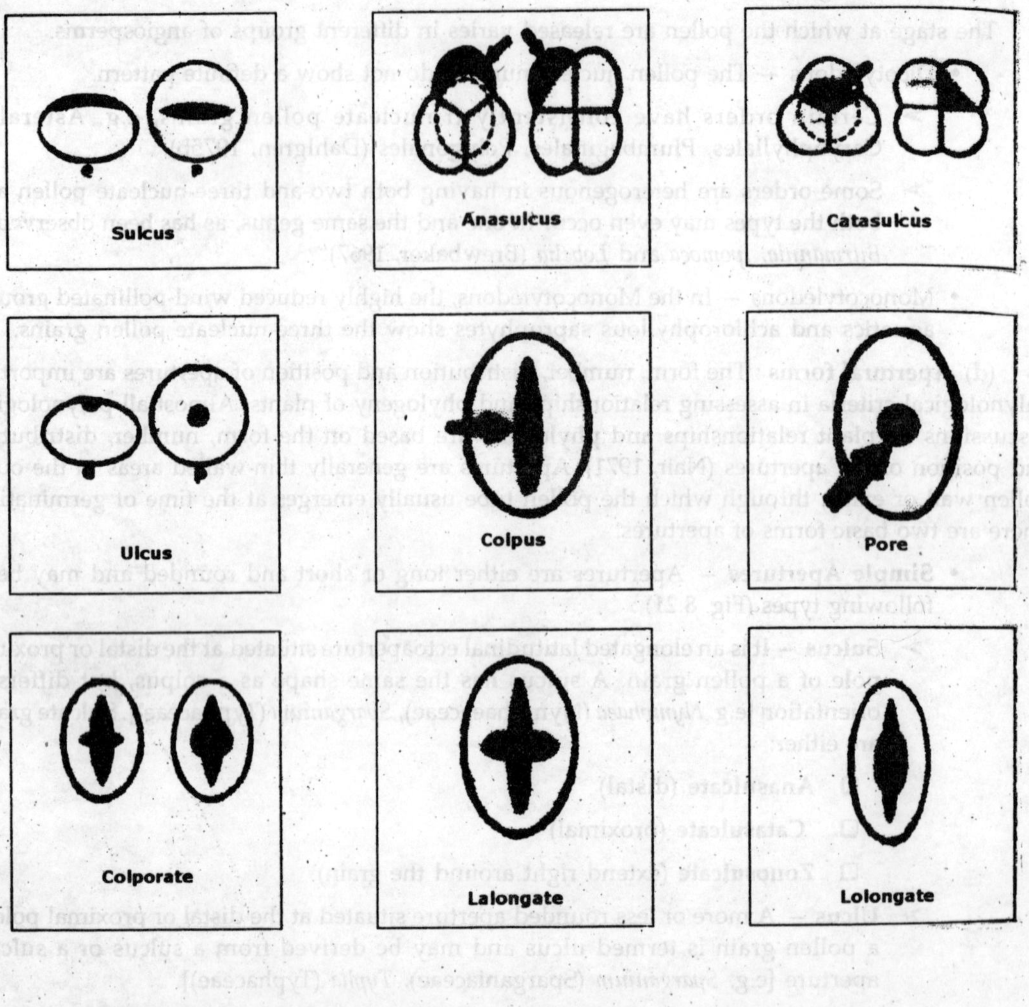

Fig. 8.21 : Types of apertures of pollen

➢ **Pore** – It is a circular or elliptic aperture is short with length and breadth ratio less than 2 and the pollen grain is called porate [e.g. *Myrica gale* (Myricaceae)].

- **Compound Apertures** – They are apertures consisting of an outer or marginal colpal (furrow-like) part and an inner or central distinctly delimited oral part. Ora are either circular or lalongate i.e., transversely elongated endoaperture [e.g. *Filipendula* (Rosaceae)] or lolongate i.e., longitudinally elongated endoaperture [e.g. *Rumex* spp. (Polygonaceae)]. In relation to ora, the furrows or apertures will be named colporate or perorate.

The position of the apertures varies in different groups of plants (Fig. 8.22) and may be :

- **Proximal** – found in uniaperturate forms and in terms of evolution are the most primitive. They are found in the gymnosperms and monocotyledons.
- **Distal** – found in uniaperturate forms and is a characteristic of pteridophytes.

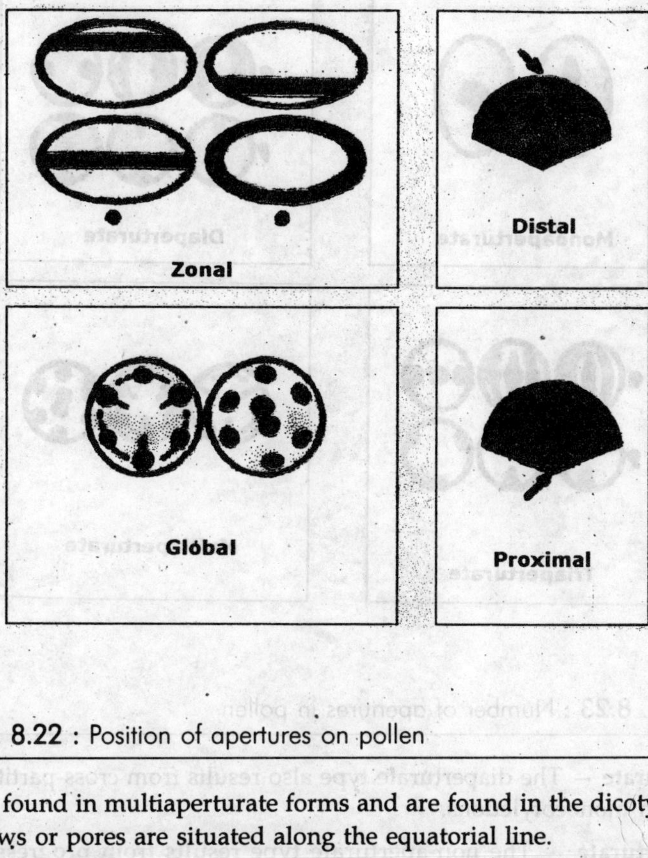

Fig. 8.22 : Position of apertures on pollen

- **Zonal** – found in multiaperturate forms and are found in the dicotyledons. In this case the furrows or pores are situated along the equatorial line.
- **Global** – found in multiaperturate forms and are found in the dicotyledons. In terms of evolution, they are the most advanced. The furrows or pores are distributed over the entire surface in this type.

Most angiospermous pollen grains fall into the following (Fig. 8.23) two general types :

- **Monoaperturate** – They are characteristic of the monocotyledons and some members of Magnoliidae. Uniaperturate pollen grains have a single germinal furrow or pore situated at the proximal or distal position.
- **Triaperturate** – The triaperturate or triaperturate-derived pollen grains occur in most of the dicotyledons and are unknown elsewhere. These pollen grains have three germinal furrows, either radiating like the lines of a trilete mark from a common point or forming a triangle at the distal position. Some less common types have been derived from the triaperturate form.

Various other forms of pollen which appears to have evolved from the uniaperturate and triaperturate forms (Fig. 8.23) , include:

- **Multiaperturate** – The multiaperturate type results from cross-partitioning of one germinal furrow into several. It is found in both monocotyledons and dicotyledons.

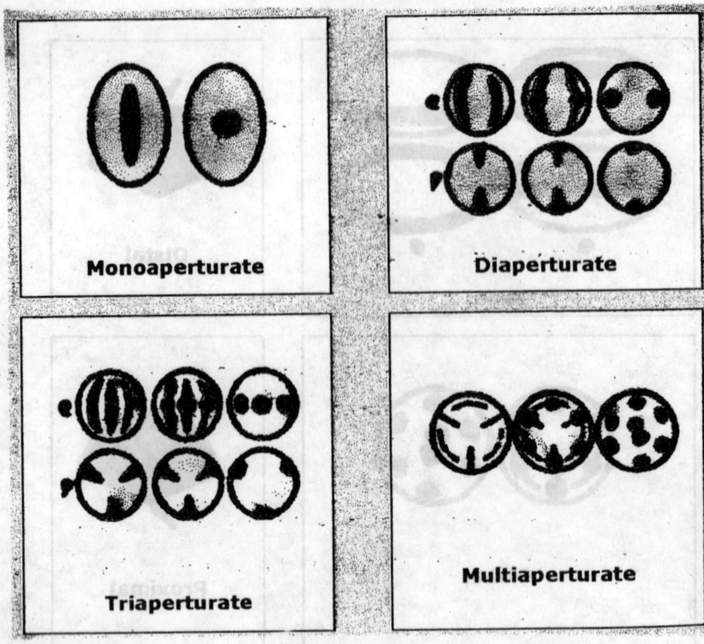

Fig. 8.23 : Number of apertures in pollen

- **Diaperturate** – The diaperturate type also results from cross-partitioning and is found in certain monocotyledons.
- **Nonaperturate** – The non-aperturate type results from progressive shortening and eventual elimination of the furrow and is found in both monocotyledons and dicotyledons.

The various types of apertural morphoforms provide one of the best taxonomic criteria in routine taxonomic work, being often constant, not very variable and easily visible under a microscope. Nair (1974 a, c) has even proposed a palynological classification of the whole plant kingdom based on such data and have classified the plants into:

(i) **Amorphosporophyta** – includes thallophytes with no apertural markings.

(ii) **Trimorphosporophyta** – includes Archegoniatae with monolete, trilete or alete pollen.

(iii) **Polymorphosporophyta** – containing angiosperms with diverse forms of apertural types.

There is a remarkable relationship between morphological types of pollen and taxonomic subdivisions in angiosperm families (Müller, 1970). An example is illustrated by the family Lecythidaceae. Based on palynological data Müller (1972) found that the pollen grains of its members fall into two groups, tricolpate and syntricolpate, and that in combination with other characters this can effectively be utilized in subdividing the family. Apertural characters have also been used in lower level taxonomy. For example in the case of *Sesamum*, the number of colpi is species-specific i.e., eleven in *S. indicum*, eight in *S. laciniatum* and nine in *S. prostratum*.

NPC-System (Erdtman, 1969)

Erdtman (1969) has even developed an elaborate terminology known as the NPC-System based on the **number (N), position (P) and character (C)** of apertures of pollen. The term treme has been used under this system for preparing keys for the classification of pollen grains.

- **Depending upon N, the pollen are of the following types :**

➢ **Atreme** —The pollen having no apertures are called atreme and represented as N_0.

➢ **Monotreme** — Pollen grains with one aperture and denoted as N_1.

➢ **Ditreme** — Pollen grains with two apertures and denoted as N_2.

➢ **Tritreme** — Pollen grains with three apertures and denoted as N_3.

➢ **Tetratreme** — Pollen grains with four apertures and denoted as N_4.

➢ **Pentatreme** — Pollen grains with five apertures and denoted as N_5.

➢ **Hexatreme** — Pollen grains with six apertures and denoted as N_6.

➢ **Polytreme** — Pollen grains with more than six apertures (seven or more) and denoted as N_7.

➢ **Anomotreme** — When the apertures, irrespective of their number, are irregular or irregularly spaced, and denoted as N_8.

Depending upon the position (P) of the apertures, the pollen are of following types :

➢ **Catatreme (P_1)** — The pollen are designated as catatreme when the aperture is on the proximal face.

➢ **Anacatatreme (P_2)** — The pollen are designated as anacatatreme when the apertures are located on the proximal as well as on the distal face.

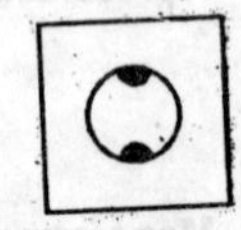

➢ **Anatreme (P_3)** — The pollen are designated as anatreme when the aperture is on the distal face.

➢ **Zonotreme (P_4)** — When the centres of apertures are located on the equator, the pollen grains are referred to as zonotreme.

➢ **Dizonotreme (P_5)** — When the centres of apertures are located on the equator in two zones, the pollen grains are referred to as zonotreme.

➢ **Pantotreme (P_6)** — The condition is said to be pantotreme, when the apertures are more or less uniformly distributed all over the pollen surface.

- **Depending upon the character (C) of the apertures, the pollen are of following types :**

➢ **Treme** — Pollen are denoted as C_0 when the character of the aperture is not known (?).

➢ **Leptoma** (C_1) — Pollen having an aperture like thin area.

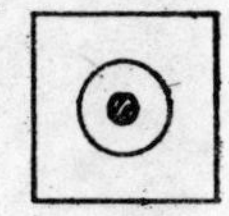

❑ **Monoleptoma** — Pollen with one leptoma.

❑ **Cataleptoma** — Pollen with leptoma on the proximal face.

❑ **Analeptoma** — Pollen with leptoma on the distal face.

➢ **Trichotomocolpate** (C_2) —The grains with a three-slit colpus.

➢ **Colpate** (C_3) — Pollen with an aperture, which is a long-furrow-like i.e. twice as long as broad.

➢ **Porate** (C_4) — Pollen with an aperture, which is short with length and breadth ratio 2:1 or 1:2.

➢ **Colporate** (C_5) — Pollen with apertures consisting of an outer or marginal colpal (furrow-like) part and an inner or central distinctly delimited oral part.

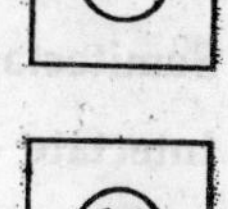

➢ **Pororate** (C_6) — Pollen with compound apertures in which, both the ectoaperture and the endoaperture are pores and the two are not congruent.

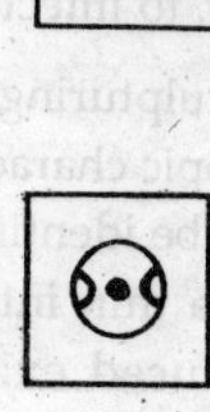

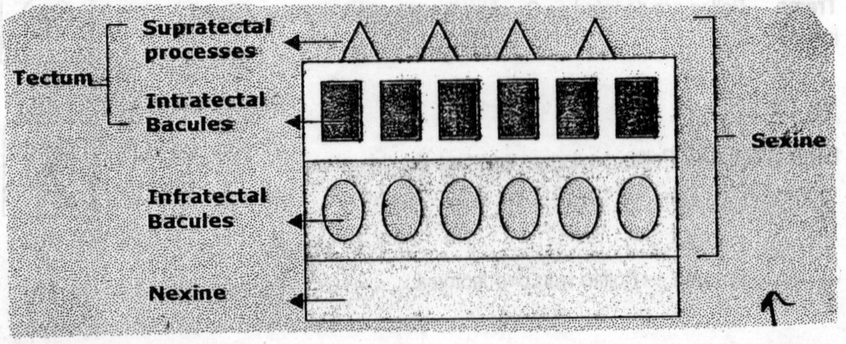

Fig. 8.24 : Sporoderm stratification (after Erdtman, 1969) of pollen

(e) **Exine stratification :** The pollen wall or **sporoderm** is generally layered (stratified). In angiosperms, it consists of two layers – the outer resistant layer composed of sporopollenin, known as **exine** and the inner more or less cellulose layer called **intine**. Topographically, the exine consists of two well-defined sub-layers – the outer sculptured **sexine or ektexine** and the inner non-sculptured **nexine or endexine** (Fig. 8.24). The sexine further consists of two main parts – a partially or wholly covering layer called **tectum** below which, are the rods or rod like elements called **bacules**. According to Walker (1974 b), there are three basic types of pollen based on the characteristic of tectum. They are:

- **Tectate** — have a continuous tectum. Tectate pollen may be either
  - ➢ **Tectate-perforate**
  - ➢ **Tectate-imperforate**
- **Semitectate** — tectum is not continuous.
- **Intectate** — without an apparent tectum.

The trend of exine evolution is from tectate-imperforate to tectate-perforate to semitectate and, more rarely, to intectate pollen.

The sculpturing of the external substance (exine) of pollen grains presents a remarkable series of microscopic characters and may vary from family to family. Commonly, the family and often the genus may be identified by the pollen grains. In general, angiosperm pollen grains have a massive exine and a thin intine. However, certain taxa among Monocotyledons are unique in having a highly reduced exine and an elaborate intine e.g. Amaryllidaceae, Cannaceae, Costaceae, Heliconiaceae, Iridaceae, Marantaceae, Musaceae, Strelitziaceae and Zingiberaceae. Further the pollen in anemophilous group is smooth and thin, whereas the ones distributed by insects or birds, are characterized by various types of sculpturing on the exine.

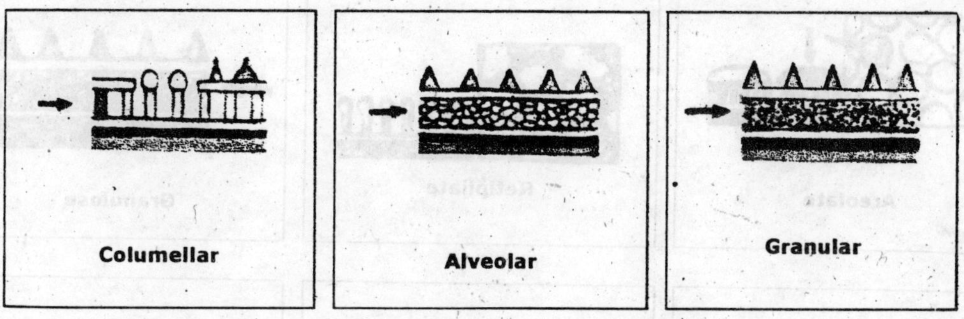

Fig. 8.25 : Types of sexine of pollen

The ektexine or sexine may be either of the following three types (Fig. 8.25) [Meyer, 1975] :

- **Columellar** — A rod-like element of the sexine/ectexine, either supporting a tectum or a caput, e.g. Angiosperms
- **Alveolar** — Describing a type of sexine/ectexine structure, in which the infratectal layer is characterized by partitions forming compartments of irregular size and shape, e.g. Gymnosperms (Pinaceae)
- **Granular** — A type of exine stratification in which the infratectal layer is composed of more or less rounded, granules rather than of columellae or other structures, e.g. some Gymnosperms and Angiosperms (Zavada, 1984b)

The endexine or nexine may be either of the following two types (Meyer, 1975) :

- **Laminated** — e.g. Gymnosperms
- **Non-laminated** — e.g. Angiosperms (Zavada, 1984b).

In recent years the pollen wall structure has become the subject of very detailed and critical study, employing phase contrast and ultraviolet microscopy as well as the more conventional light microscopy (Erdtman, 1954, 1956, 1959). Ultrafine sections of grains have also been studied under the electron microscope (Fernandez-Moran and Dahl, 1952; Muhlethaler, 1953). Ultraviolet microscopy has proved useful in the study of fine details of the intine and cytoplasm. With the improvement of techniques (Afzelius, 1954, 1956) it has been possible to demonstrate fine details of the structure of the walls (sporoderms) and their stratification.

The exine ornamentation may be of either of the two types (Fig. 8.26) :

- **Depression type** — In this case the exine is smooth. This includes many subtypes, such as
  - **Areolate** — luminoid network surrounds islands of raised areas i.e., a feature of ornamentation in which the exine is composed of circular or polygonal areas separated by grooves.

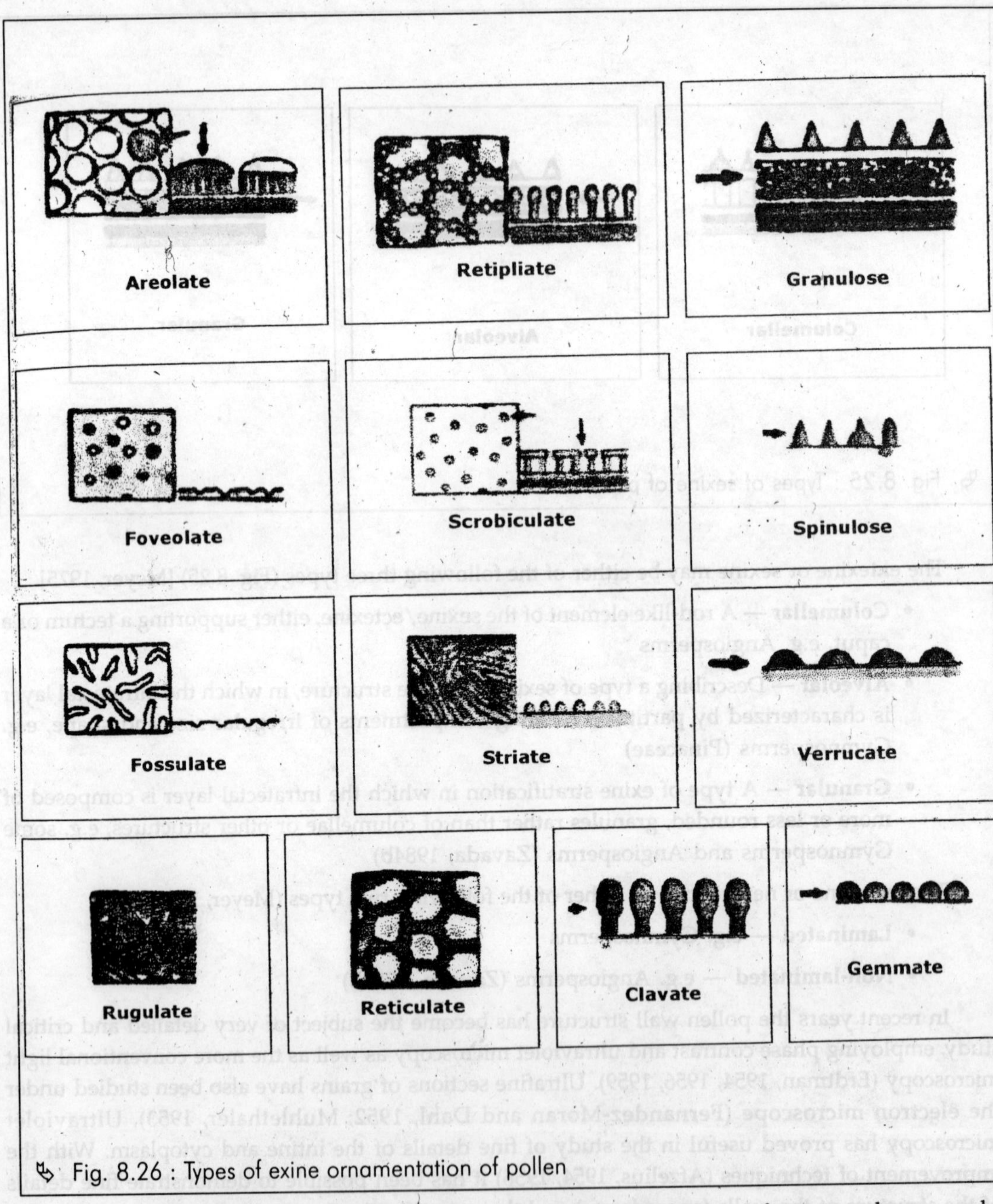

Fig. 8.26 : Types of exine ornamentation of pollen

- **Foveolate** — A feature of ornamentation consisting of more or less rounded depressions or lumina more than 1μm in diameter. The distance between foveolae is greater than their breadth.
- **Fossulate** — A feature of ornamentation consisting of an elongated, irregular groove in the surface.
- **Rugulate** — a type of ornamentation consisting of elongated sexine elements more than 1μm long, arranged in an irregular pattern that is intermediate between striate and reticulate.

➢ **Reticulate** — lumina in the form of a network-like pattern consisting of lumina or other spaces wider than 1μm bordered by elements narrower than the lumina.

➢ **Retipliate** — incomplete fusion of columella i.e., describing a reticulum formed by rows of pila instead of muri.

➢ **Scrobiculate** — A rounded or elongate tectal perforation, less than 1μm in length or diameter.

➢ **Striate** — elongated, generally parallel elements separated by grooves.

- **Excrescences type** — In this case the exine carry processes or excrescences of various kinds. This includes many subtypes, such as

➢ **Baculate** — cylindrical or rod shaped free-standing excrescences, more than 1μm in length and less than this in diameter.

➢ **Clavate** — club shaped excrescences that is higher than 1μm, with diameter smaller than height and thicker at the apex than the base.

➢ **Gemmate** — excrescences are in the form of rounded warts with constricted base, higher than 1μm, and that has approximately the same width as its height.

➢ **Granulose** — A type of exine stratification in which the infratectal layer is composed of more or less rounded, granules rather than of columellae or other structures.

➢ **Spinose** — long pointed excrescences

➢ **Spinulose** excrescences are as long and tapering pointed elements, exceeding 1μm.

➢ **Tuberculate** — excrescences in the form of knobbly projections.

➢ **Verrucate** — excrescences are rounded but base is not constricted i.e., wart-like excrescences, more than 1μm wide, that is broader than it is high and is not constricted at the base.

8.5.2 Role of palynology in taxonomy

The taxonomic and evolutionary importance of pollen morphology may be at specific, generic or higher levels. Pollen characters have helped in differentiation of various taxa (e.g., families, genera and species), classification of newly discovered taxa and in phylogenetic considerations.

(a) **Stenopalynous and Eurypalynous taxon**

- **Stenopalynous** — In many taxon, the type of pollen is characteristic and constant. Such a taxon is termed stenopalynous or unipalynous e.g., Asclepiadaceae, Cruciferae, Poaceae, Labiatae, etc. Stenopalynous taxa are generally considered to be very natural.
- **Eurypalynous** — The taxon in which the types of pollen may vary considerably in size, aperture, stratification of exine, etc., are termed eurypalynous or multipalynous e.g., Acanthaceae, Rubiaceae, Verbenaceae. Eurypalynous taxa are heterogeneous, at least in certain instances. Palynological data are particularly useful in the delimitation of eurypalynous taxa e.g. Acanthaceae (Bhoj Raj, 1961), Asteraceae (Stix, 1960), Euphorbiaceae (Punt, 1962; Kohler, 1965), Gentianaceae (Nilsson, 1967), Saxifragaceae (Erdtman, 1954; Nair, 1970), etc.

Like families, genera may be stenopalynous or eurypalynous. For example, the acanthaceous genera such as *Strobilanthus*, *Justicia*, etc. have been split up by Bremekamp (1944) and more natural groupings have been proposed.

(b) **Family and tribal level** : The families may be stenopalynous or eurypalynous. A few examples of the use of pollen characters in the taxonomy of some families and tribes are given below:

- The family Berberidaceae has been variously circumscribed by different taxonomists. Recently based on pollen characters, *Podophyllum* where the pollen grains remain united, has been removed to a separate family Podophyllaceae. The pollen grains are free in the other members of Berberidaceae.
- Several families have distinctive pollen types. For example, Poaceae have smooth sulcate pollen, Malvaceae and Asteraceae have typically spinulose exine, and Plumbaginaceae have verrucate exine.
- The two families, Araceae and Lemnaceae under the same order Arales of Hutchinson are characterized by stenopalynous, 1-2-4-colpate, 3-porate or inaperturate and Arecaceae exine sculptured pollen in the former, while the latter is characterized by eupalynous, 1-porate and spinose pollen.
- The palynological studies support the treatment of the tribe Bombaceae of the family Malvaceae as a separate family Bombacaceae. Palynological studies reveal that the exine is reticulate in the Bombaceae, whereas it is spinose in most of the Malvaceae.
- Palynological evidences have also supported the separation of Faeoniaceae from Ranunculaceae, Fumariaceae from Papaveraceae, and Nelumbonaceae from Nymphaeaceae.

(c) **Generic level** : Following are few examples of the taxonomic significance of palynology at the level of genus:

- Pollen characters have been extensively employed in classifying the genera of the Acanthaceae and the Primulaceae.
- The two genera *Phytolacca* and *Rivinia* of the Phytolaccaceae, can be recognized on the basis of palynological characters. The pollen of *Phytolacca* is 3-zonocolpate, whereas that of *Rivinia* is pantocolpate.
- Hedberg (1946) has suggested the division of the genus *Polygonum* into seven genera based on pollen morphology viz. *Koenigia, Persicaria, Polygonum, Pleuropteropyrum, Bistoria, Tiniaria* and *Fagopyrum* which are distinct in their pollen types, and is accepted by several recent taxonomists.
- In the family Betulaceae, the thickening of exine around the pores make a distinguishing character for different genera, which is knob-like in *Betula*, club-shaped in *Corylis*, unexpanded in *Caprinus*, and an arcus is present between adjacent pores in *Alnus*.
- The genera *Salix* and *Populus* under Salicaceae can be distinguished on the basis of pollen characters. *Salix* has long and narrowed 3-furrowed pollen, while *Populus* has spherical pollen without distinct apertures.

(d) **Species and infraspecific level** : Pollen characters have also proved helpful in distinguishing the species within a genus. Following are a few examples :

- Sharma (1987) investigated the pollen morphology of 16 Indian species of *Cyperus*, and prepared a key to differentiate all of them on the basis of pollen characters. Of the 16 investigated species, the pollen grains of six species (*C. iria, C. difformis, C. squarosus, C. triceps, C. flabelliformis* and *C. paniceus* var. *roxburghianus*) are predominantly 1-colpate, two species (*C. exaltatus* and *C. pumilus*) are predominantly 2-colpate, four species (*C. rotundus, C. laevigatus, C. alulatus* and *C. bulbosus*) are predominantly 4 aperturate with 3 colpi and 1 pore, three species (*C. compactus, C. kyllingia* and *C. globosus*) are predominantly pantoaperturate, and one species (*C. digitatus*) is predominantly 1-porate.
- Species of *Anemone* can be distinguished on the basis of germinal aperture of pollen. It is 3-zonocolpate in *A. obtusiloba*, pantoporate in *A. alchemillaefolia*, pantocolpate in *A. rivularis* and spiraperturate in *A. fulgens*.
- The exine pattern has been useful in recognizing different species of *Bauhinia*. According to Nair (1974), pollen are psilate in *B. acuminata*, striate in *B. krugii*, spinulate in *B. malabarica*, reticulate tuberculate in *B. purpurea*, reticulate in *B. racemosa*, and verrucate in *B. retusa*.
- Pollen size is helpful in distinguishing two species of *Malva*, *M. rotundifolia* (pollen 74-84 μm) and *M. sylvestris* (pollen 105-126 μm).

(e) **Phylogenetic considerations** : Palynological studies have also been helpful in the elucidation of phylogenetic relationships. Following are a few examples :

- Pollen morphology does not support sharp demarcation between the dicotyledons and the monocotyledons as suggested by most systems of classification, because dicotyledonous pollen characters occur in some monocotyledons and vice versa.
- Palynological data suggest two distinct phylogenetic stocks in the dicotyledons – monocolpate, represented by the Magnoliaceae and tricolpate represented by the Ranunculaceae. Due to the presence of monocolpate element, the monocolyledons, are considered to be more closely related to the magnolian stock. Furthermore, both the monocotyledons and the magnolian dicots have monocolpate elements characteristic of the preangiospermous archegoniates and are considered more ancient palynologically than the ranalian dicots where monocolpate elements are completely absent and new apertural forms are present.
- Based on pollen morphological studies in the monocotyledons, Kuprianova (1948), proposed that the Helobiae are not related to the other monocotyledons but are specialized Polycarpiceae with ranalian affinities. Kuprianova is also of the view that most monocotyledonous families could be considered to have evolved from Arecaceae or Liliaceae.

PLATE 17

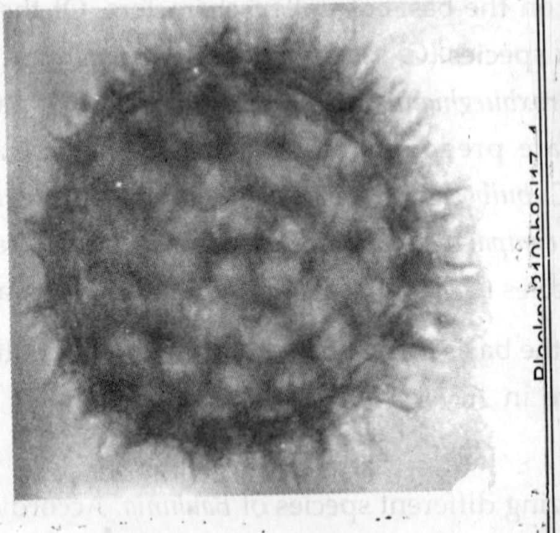

Pollen of *Ipomoea fistulosa*

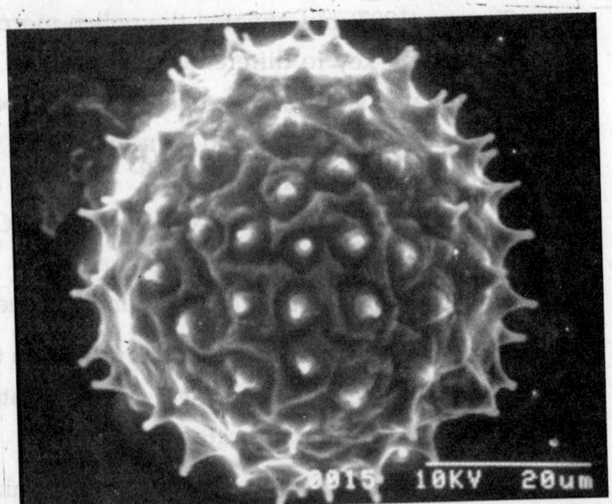

Scanning electron microscopic photograph of pollen of Convolvulaceae showing the beautiful surface ornamentation

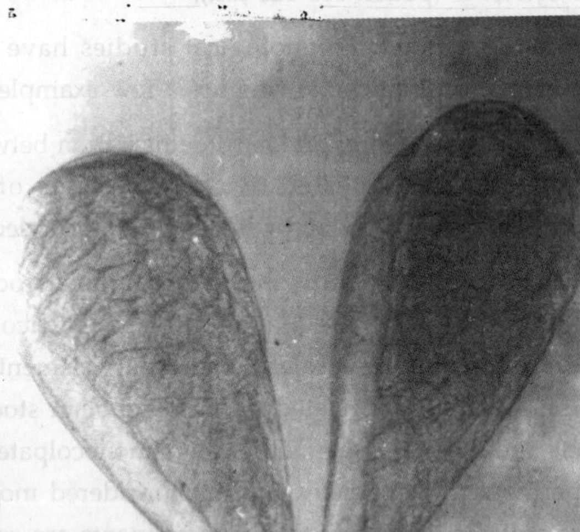

The pollinia of *Calotropis procera* (Asclepiadaceae) under the Light Microscope (40 X)

8.6 Phytochemistry in Relation to Taxonomy

Chemical systematics is the study of chemical variation in different organisms and their relationships (Naik, 1984) and this approach of taxonomy in which chemical features of plants are used in developing classifications or in solving taxonomic problems is called Chemotaxonomy. Data from chemistry of plant products is termed phytochemistry. The principles, procedures and results of investigations of chemical variation of plant groups can be applied mainly for two purposes:

- To provide taxonomic characters which may help in plant classification and may improve the existing systems of plant classification.
- To have additional knowledge of phylogeny or evolutionary relationships of plants.

Chemotaxonomy has been used in all the groups of the plant kingdom starting from the simple organisms, such as fungi and bacteria, up to the most highly advanced and specialized groups of angiosperms and at all levels of the hierarchy of classification in plants, starting from the rank of Variety up to the rank of Division. Chemical characters are particularly of high taxonomic value when they are stable, unambiguous and not easily, if at all, changeable.

The use of chemical characters in plant classification has a long history. Since the early 1960s, phytochemical characters started to attract the attention of plant taxonomists. However recently, due to the development of new and powerful analytical techniques and the speed and simplicity of these techniques (Alston, 1967), it has been possible to screen a large number of individuals in a very limited time and utilize such information in plant taxonomy. According to Smith (1976), a paper published by Abbott (1886) on chemical constituents of plants can be taken as the beginning of the modern phase of chemotaxonomy. The innumerable papers that have been published later in various journals of the biological sciences [Alston & Turner (1963), Bendz & Santesson (1973), Cagnin et al. (1977), Gibbs (1974), Harbome, Boulter & Turner (1971), Heywood (1973a), Swain (1966), etc.] are an indication of the interest that has been generated among biologists working in this area of systematics and have put this promising sub-discipline of taxonomy on a sound footing. The various data from these studies confirms early ideas that phytochemical characters correlate quite well with other types of characters and should be considered seriously in taxonomy. However, at the same time, chemotaxonomy should not be considered as more indicative of relationship than other characters such as external morphology, anatomy, cytology, etc. and as a replacement of other taxonomic characters, but at best a major source of new characters and information. The new ideas and facts that are still being obtained from all other existing sources of taxonomic evidence should not be ignored. The chemical characters are considered more important, only when they show a high degree of correlation with other features.

8.6.1 Stages in chemotaxonomic investigation

Hegnauer (1967) recognized six chief stages in a phytochemical study, reconciling the need for informed chemical procedure and a project design satisfying formal taxonomic requirements:

(i) Choice of a group, taxonomic survey and sound sampling.

(ii) Choice, mastery and modification of suitable chemical techniques in a pilot survey.

(iii) Full analysis of all material.

(iv) Interpretation and comparison with data from all other sources.

(v) Adjustment of classification as necessary, and

(vi) Treatment of any evolutionary relationship as indicated by the new data.

8.6.2 Phytochemical characters of taxonomic significance

The occurrence and distribution of the various types of chemical substances present in plants prove to be of taxonomic significance. However, it should be noted that, all kinds of chemical substances present in plants do not reveal information useful to the taxonomist. Phytochemical characters of taxonomic significance have been classified into three types (**Mentzer, 1966**). These include:

- **Primary constituents** — These include the macromolecular compounds directly taking part in metabolism and include proteins, nucleic acids, chlorophyll and polysaccharides. All chemical materials synthesized by an organism reflect the information in DNA, RNA and proteins. These latter molecules have been termed as semantides (Zuckerkandl & Pauling, 1965). Semantides, thus contain useful information of taxonomy and phylogeny.
- **Secondary constituents** — They include compounds lacking nitrogen and not involved directly in plant metabolism i.e., simple phenolic compounds like caffeic, benzoic and nicotinic acids and polyphenolic compounds like flavonoids, terpenes, coumarines, alkaloids and pigments of which flavonoids are most widely studied with respect to plant systematics (Gardner, 1977; Cronquist, 1977).
- **Miscellaneous substances**

However, no suitable classification of the chemical characters and their use in taxonomy is developed so far. On the basis of their molecular weight, **Jones and Luchsinger (1987)** has divided the natural chemical plant products useful in taxonomy, into two major groups :

- **Micromolecules** — they are low molecular weight compounds with a molecular weight of 1000 or less, e.g. amino acids, alkaloids, fatty acids, terpenoids, flavonoids, etc.
- **Macromolecules** — they include the high molecular weight compounds with a molecular weight of over 1,000, e.g. proteins, DNA, RNA, complex polysaccharides, etc.

Some taxonomically important chemical compounds, along with a few examples of their systematic value is given below :

(a) **Proteins** — Among the various semantides, proteins serve as the most important tool in chemotaxonomy. The importance of proteins in chemotaxonomy has several reasons behind it.

- Firstly, these are large, complex molecules showing little qualitative variation with changing environmental factors.
- Secondly, they are universally distributed.
- Thirdly, they are relatively simple to extract and handle and present in appreciable amount.

- Lastly, numerous cheap, simple and rapid methods have developed in the recent past, for protein analysis and comparison by taxonomists, of which electrophoresis is particularly an important method of protein separation.

The variation in protein structure between plants effectively provides an inner view into the cell's genome, as the amino acid sequences of proteins are encoded by the nucleotide sequences of a cell's DNA. Proteins have been used in taxonomic investigations in the following ways:

- **Comparison of the protein banding patterns** — In recent times protein banding patterns obtained by gel electrophoresis, have been focused on the problem of identification of critical taxa, their relationship and taxonomic status. Taxonomic interpretations should be based on comparison of proteins from homologous organs of the same age in order to avoid any confusion due to variability of proteins. Considerable variation in protein complements has been recorded at the level of species and genus, and even between the same plants in different populations. These form evidence upon which taxonomic systems may be founded, tested or demolished.

  Electrophoresis is one of the most extensively used techniques in protein investigations. Because of the presence of ionizable molecules on the surface of proteins, they will migrate when subjected to an electrical field in a solution of suitable pH. Soluble proteins thus migrate within an electric gradient at a rate that depends on their net electric charge and on their molecular size and shape. The rate of migration of each protein is constant under identical conditions and hence can be used as a reliable character for the detection of homologous proteins. Separation of proteins can be done either on paper (paper-electrophoresis) or on a gel medium (gel-electrophoresis). The commonly employed gel media include starch, polyacrylamide and cellulose acetate.

  It has been possible to obtain useful systematic information from electrophoretic analysis of crude biological protein samples. For example :

  - Johnson and Hall (1965) have demonstrated the phylogenetic affinities in Triticinae by the process of protein electrophoresis.
  - This process also helped in establishing a close relationship between *Vicia* and *Lathyrus* (Jackson et al., 1967).
  - Interspecific variations among eight species of *Cassia* (Caesalpiniaceae) were evaluated on the basis of seed protein and mitochondrial DNA RFLP by gel electrophoresis (Mondal et al., 2000) as well as pollen protein patterns (Mondal et al., 1998).

- **Amino acid sequence studies of homologous proteins** — Comparison of amino acid sequences of homologous proteins from different taxa has also provided a powerful tool for evolutionary and systematic studies (King, 1977). The degree of similarity in the amino acid sequence is presumably proportional to the degree of genetic relationship (Sivarajan, 1984). For example, based on amino acid sequence studies, Martin and his associates have tried to trace the phylogeny of various taxa like Malvaceae, Ranunculaceae (Martin & Dowd, 1984a), Magnoliaceae, Polygonaceae (Martin & Dowd, 1984b), Myrtales (Martin & Dowd, 1986b) and some Solanaceae (Martin et al., 1986), and have elucidated the phylogeny and taxonomy of angiosperms (Martin et al., 1985; Martin & Dowd, 1986a; Martin & Jennings, 1983)

- **Analysis of isoenzymes and alloenzymes** — Electrophoresis of enzymes can reveal two distinct types of genetically controlled variation in enzyme phenotype.
    - ➢ Alloenzymes — They are different forms of a particular enzyme that are coded by a single gene locus but by more than one different allele, each coding for a slightly different amino acid sequence, which have difference in mobility during electrophoresis.
    - ➢ Isoenzymes — They are different forms of the same enzyme formed from genes at different loci.

Nowadays, a large number of such different enzymes are being exploited electrophoretically for taxonomic purposes, particularly to reconstruct phylogenies either within or between species. Most of these include enzymes involved in fundamental cellular processes such as those involved in glycolysis (Hexokinase, phosphofructokinase, aldolase, etc.), Krebs cycle (isocitrate dehydrogenase, malate dehydrogenase, etc.), enzymes involved in protecting cells from free radicals (catalase, superoxide dismutase, etc.), etc. Comparative studies of enzymes may yield data about differences in primary structure, which can be traced to evolutionary adaptations. For example, Natarella and Sink (1975) studied the peroxidases and proteins of four species of the genus *Petunia* (viz. *P. axillaris, P. inflata, P. violaceae,* and *P. parodii*) and some 11 cultivars of *P. hybrida* Hort., by electrophoresis. Electrophoretic patterns revealed that all the species are closely related and that *P. inflata, P. axillaris* and *P. violacea,* are involved, in *P. hybrida.* Similarly based on peroxidase, esterase and acid phosphatase isoenzyme pattern, Symenoidis and Tsekos (1984) suggested that the genus *Taenianthерum,* formerly considered as a part of the genus *Horedum,* should be treated as an independent genus.

(b) **Nucleic acids** — The potentiality of the huge amount of phylogenetic information comprising the base sequences of cellular nucleic acids has been recognized relatively recently. Lack of the necessary expertise and equipment, required for the nucleic acid technique, to most of the taxonomists, hindered the progress in this direction. At present this field is expanding at such a prodigious rate that it is now within the reach of most laboratories to obtain informative sequence data from a wide range of plants, and recent advances in DNA technology have created a wealth of new opportunities for taxonomy. It has now become possible to sequence any form of DNA from any plant. However, this is still a new technology, and we are only beginning to understand its problems as well as its potential.

- **Nuclear DNA and RNA** — The relative homology of DNA or RNA of various plants is useful in a taxonomic study (Mc Carthy & Bolton, 1963; Hoyer et al., 1964) and as a possible screening method for interfertility of species (Bendich & Bolton, 1967).

  The nuclei of the cells of higher plants contain relatively enormous amount of DNA ranging between 1000-10,000 megabase pairs. However, some plants like *Aesculus hippocastanum, Arabidopsis,* etc. have smaller genomes of the size in the region of 100 megabase pairs, while some lilies have DNA with almost 100,000 megabase pairs in each nucleus.

  Ribosomes, which are essential for protein synthesis, are ubiquitous and virtually abundant in the vast majority of plant cells. Their constituent RNA and proteins are encoded by multiple gene copies, making them particularly amenable for sequencing. Although ribosomes show some variation between organisms, on the whole they are

PLATE 18

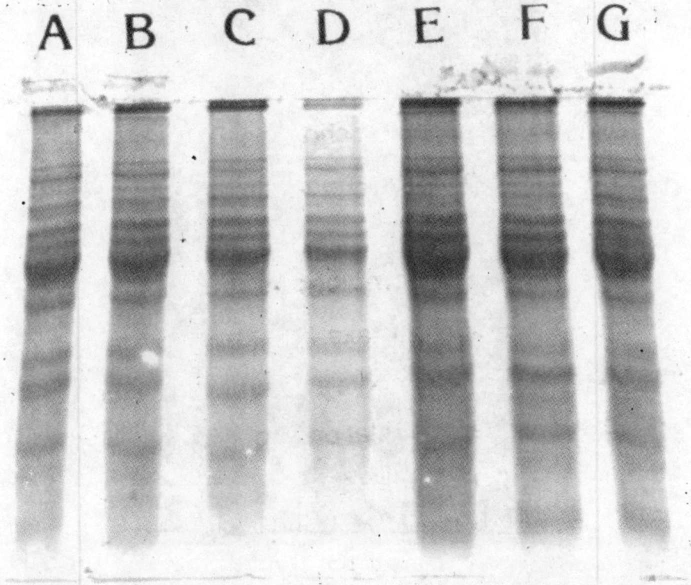

↳ The protein profile of the seeds of *Phaseolus* sp. (Fabaceae) as revealed by gel electrophoresis

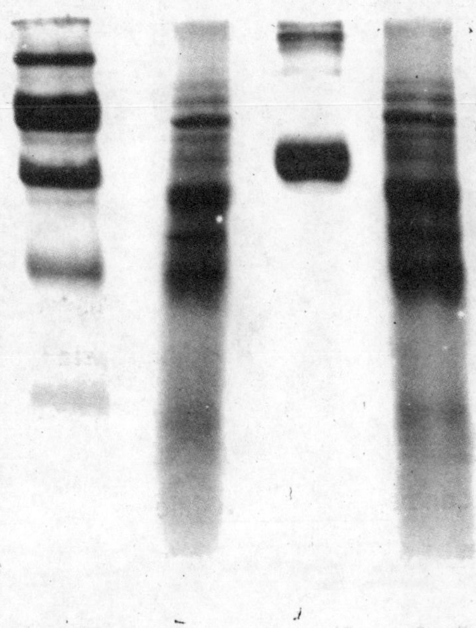

↳ The protein profile of the pollen of *Cassia siamea* (Caesalpiniaceae) as revealed by gel electrophoresis

PLATE 19

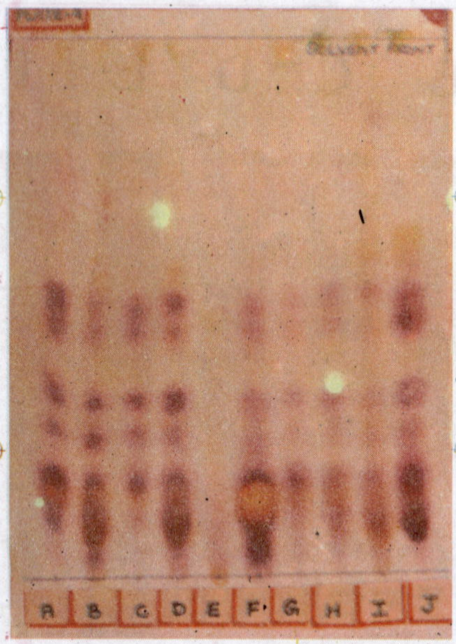

Thin Layer Chromatography (TLC) of the free amino acids of the pollen of several species of plants under the same genus showing the homology

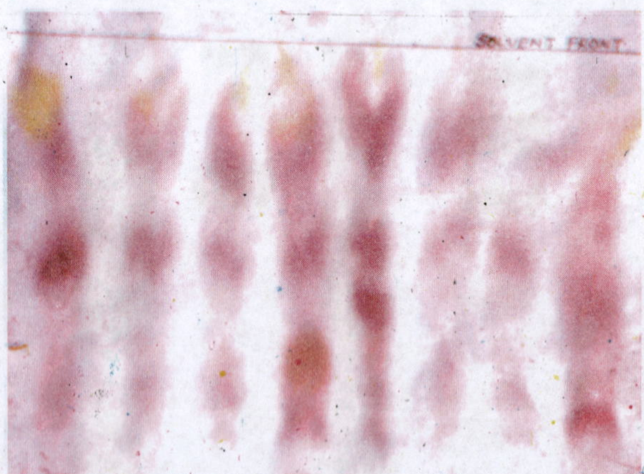

Paper Chromatography of the free amino acids of the pollen of some plants.

PLATE 20

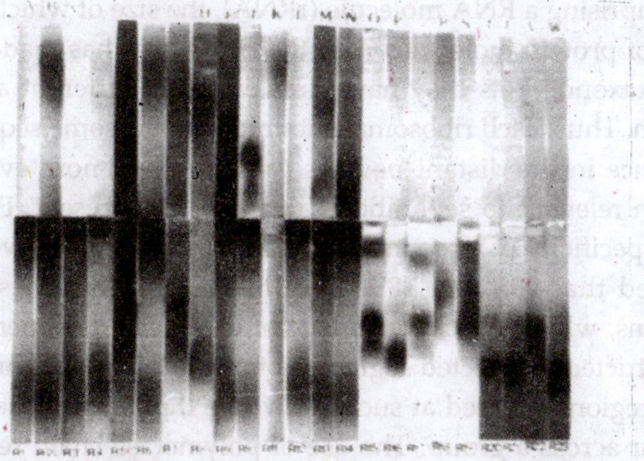

↳ Peroxidase isoenzyme profile of the leaves of some plants

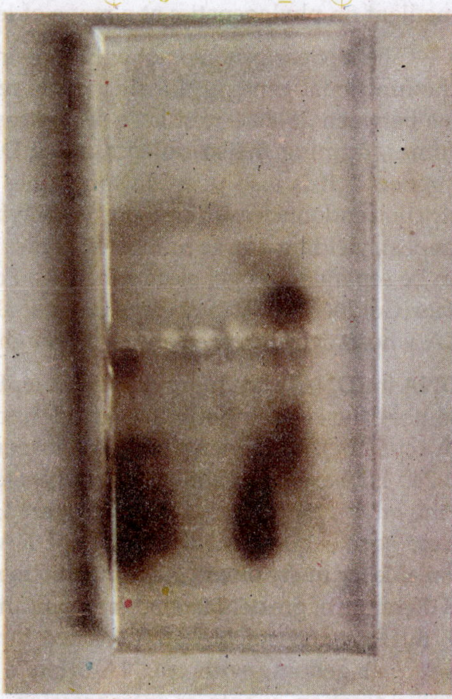

↳ Peroxidase isoenzyme profile of the leaves of some plants

very much similar in both their gross structure and in sequences, even in the most disparate members. All ribosomes are composed of a larger and a smaller subunit, each in turn comprising a RNA molecule (rRNA), the size of which, vary between plants, and a number of protein molecules. It is the rRNA that has made a considerable impact on molecular taxonomy as they have domains with different average rates of nucleotide substitution. Thus, each ribosomal subunit contains some sequence information relevant to divergence in the distant past, as well as other more evolving sequences carrying information relevant to separations almost to the present. For example, Pollard (1964) noted the specificity of ribosomal RNA in cabbage, cauliflower, celery, corn and parsnip and showed that 28S and 18S ribosomal RNA from these taxa have distinct base compositions, which are characteristic of the species. Sequencing of rRNA presently is usually restricted to selected segments of one or other subunits. This is based on the fact that some regions evolved at such a low rate that their sequences would be identical or very similar across a group. The use of polymerase chain reaction, along with suitable primers, has greatly enhanced the usefulness of rRNA sequence data at all levels of phylogenetic analysis.

Transfer RNA (tRNA), like ribosomes are ubiquitous and highly conserved and forms an integral part of the protein synthesis machinery, coding for the various amino acids. Unfortunately so far there have been few comparative studies of codon usage between higher taxa, but as the pattern of codon usage is a genetically controlled character, it might throw some light on phylogenetics at a range of levels.

- **Repetitive DNA** — The nuclear genome not only contains the largely single copy DNA that comprises the functional genes and their various regulatory units, etc., but it also typically contains between 25-80% nonsense sequences of varying length which are repeated many times, forming the so called satellite DNA, which are spread widely throughout the genome. It is also possible sometimes to isolate tandemly repeated sequences by restriction endonuclease digestion, for cleaving the DNA at the same relative position in each repeat, yielding a large number of copies of the repeat from each nuclear digest, which can be separated by gel electrophoresis. For example, studies on thirteen species of *Cucurbita* (Goldberg et al., 1972) revealed that all the species possess mainband and satellite DNA whose densities are constant throughout the genus.
- **Mitochondrial DNA (mtDNA) and chloroplast DNA (cpDNA)** — Apart from the nuclear DNA, plants also have the circular double-stranded mitochondrial DNA (mtDNA) and chloroplast DNA (cpDNA). In higher plants (angiosperms), the total size of mtDNA ranges from 100-240 kilobase pairs and therefore considerably difficult to work with. The cpDNA on the other hand are generally between 1,35,000 and 1,60,000 base pairs long. The complete gene sequences for several cpDNAs are now known and partial sequences are available for many others. As compared to mtDNA, cpDNAs are relatively conservative and therefore of greater taxonomic significance at the level of species, genera and family. Both mtDNA and cpDNA are inherited through the maternal line. Thus, comparision of nuclear genome with cpDNA offers a potentially powerful tool for revealing past hybridization events or introgression. Further, restriction fragment analysis of cpDNA can help distinguish between potentially single or multiple origins of hybrid taxa (Ogihara & Tsunewaki, 1988; Sytsma, 1990). For example, presence of two quite different chloroplast genomes in the hybrid species *Aegilops triuncalis*, indicates that it has arisen through two hybridization events, one with *A. caudate* as the female parent and one with *A. umbellulata*.

PLATE 21

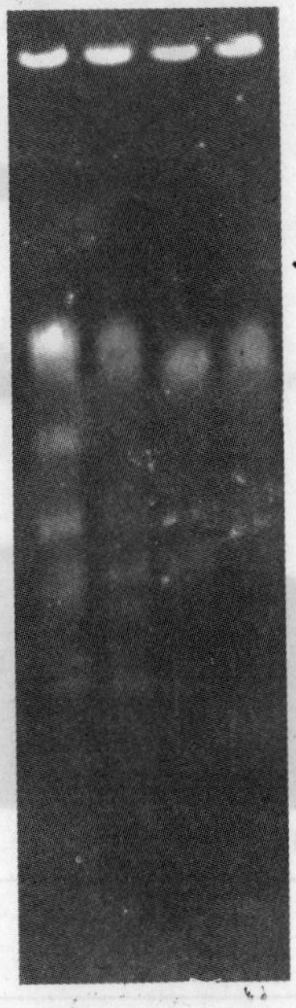

↳ Agarose gel electrophoretic profile of DNA of pollen — The first lane showing marker DNA; the last lane showing total DNA; the middle two lanes showing DNA cut with restriction enzymes

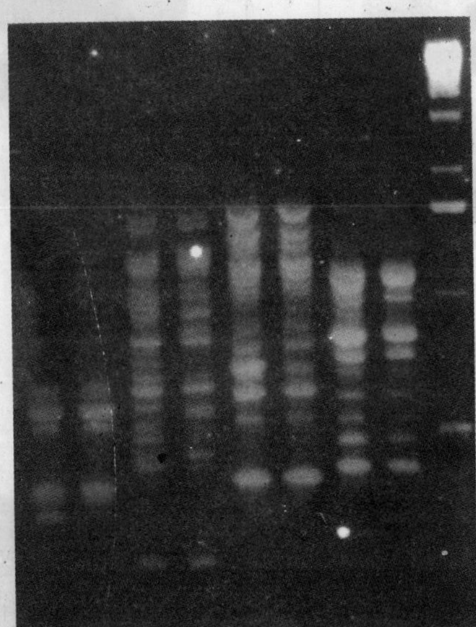

↳ The RFLP profile of the mitochondrial DNA of the seed of some species of *Cassia* cut with the restriction enzyme Eco RI

PLATE 22

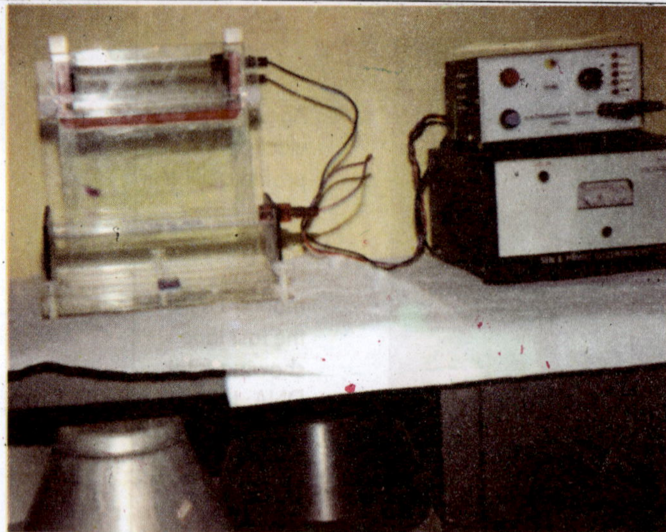

The horizontal protein gel apparatus

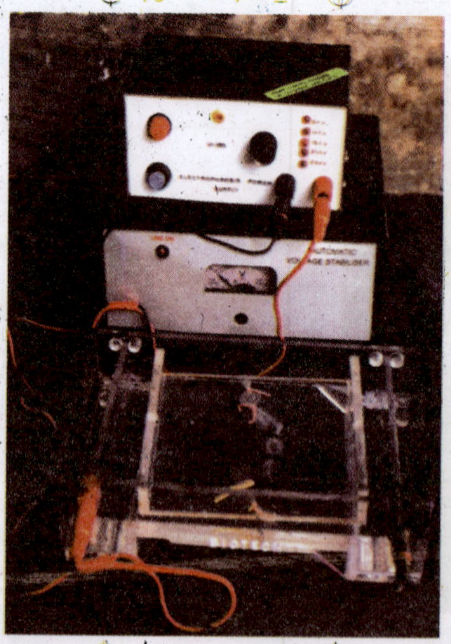

The submarine gel apparatus for DNA

Electrophoresis of DNA being performed with the DNA gel apparatus

The various techniques used for sequencing and characterizing nucleic acids are as follows:

➢ **Gel electrophoresis** — It is now possible to study DNA or RNA molecules in a routine manner by gel electrophoresis. This method is based on the principle that dissolved molecules in an electric field move at a speed determined by their charge-mass ratio, which allows the separation of molecules in a mixture according to size. Since the phosphate groups in the nucleic acids are ionized, they have a negative charge in solution and hence migrate towards a positive electrode. However, nucleic acid molecules consisting of long chains have almost identical charge-mass ratios, as each residue contributes about the same charge and mass, whatever may be the length of the chain. Therefore, little or no separation of molecules of varying lengths would occur if the electrophoresis of nucleic acids were simply carried out in solution. Hence, instead of using a liquid solution, nucleic acids are now separated by electrophoresis in an agarose gel, in which the rate at which molecules can move depends on the size of the pores. Nucleic acids with identical charge-mass ratios separate according to length, with the longer ones moving more slowly. Even very long nucleic acids, with chains containing 10,000-20,000 residues, that differ by only a few percentage points in their length can be separated. Nucleic acid mixtures containing chains of 500 nucleotides or less can be separated on polyacrylamide gels, in which each chain length can be resolved making DNA sequencing possible. The location of DNA bands within the gel can be determined directly by staining the gel with the intercalating dye ethidium bromide and very low concentrations of DNA i.e., as little as 1ng can also be detected by direct examination of the gel in ultraviolet light. Thus this technique is simple, fast, and capable of resolving mixtures of DNA fragments that cannot be separated adequately by other sizing procedures.

➢ **The Polymerase Chain Reaction (PCR)** — This method, was invented by Kary Mullis and it is an *in vitro* method for the enzymatic synthesis of specific DNA sequences using two oligonucleotide primers that are hybridized to opposite strands flanking the region of interest in the target DNA. It involves a repetitive series of cycles (Fig. 8.27), each cycle involving the steps of template denaturation, primer annealing, and the extension of the annealed primers by DNA polymerase, which results in the exponential accumulation or multiplication of a specific fragment whose termini are defined by the 5' ends of the primers. The primer extension products synthesized in one cycle in turn can serve as a template in the

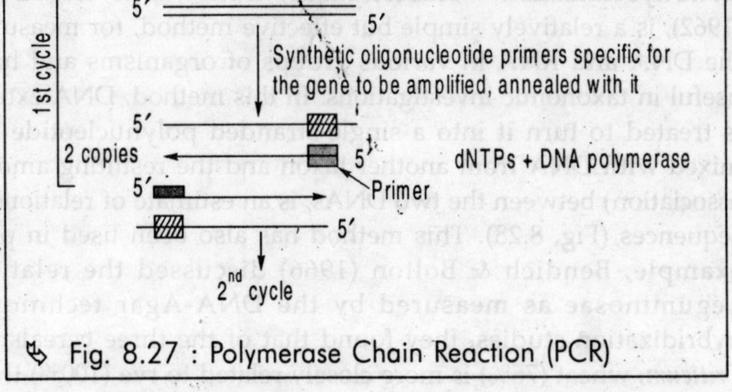

Fig. 8.27 : Polymerase Chain Reaction (PCR)

next, thus doubling the number of target DNA copies approximately at every cycle. Initially, the PCR technique used the Klenow fragment of *E. coli* DNA polymerase for extention of the annealed primers. But since this enzyme was inactivated by the high temperature required to separate the two DNA strands at the outset of each PCR cycle, the thermostable DNA polymerase (*Taq* polymerase) isolated from *Thermus aquaticus* was introduced, which transformed this technique into a simple and robust reaction which can now be automated by a thermal cycling device. The higher temperature optimum for the *Taq* polymerase allows the use of higher temperatures for primer annealing and extension, thereby increasing the overall stringency of the reaction. Various reaction parameters such as enzyme, primer, Mg^{2+} concentration, the temperature cycling, etc. are required for this process, optimal set of conditions varying for any given pair of oligonucleotide primers. PCR is thus an important technique when it comes to dealing with small samples. It is also of enormous value as it allows the amplification of particular DNA sequences without the need of cloning.

➢ **Random Amplification of Polymorphic DNA (RAPD)** — This is a recently developed technique involving PCR, in which random primer sequences are employed to amplify sections of genomic DNA of unknown function. In this method oligonucleotides are used as primers, but unlike PCR which is used to amplify a specific gene sequence, only one primer is added to the reaction mixture and it is a matter of chance as to the number of short sequences, which are bordered at either end by an inverted pair of that particular primer sequence, so that they and the intervening sequence can be amplified. Generally an RAPD amplification with 9 or 10 primers may yield 1-20 or so distinct amplified DNA sequences, which can be separated electrophoretically. RAPD has the advantage over PCR in being simpler and cheaper. This technique allows to determine relationships between closely related species, assessing kinship relationships to the production of species-specific probes, to detect hybridization and gene flow between two samples.

➢ **Restriction Fragment Length Polymorphism (RFLP)** — This technique involves the cutting of DNA molecules by specific enzymes called restriction endonucleases, which cut the DNA within their own particular recognization sequences, the length of which may vary between 4,6 and 8 bases. For example the restriction enzyme BamHI cuts between the first two Gs of its recognization sequence GGATCC. The restriction fragments are then seperated electrophoretically. The significance of restriction enzymes for phylogenetic research lies in the fact that in many plants the distribution of sites in the genome where a particular enzyme can cut is relatively fixed, generally showing variations between species or genera.

➢ **DNA hybridization** — This technique, which was developed by Bolton & McCarthy (1962), is a relatively simple but effective method, for measuring the homology of the DNA and RNA in various groups of organisms and has been shown to be useful in taxonomic investigations. In this method, DNA extracted from one plant is treated to turn it into a single-stranded polynucleotide chain, which is then mixed with DNA from another taxon and the resulting amount of annealing (re-association) between the two DNAs, is an estimate of relationship of the nucleotide sequences (Fig. 8.28). This method has also been used in plant systematics. For example, Bendich & Bolton (1966) discussed the relationships among the Leguminosae as measured by the DNA-Agar technique. Based on DNA hybridization studies, they found that of the three cereals; *Secale, Hordeum* and *Triticum*, wheat (75%) is more closely related to rye (100%) than is barley (57%).

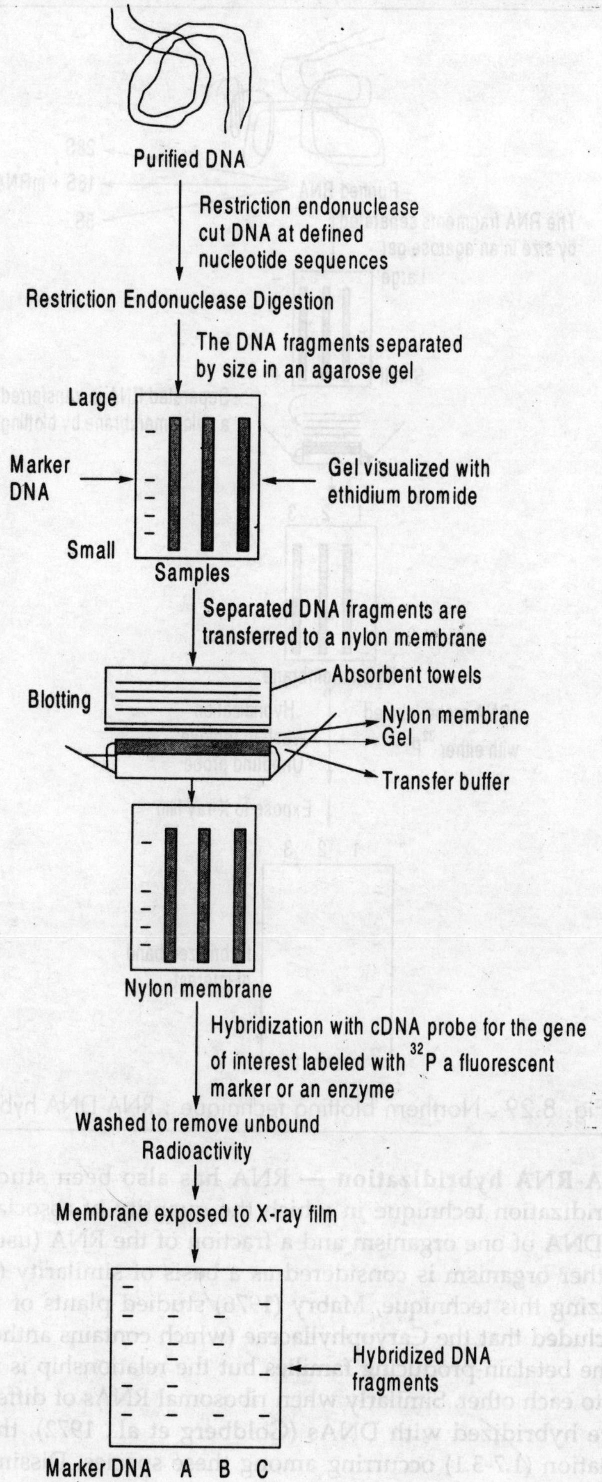

Fig. 8.28 : Southern blotting technique : DNA-DNA Hybridization analysis

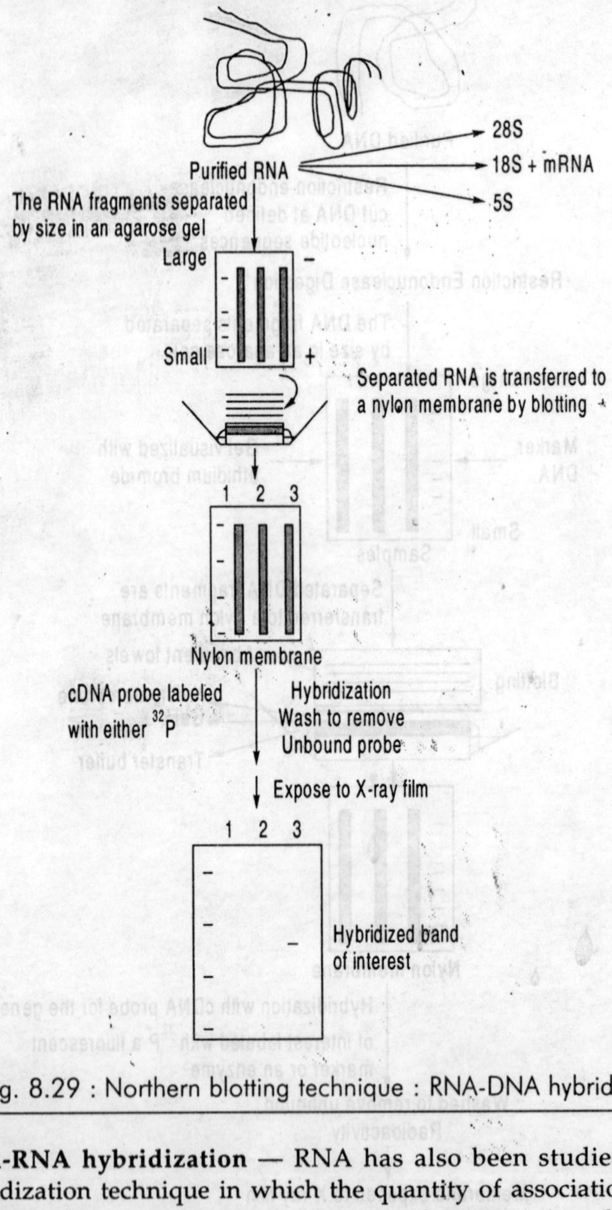

Fig. 8.29 : Northern blotting technique : RNA-DNA hybridization

➤ **DNA-RNA hybridization** — RNA has also been studied by the DNA-RNA hybridization technique in which the quantity of association occurring between the DNA of one organism and a fraction of the RNA (usually ribosomal RNA) of another organism is considered as a basis of similarity (Fig. 8.29). For example utilizing this technique, Mabry (1976) studied plants of the Centrospermae, and concluded that the Caryophyllaceae (which contains anthocyanins only) is related to the betalain-producing families but the relationship is not so close as the latter are to each other. Similarly when ribosomal RNAs of different species of *Cucurbita* were hybridized with DNAs (Goldberg et al., 1972), there was a considerable variation (1.7-3.1) occurring among these species. Dissimilar species C. *andreana* and C. *sororia* had the same proportion (1.7) of ribosomal DNA whereas similar species C. *andreana* and C. *maxima* had different proportions with 1.7 and 3.1 respectively.

(c) **Amino acids** — There are two groups of amino acids present in plant cells. They are :

- **Protein amino acids** — They are the building blocks of proteins and are released on hydrolysis of proteins. They are universally distributed in plant tissues. Amino acids present in the free form in various plant cells has also been used as a taxonomic factor in some cases. For example, Shellard and Jolliffe (1969) studied the free amino acid composition of the pollen of 11 different grass species, and found it to be similar in all the species. Generally total free amino acids are usually higher in pollen than in leaves or other plant tissues (Virtanen & Kari, 1955). Hence, free amino acid composition of pollen has been used to study the homology between various species, e.g. Asteraceae (Mondal et al., 1998), *Cassia* (Mondal et al., 1998), etc. However, it is very difficult to draw any conclusion on evolution based upon the data on free amino acid composition alone, as amino acid composition greatly varies with climatic and nutritional conditions, as well as with storage and handling patterns.

- **Non-protein amino acids** — They are not found in combination in proteins and are more numerous, estimated to be about 300 in number. The non-protein amino acids are discontinuous in distribution and less susceptible to rapid change, which increase their taxonomic value. They occur in plants, which are only distantly related. The occurrence of such compounds may be explained by the fact that different plants can probably produce the same amino acid by different biochemical routes. Several workers have studied the taxonomic potential of such amino acids. Following are a few examples of the taxonomic significance of specific non-protein amino acids present in different groups:

  ➢ Acetyl ornithine has been identified as the main non-protein amino acid in Fumarioideae under the Fumariaceae. It has also been found in ferns and grasses.

  ➢ Turner and Harborne (1967), detected canavanine only in 60 percent of the species under Lotoideae of the Leguminosae. The prevalence of canavanine was considered as an advanced character by Birdsong et al. (1960), as it was unknown from the primitive tribes Podalyrieae and Sophoreae. However, this amino acid was subsequently discovered in both the tribes.

  ➢ The amino acid azetidine-2-carboxylic acid is extremely restricted in its distribution, being present in the Agavaceae, Amaryllidaceae and Liliaceae. This amino acid is also reported from the legumes *Delonix* and *Peltophorum*.

  ➢ *Lathyrus* species could be grouped under seven infrageneric groups on the basis of the association of amino acids within the seeds (Bell, 1962). Each group is characterized by a different amino acid or group of amino acids.

  ➢ Species of the section Gummiferae of the genus *Acacia* can be recognized easily on the basis of the amino acid contents of their seeds.

  ➢ Four infrageneric groups have been recognized in the genus *Vicia* on the basis of distribution of amino acids.

  ➢ Free amino acid, lathyrine is so far known only in the genus *Lathyrus* and this supports the present circumscription of the genus.

(d) **Flavonoids** — Of all the various secondary constituents, flavonoids have been the most widely exploited phytochemical constituent in chemotaxonomic studies (Gardner, 1977; Cronquist, 1977). They are phenolic glycosides consisting of two benzene rings linked together through a heterocyclic pyrane ring. Flavonoids are further classified into various types, such as:

- Flavones — e.g. apigenin, levtolin
- Flavanones — e.g. naringenin
- Isoflavones — e.g. orobol
- Isoflavonoids — e.g. ferreirin
- Flavanols — e.g. kaempferol, quercetin
- Anthocyanidins — e.g. cyanidin, delphinidin
- Chalcone — e.g. butein
- Aurone — e.g. sulphuretin

Flavonoids are present in leaves, flowers and fruits. They function not only as insect repellents, but also play a vital role in resisting chronic pests and pathogens and by interaction with plant hormones, they regulate certain growth processes (Harborne et al., 1975). These chemically complex compounds are easily and quickly identified, have great structural variation, widespread in distribution and physiologically relatively stable, which makes them taxonomically significant. The use of flavonoids in evaluating contemporary classificatory systems has been mainly based on their distribution patterns, i.e. their presence or absence. Phytochemical taxonomists have considered them as evolutionary markers due to the many positive correlations that they display. The primitive and advanced types of flavonoids in angiosperms were, deduced by Harborne (1977). According to Swain (1975), during the course of evolution flavonoids increased in their structural complexity from the primitive glycoflavones (in *Chara*) to the proanthocyanidins (in ferns, gymnosperms and putative angiosperm progenitors). The content of flavonoids have been found to vary in different plant organs and have also been found to vary in response to light and growth (Galston, 1969). The flower pigments, which are usually anthocyanins and anthoxanthins, vary greatly and have been shown to be under genetic control (Moncrieff, 1931). It is also reported that the anthocyanin production increases with the deficiency in nutrients such as nitrogen and phosphorus. However, all variation in pigment production is not genetic in origin. In general, vegetative tissue extracts are more reliable and convenient in chemotaxonomical studies. Flavonoids from vegetative parts and seeds provide more reliable taxonomic evidence than flower pigments, which are too variable.

Many phylogenetic inferences have, been based on these characters in various groups. Following are a few examples:

- The flavonoid systematics of the genus *Perideridia* (Umbelliferae), was studied by Giannasi and Chuang (1976), who found that although the 16 species studied by them had the same flavonoids, they could be categorized into three discrete groups. They are:
  - ➢ those producing only flavonols.
  - ➢ those with flavonols mainly and a few flavones.
  - ➢ those which produce flavones predominantly.

  They also tried to correlate between these three different types and their respective geographical distributions in America. The flavonol-producing group was centered in California and the second and third groups in the Pacific North West and eastward to mid-western USA.

- An extensive survey of the families Umbelliferae and Araliaceae by Hegnauer (1971) revealed that, the two families are closely related. Hegnauer also concluded that they may be regarded as constituting the order Urabellales, which in turn may have evolved from a Rutalean stock.

- Analysis of flavonoids of leaf of Liliaceae, Juncaceae, Cyperaceae and Poaceae by Williams (1976) suggest that all these families have arisen from Liliaceous ancestors.
- The flavonoid pattern of 5 species of *Tragopogon* of Asteraceae, and their F_1 and F_2 hybrids were studied, by Belzer and Ownbey (1971), who noted that all the species and their population showed different chromatographical results.
- Flavonoids have been used in the estimation of generic affinities in Ulmaceae (Giannasi, 1978a) and of species relationships in *Chenopodium* (Crawford & Evans, 1978).
- The evolutionary pathways in many taxa, e.g. Lemnaceae (Zennie & McClure, 1977) and *Ruellia* (Bloom, 1976) have been traced based on the data of flavonoids.
- Flavonoid chemistry has been used to distinguish species of *Spirodella*, which cannot be distinguished on the basis of their morphology.

(e) **Betalins** — Betalins differ from flavonoids and other phenolic compounds in that they contain nitrogen in them. They are however functionally equivalent to phenolics. The betalins are popularly known as nitrogenous anthocyanins and comprise of:

- Red to violet betacyanins.
- Yellow betaxanthins.

Plants containing betalin do not contain anthocyanins, which are the normal pigments found in other angiosperm families. The two groups are unrelated both chemically and biosynthetically.

The distribution of betalin has proved to be a notable chemotaxonomic criteria. Some examples of the taxonomic value of betalins are mentioned below :

- The betalins are confined to ten families of angiosperms (which contain only betalin but no anthocyanin) i.e., Chenopodiaceae, Portulacaceae, Aizoaceae, Cactaceae, Nyctaginaceae, Phytolaccaceae, Stegnospermaceae, Basellaceae, Amaranthaceae and Didieraceae, which have been placed under a single order Centrospermae.
- Centrospermae also includes two anthocyanin-containing families, Molluginaceae and Caryophyllaceae. Mabry et al. (1963) had suggested the transfer of the anthocyanin-containing families, Molluginaceae and Caryophyllaceae, to a distinct order Caryophyllales, restricting the Centrospermae to betalain-containing groups alone. However, other workers did not support this suggestion. Mabry again re-examined his earlier view in 1977, and concluded that the Centrospermae including both betalain and anthocyanin-containing families, is a monophyletic group, and suggested the recognition of two Centrospermous suborders, based on the presence of the respective pigments.
- The systematic position of the family Cactaceae under Centrospermae, which contain betalin has been a matter of dispute in the past. It was often placed in the order of its own i.e., the Cactales or Opuntiales. However the presence of betalin in the Cactaceae establishes its position in the Centrospermae.
- The genus *Gisekia*, based on morphological similarities was traditionally placed under Molluginaceae. However, it has been found to be anomalous in that family because it contains betalains instead of anthocyanins. Takhtajan (1980) has recently transferred this genus to Phytolaccaceae, where it seems to belong more naturally than in any other family of Centrospermae.
- Mabry and Behnke (1976) found that the genus *Dysphania*, assigned to various families (Chenopodiaceae, Caryophyllaceae and Illecebraceae) by various authors, possessed

betalain pigments and sieve tube plastids, characteristic of the Centrospermae. They have strongly supported its allocation to the Chenopodiaceae.

(f) **Alkaloids** — Alkaloids are a heterogeneous group of organic nitrogen containing bases, often with a heterocyclic ring. The true alkaloids have a nitrogen-containing heterocyclic nucleus derived from a biogenetic amine and they can be related structurally to parent bases such as isoquinoline, pyridine, piperidine and tropane. They are by-products of plant metabolism and are synthesized from amino acids or their derivatives. Three main categories of alkaloids are distinguished:

- **True alkaloids** — They contain a nitrogen-containing heterocyclic nucleus derived from a biogenetic amine.
- **Protoalkaloids** — They are derived from amino acids but lack any heterocyclic ring.
- **Pseudoalkaloids** — They are biogenetically unrelated to amino acids, and are derived from terpenes, sterols, aliphatic acids, nicotinic acids or purines.

They are distributed throughout the plant tissues, and are present in the vacuoles in the form of salts. However they are not essential for the growth of plant, but have striking properties of affecting the nervous system of animals. They are well known for their medicinal, poisonous and systematic viewpoints.

Alkaloids are not universally accumulated by plants, but are unexpectedly widely distributed in flowering plants and ferns. They are somewhat characteristic of particular families. According to Manske (1950), there are about 38-39 families that can be regarded as alkaloid-containing families. On the other hand, Hegnauer (1963) has suggested that none of the larger families is free from alkaloids. Over 5000 alkaloids have been reported from angiosperms mostly from the Dicotyledons, as for example:

- Families like Berberidaceae, Fabaceae, Ranunculaceae and Solanaceae are especially rich in alkaloidal species. Other families include Annonaceae, Fumariaceae, Hydrastidaceae, Menispermaceae, etc.
- The members of the Papaveraceae synthesize isoquinoline alkaloids always including protopine, those of the Fabaceae lupin alkaloids, while Solanaceae have tropane derivatives. Asteraceae and Poaceae, on the other hand, produce many different types of alkaloids (Hegnauer, 1966).
- Sometimes alkaloids have a very narrow distribution. For example morphine is restricted to *Papaver somniferum*, coniine to a few Apiaceae and strichnine to a few species of *Strychnos.*

Alkaloid content can be considered as a source of taxonomic evidence as alkaloids characterizing species of a particular taxon are frequently of the same chemical or biogenetic group. This suggests that related plants share the same pathways of alkaloid synthesis. Therefore, alkaloid content can be considered as a source of taxonomic evidence. However, mere accumulation of alkaloid does not imply taxonomic relationship at family level (Smith, 1976). Some of the taxonomic aspects of alkaloids are mentioned below:

- Distribution of alkaloid has proved useful in the taxonomy of the Fabaceae. Of the three genera viz. *Genista, Ammodendron* and *Adenocarpus* under this family, *Genista* and *Adenocarpus* were included in the tribe Genisteae, whereas *Ammodendron* was placed in the tribe Sophoreae, which is characterized by the presence of matrine alkaloids. However, phytochemical studies have shown that all the three genera contain ammodendrine-hystrine alkaloids, which suggests that *Ammodendron* should be transferred to the tribe Genisteae.

- Lupin alkaloids have been found in three tribes under the subfamily Lotoideae of the Fabaceae viz. Sophoreae, Genisteae and Podalyrieae, which suggest that these tribes may have originated from a common ancestral stock.
- According to Jones and Luchsinger (1987), alkaloids are useful in taxonomic studies in *Papaver* and *Argemone* (Papaveraceae), *Veratrum* (Liliaceae), as well as in *Lycopodium*, *Lupinus* and Caryophyllales.
- The presence of isoquinoline alkaloids in the families Fumariaceae and Papaveraccae indicates very close relationship between the two families.
- On the basis of the distribution of benzylisoquinoline alkaloids, several modem taxonomists, including Takhtajan, Cronquist, Thorne and Dahlgren, rearranged the families with apocarpous gynoecium in Magnoliidae.

(g) **Terpenoids :** These are a biogenic group of volatile compounds, which are mostly polymerized isoprene residues (isoprene unit -2-methyl 1,3 butadiene). According to the number of isoprene units present in a terpenoid molecule, they may be of following categories:

- **Hemiterpenes** (single isoprene unit) — e.g. tiglic acid
- **Monoterpenes** (two isoprene units) — e.g. menthol
- **Sesquiterpenes** (three isoprene units) — e.g. farnesol
- **Diterpenes** (four isoprene units) — e.g. phytol
- **Triterpenes** (six isoprene units) — e.g. squalene
- **Tetraterpenes** (eight isoprene units) — e.g. carotenoid
- **Polyterpenes** – e.g. rubber

Terpenoids are almost universal, though not as widespread as flavonoids and have been used extensively in the chemotaxonomy of mints, umbellifers, *Citrus* plants, and gymnosperms. Unlike flavonoids, which are more difficult to work with at the technical level, monoterpenes can be rapidly identified by combined gas-liquid chromatography (GLC) and mass spectrometry. Some specific terpenoids are found in certain families, e.g. sesquiterpene lactones in Asteraceae, cucurbitacins in Cucurbitaceae and asperuloside in Rubiaceae.

Some of the chemotaxonomic applications of terpenes are mentioned below:

- Sesquiterpene lactones are common in the family Asteraceae, in which the oxidation level of these sesquiterpenes is sometimes specific to a tribe, subtribe or even a genus. The absence of sesquiterpenes in the Astereae may be rare. It has been found that if *Ambrosia* and its allies viz */va, Franseria* and *Xanthium* are removed from the Heliantheae, then the residual elements are similar to the Helenieae in terms of lactone distribution. Thus, there is a need to construct a separate tribe or a distinct family to accommodate allied forms of *Ambrosia,* which exhibit diversity in lactone production.
- Geographical races of *Pseudotsuga menziesii* were distinguished based on the terpenoids of their conical oleoresins.
- The classification of *Eucalyptus* has been very difficult on the basis of gross morphology. Baker and Smith (1920) divided this genus into larger groups containing different oil-combination, which they correlated with leaf venation and bark structure:
  - ➢ Butterfly venation, shows high yield, with phellandrene and piperitone as main components.
  - ➢ Obtuse feather venation, shows a low yield, with pinene as the dominant compound.

> More acute venation, having marginal veins, shows a slightly higher oil yield, with cineole and pinene as chief constituents.

- Origin of certain *Citrus* cultivars was determined by studying their rind and leaf terpenoid pattern.
- A survey of monoterpenes in 19 species of *Salvia* by Emboden and Lewis (1967), indicates that terpene composition is a valid morphological trait in the analysis of introgression within a group.
- Gum terpentines, have been used by Mirov (l96l) in the taxonomy of *Pinus*.
- Irwin (1968) found that there are three sympatric taxa of *Hedeoma* (Labiatae) coexisting without any sign of natural hybridization, and used their terpenoids in identifying them.

(h) **Iridoid compounds** — They are monoterpenoid cyclopentanoid lactones, which represent a separate class of taxonomically significant compounds. The distribution of iridoid compounds has attracted the attention of plant taxonomists as a character of systematic importance in recent times and is considered to be an evolutionary marker. The significance of the monoterpenoid cyclopentanoid lactones called iridoids, has been reviewed by Bate-Smith and Swain (1966). Among these compounds, asperuloside is particularly common in the Rubiaceae. Aucubin is frequently noted in the Cornaceae, Scrophulariaceae, Orobanchaceae and some closely related families. Some of the aspects of their taxonomic value are mentioned below:

- On the basis of iridoid distribution, some taxonomic changes have been suggested in the 12th edition of Engler's syllabus. The aucubin-containing genus *Buddlela* must be removed from the Loganiaceae to the Buddleiaceae, with a position near the Scrophulariaceae.
- Removal of Garryaceae from apetalous orders to the vicinity of the Cornaceae is suggested in Engler's syllabus on the basis of morphological evidence and this treatment receives support from chemical evidence as both Garryaceae and Cornaceae produce acubin.
- An extensive survey of the iridoids among angiosperms, was made by Jensen et al. (1975), who discussed their systematic importance. From their studies, they concluded that the sympetalous subclass Asteridae, is not natural as defined by Takhtajan (1969), but polyphyletic. The orders Dipsacales, Gentianales and Scrophulariales (excluding Solanales) are characterized by the presence of iridoids while the others lack it. They have proposed the splitting of Scrophulariales and also the separation of Goodeniales from Campanulales and the Loasales from Violales. All these studies thus, suggest a double or even multiple ancestry for the Sympetalae (Jensen et al., 1975), which has to be rejected as an unnatural group.
- Kubitzki (1969) connected the Rosalean and Guttiferalian complexes on the basis of the presence of iridoid compounds. According to Cronquist (1968b) both these groups came from Magnoliidae, which, however, is completely devoid of iridoids.
- Meeuse (1970) pointed to the presence of iridoids in some taxa of the traditional Rosiflorae and their absence from others to support his concept of the polyphyletic origin of angiosperms.
- The restriction of iridoid compounds mainly to groups with unitegmic ovules, suggests that they developed along an evolutionary line where the ovules were just about to evolve from bitegmic to unitegmic condition. According to Jensen et al. (1975), Altingiaceae and Daphniphyllaccae (Hamamelidales) are perhaps relicts of primitive iridoid-bearing groups with bitegmic ovules.

- Studies on iridoids in correlation with other traditional taxonomic characters of the monogeneric Fouquieriaceae and related orders Tamaricales (in which the family is usually included), Ericales, Cornales and Solanales by Dahlgren et al. (1976), suggests that the Fouquieriaceae is quite distinct and that it should be placed in an order of its own.
- A minor modification in Takhtajan's classification can lump all iridoid-producing orders into a single group, indicating a possible common origin.

(i) **Oils, fats and waxes** — Along with the proteins and carbohydrates, lipids or fats form the bulk of the organic matter of plant tissue and are therefore a potential source of taxonomic evidence. Lipids are the esters of fatty acids with glycerol, and are mostly made up of carbon, hydrogen and oxygen (simple lipids). The fatty acids present may be saturated or unsaturated. The greater the proportion of saturated fatty acids in a lipid, the higher is the melting point. Saturation is usually measured by the iodine number, i.e. the number of grams of iodine absorbed by a hundred grams of fat, oil or fatty acid. Some lipids may contain nitrogen, a carbohydrate group, phosphorus or some other group, in addition to carbon, hydrogen and oxygen. They are called conjugated or complex lipids. The waxes on the other hand are esters of long-chain alcohols with long-chain fatty acids and may contain free alcohols, free fatty acids, aldehydes, ketones or hydrocarbons. Lipids and waxes are somewhat heterogenous group and are completely or partially soluble in organic solvents such as ethanol, ether or chloroform. Fats and oils can be distinguished by their physical state at normal temperatures i.e., fats are solids while oils are liquids. Lipids are found in all parts of plants but are dominant in the storage organs, seeds and fruits, forming droplets suspended in the cytoplasm. Waxes occur in the cuticular layers of plants.

The taxonomic significance of lipids was assessed for the first time, by McNair (1929). According to him, the iodine number of lipids was higher (i.e. lipids are more saturated) in more advanced plant groups. However, since the iodine number is a rather unstable character, it must be used carefully. Fatty acids specific to particular plant groups may be of taxonomic significance. For example, ximenyric acid is found in the Olacaceae and Santalaceae, petroselinic acid is almost completely restricted to the Umbelliferae and erucic acid to Cruciferae. Certain species of the Flacourtiaceae are characterized by the presence of chaulmoogric acid and some related acids. Bacterial lipids contain fatty acids, which are not found in other plants. Similarly, the highly evolved Asteraceae do not contain lipid with unsaturated fatty acids, while the algae, possess very unsaturated lipids. Thus, keeping in mind the fact that lipids are not constant in absolute composition, characters relating to their intact structure would not be of any value in taxonomy.

Some of the taxonomic applications of lipids and waxes are undermentioned:

- The fatty acid composition of lipids has been used by Shorland (1963) as a possible source of taxonomic importance, and on the basis of the major fatty acids released by hydrolysis of their lipids, he recognized groups of families containing:
  - ➢ Linolenic acid - rich seed fats.
  - ➢ Linoleic acid - rich / oleic acid - rich seed fats.
  - ➢ Linoleic acid and oleic acid rich seed fats rich, with linolenic acid or a conjugated polyethanoic acid as principal component.
  - ➢ Seed fats rich in palmitic, oleic and linoleic acids.
  - ➢ Seed fats with other characteristic acids in addition to oleic, linoleic and palmitic acids.

However, he found that, almost all fatty acids are present in all groups at low concentrations, but oleic acid is the main fatty acid in most groups. Simultaneously, distantly related families sometimes occur under the same group and a family may appear in more than one group. However, the family Palmae is an exception, where the proportions of fatty acids in lipids are constant in both species and genera (Hilditch, 1952).

- The presence of unusual fatty acids in groups has also been of some taxonomic significance. For example :
  - ➢ Linolenic acid and the unusual fatty acid, octadecatetranoic acid, are present in the lipids of both leaf and seeds of eight members of the Boraginaceae. At the same time, the same unusual fatty acids are present in ten species of the unrelated Caryophyllaceae.
  - ➢ Capric acid, which is present in *Ulmus*, is also dominant in seed lipids of *Zeikova* (Ulmaceae) as well as members of the Lauraceae and Lythraceae.
  - ➢ Malvalic acid is present in members of the Malvaceae.
- Proportions of fatty acids in lipids are constant in both genera and species in Palmae (Hilditch, 1952).
- The *Eucalyptus* species of the Australian arid zone form a homogeneous group on the basis of wax characters.
- Wax alkanes also appear to be attractive taxonomic characters. However according to Martin-Smith et al. (1967), a systematic investigation into the possible influence of season, climate, geographical distribution and the kind and age of organs on the composition of plant surface waxes is essential before the method can be accepted.
  - ➢ Alkane variation in cuticular waxes serves as a useful source of taxonomic information in the Poaceae.
  - ➢ There is a general uniformity in proportions of alkane hydrocarbon constituents within the genera of the Crassulaceae.
  - ➢ Wax alkane variation can be related to morphology, cytogenetics and protein analysis in 22 tuber-bearing species of *Solanum*.

(j) **Steroids** — Steroids may be considered as derivatives of a fused and fully saturated ring system called cyclopentanoperhydrophenanthrene or sterane, which is formed by the fusion of 3 cyclohexane rings in non-linear or phenanthrene manner and a terminal cyclopentane ring. True steroids possess two methyl groups and are mostly alcohols or esters. In plants they serve the role of water-proof, being located in the plant cutins. Steroids have also proved to be of some taxonomic significance in some taxa. As for example, their distribution has proved helpful in the taxonomy of the genera of tribe Veratreae of family Liliaceae. According to Kupchan et al. (1961), these genera contain the steroid veratum.

(k) **Polysaccharides** — Polysaccharides perhaps offer the greatest hope for taxonomic evidence because of their complexity and diversity. But so far they have been examined from a systematic viewpoint only, particularly due to the difficulties in the procedures for isolation and fractionation of these compounds.

Following are some of the examples of the use of polysaccharides in taxonomic evaluation:

- MacLeod and McCorquodale (1958) studied the water-soluble polysaccharides in the seeds of 22 grass species, and observed considerable taxonomic variations. Lesser or greater amounts of β-glucosan was present in all the Festuceae, but *Festuca* and *Lolium* are distinctive in having an unusual trisaccharide. On the other hand, due to the presence of fructosans and absence of raffinose, the Bromeae forms a very natural tribe.

- The sugar alcohols have been studied by Plouvier (1963), from a taxonomic standpoint. He found that:
  - ➢ Quercitol is exceptionally common in the Menispermaceae, although it is found elsewhere. Though it is present in all 35 species of *Quercus*, it is absent from *Castanea* and *Fagus* of the Fagaceae.
  - ➢ Sorbitol distribution is in accordance with the taxonomy of the Rosaceae, supporting the transfer of *Ulmaria* to the Rosoideae (usually without sorbitol) from the Spiraeoideae (which contains genera with sorbitol).
  - ➢ Similarly, Pinitol is widespread but is particularly common in the Caryophyllaceae. It is found only in the genera *Magnolia* under the Magnoliaceae, but absent in closely related families.
- Among storage polysaccharides, the distribution of seed amyloids, which consist of a principal chain of glucose with side-chains of galactose and xylose, has also proved to be of some taxonomic significance.
  - ➢ Amyloid was detected in 16 families of the Dicotyledons and none of the Monocotyledons by Kooiman (1960) in a test involving over 2,500 species.
  - ➢ Structures of these compounds have been fully worked for *Annona muricata* and *Tamarindus indicus*, being basically similar.
  - ➢ Although Leguminosae and Acanthaceae are not closely related, yet both of them contain amyloid and the amyloid distribution follows taxonomic lines within these two families.
  - ➢ *Paewia* is the only amyloid-producing genus among the 30 tested genus from Ranunculaceae, although some taxonomists would prefer to transfer it to a monotypic family.
  - ➢ Amyloid is present in a few species of the Lotoideae.
  - ➢ Amyloid is found in many genera of the Caesalpinioideae. However, all compounds are restricted to the Amherstieae, Cynometreae and Sclerobiae.
- Linear and branched hemicellulose fractions of species of the Leguminosae and Poaceae were compared by Gaillard (1965). He found prominent differences between the two families. While branched polymers of Leguminosae contain a high proportion of arabinose, galactose and uronic acid, those of Poaceae have comparatively more of xylose. Uronic acid is attached to arabinose in the legume polymer and to xylose in the grass polymer.

(l) **Ellagitannins** — The systematic importance of ellagitannins have been focussed by Bate-Smith (1972), Harborne (1973) and Spome (1975). Extensive surveys of the flowering plants have revealed that these substances are exclusive to the Dicotyledons, and absent from the Monocotyledons. They have indicated a fundamental cleavage between the Magnoliidae, which lack ellagitannins and the Hamamelidae-Dilleniidae, which possess them, which in turn probably points to the origin of the Monocotyledons from a Magnolian stock.

(m) **Cyanogenic compounds** — Cyanogenic compounds are poisonous compounds produced by plants after injury to their cells, such as hydrocyanic acid, amygdalin, etc. The ability of these plants to release such poisonous compounds is called cyanogenesis and plants containing these compounds are called **cyanophoric plants**. Hegnauer (1977) defined the term **cyanogenesis** as the ability of certain plants to release hydrocyanic acid (Prussic acid) after injury of cells. Cyanophoric plants usually contain one or several cyanogenic glycosides. Most cyanogenic activity is located in leaves and seeds of plants. Cyanogenic glucosides are known to be transported from one tissue in a plant to another e.g. in cassava from the young leaves to the tubers. In the cell, cyanogenic glucosides are thought to be stored in the vacuole. The same plants also contain degradative enzymes, which upon cellular disruption of the plant tissue get in contact with the cyanogenic glucosides thereby causing the rapid release of hydrogen cyanide. This binary system - two sets of components, which are inert individually, comprises the "cyanide bomb" and plays a role in the chemical warfare of plants against herbivores, pests and pathogens. Cyanide, or prussic acid, is a naturally occurring glycoside in certain plants. Corn, sudan grass, sorghums, cherry, apple, and peach may accumulate large quantities of cyanide. This accumulation occurs when plants are injured or drought stressed, which can result in release of large amounts of cyanide.

It was Bohm, who first detected HCN and amygdalin (the first plant glycoside) in seeds of *Prunus amygdalus* (Rosaceae) in the early 1800s. Till date about 2,056 species of vascular plants are known to be cyanophoric (Nahrstedt, 1985), including ferns, monocots and dicots. The taxa belonging to Araceae, Poaceae, Juncaceae, Juncaginaceae and Scheuchzeriaceae are common. Several dicotyledonous families belonging to Asteridae, Rosidae and Dilleniidae are also cyanogenic. In angiosperms these compounds have been found to occur erratically.

The most frequently occurring cyanogenic glycosides fall into five groups:

- The prunasin group
- The taxiphyllin group
- The linamarin group
- The gynocardin group
- The cyanolipid group

The type of biogenesis of cyanophoric compounds has been found to differ in the major groups of tracheophytes viz. Pteridophytes, Gymnosperms and Angiosperms

- Phenylalanine acts as the precursor of cyanophoric compounds in the pteridophytes.
- In Gymnosperms, tyrosine is the only precursor.
- Angiosperms have valine, leucine and cyclopentenyl glycine, as well as phenylalanine and tyrosine as the precursors of cyanophoric compounds.

According to Hegnauer (1977), cyanogenesis, may prove to be of considerable value in the classification of plants, if adequately and carefully used as a character in plant systematics, even at the higher systematic levels. More than one biosynthetic group of cynophoric compounds occurs only in very large genera or families, and in such cases cyanogenic compounds are of systematic value at subfamily levels (Hegnauer, 1977).

Following are a few examples of the role of cyanogenic compounds in plant systematics:

- The theory that the Liliopsida (Monocotyledons) evolved from ancestors resembling present day Magnoliidae is well supported by the study of the distribution of individual cyanogenic compounds in the tracheophytes. In both these groups, cyanogenesis proceeds in exactly the same way, i.e., with only tyrosine as the precursor (Hegnauer, 1973; van Valen, F., 1978a).
- The tribe Calendulae of Asteraceae and Trifolieae and Phaseoleae of Fabaceae, are characterized by the presence of the cyanogenic compound linamarin, while amygdalin

or prunasin are characteristic of the many cynophoric taxa of Rosaceae having a basic chromosome number 9 (Hegnauer, 1977).

- The taxonomic evaluation of cyanogenesis in vascular plants based on chemical characterization of the cyanogenic constituents, attributable to distinct biosynthetic pathways, is much more meaningful for plant systematics than cyanogenesis itself (Fat, 1977). For example, the family Poaceae is biochemically homogeneous with regard to cyanogenesis, as although the subfamilies of Poaceae are characterized by three different cyanogenic compounds, namely dhurrin (Andropogonoideae), triglochinin (Festucoideae and Eragrostioideae) and taxiphyllin (Bambusoideae) (van Valen, F., 1978b), all are derived from tyrosine, and these compounds only differ in the position of attachment of glucose and not in biogenesis.
- In certain cases, cyanogenesis is however quite variable. For example:
  - ➢ The genus *Glyceria* comprises species, which may be cyanogenic, facultatively cyanogenic or non-cyanogenic.
  - ➢ Extreme polymorphism has been revealed in the cyanogenesis of 12 species of *Lotus* (Leguminosae) from Israel. Some species are cyanogenic and others acyanogenic, while still others contained both cyanogenic and acyanogenic populations (Band et al., 1981).

8.7 Serology in Relation to Taxonomy

Serology is defined as that portion of biology, which is concerned with the nature and interactions of antigenic material and antibodies (Boyden, 1964). Smith (1976) defined it as "the study of the origins and properties of antisera." When foreign cells or particles (antigens) are introduced into an organism, antibodies are produced in the blood (antiserum). The substance capable of stimulating the formation of an antibody is called **antigen** and the highly specific protein molecule produced by plasma cells in the immune system in response to the antigen is called **antibody**. Proteins most widely used as antigens in serotaxonomy are those, which carry useful taxonomic information and are easy to handle. Both structural and reserve proteins can be used in the field of systematics, as long as they belong to the same group and the same organs are always compared. Generally, storage proteins are most amenable for taxonomic studies followed by pollen proteins. Stem tubers, algal cells, fern spores, fruits and leaves can also be employed as satisfactory antigenic material for systematic investigations.

8.7.1 Brief history of serotaxonomy

Phytoserology, which deals with immunochemical reactions, between serum antibodies and antigens (Knoz, 1971), has also established itself as a valid method in systematics (Frohne & Jensen, 1984; Moritz & Jensen, 1961), because it helps to detect homologous proteins. It uses the specific properties of antisera produced by animals against plant proteins as characters to assess plant relationships. Serotaxonomy developed and became popular in Germany, which has been an active center since the beginning of this century. Nuttal (1901) was the first biologist to compare the immunochemical specificity of serum proteins for systematic purposes. Kowarski (1901), Bertarelli (1902) and Magnus (1908) were the other early notable serologists, who compared proteins from various grass and legume species, showing similarities and differences. Their experiments gave phytoserology a hopeful start, but unfortunately due to inadequate methods and extraordinary claims of the Königsberg school, founded in 1914 and headed by Gohlke, spoiled this momentum. Based largely on serological studies, Mez and Ziegenspeck (1926) produced a "Stammbaum" or "family tree" for the whole plant kingdom. But the results of the Köenigsberg school was unaccepted by another school, that of Gilg and Schurhoff, in Berlin. As a result of this conflict, there was a decline of serology in Germany. Later, Otto Moritz gave a critical new start to phytoserology in the 1950s, and plant serology has now been established as a valid method in systematics (Frohne & Jensen, 1984). Rives (1923), Nelson and Biskeland (1928) and Moritz (1933) are some of the other early workers, who have made significant contribution to the application of serology to systematics. The role of the school of serology at Rutgers University, New Jersey, headed by Boyden, has also been very significant in the theoretical and technical progress of serotaxonomy in recent years.

8.7.2 Methods used in serotaxonomy

In this method, a crude protein extract of a particular plant taxa (antigen), is injected into the blood stream of an experimental animal, usually a rabbit or a rat, to develop antibodies. In response to the specific antigen injected, a specific antibody is produced in the blood of the animal. The serum (termed the antiserum) containing the antibody is then collected and made to react *in vitro* with the antigenic proteins as well as proteins from other related taxa, of which the affinities are in question. Serological reactions between antibodies and antigenic material results in the formation of a precipitate. This is called **precipitin reaction.** Kraus (1897) showed that this reaction indicates similarity of antigens. The degree of protein homology is determined by the amount of precipitation and hence is taken as a phylogenetic marker and taxonomic character.

However, according to Moritz and Jensen (1961), crude protein extracts should not be used as antigens, as has been the normal practice in serotaxonomic work. This is because, crude protein extracts contain a large number of proteins, which stimulates the production of a vast range of antibodies, which differ in their specificity and reactivity. Some are produced in abundance while others are hardly detectable. Further, each protein in turn carries a number of **antigenic determinant** sites (the portion of the antibody molecule that reacts with a portion of the antigenic material) or **epitopes,** each of which is capable of eliciting the production of distinct antibody molecules with a specific antibody determinant site or **paratope**, specific to that epitope. To overcome these problems and to make highly refined serotaxonomic comparisons possible, Lester (1984) has developed a novel method of pre-absorption i.e., an antibody system induced by immunization with a crude protein extract (the antigen system) of one species is 'pre-absorbed' by the antigen system of a second species and then tested with the antigen systems of other species. The recent development of powerful analytical techniques and instrumentation, has also made possible the use of monoclonal antibodies, i.e. single kinds or species of antibody directed against single epitopes, although such techniques may not be of great importance in taxonomic studies.

Several serological methods are now available. The most common methods used are as follows:

(a) **Immunodiffusion in agarose gels** — In this method the antigen-antibody reaction is carried out in gels, mostly of agarose, in petridishes. The antiserum containing antibodies is filled in a well at the center of the gel and the antigens from related taxa are placed in outer or radial wells. The antigen and antibody react to produce the insoluble antigen-antibody complex, forming a thin immobile band of precipitin (protein) at equilibrium, which can be visualized either directly or after protein staining for interpretation. Immunodiffusion in gels can be further of two types:

- **Single radial immunodiffusion** — In this technique, the antigen is usually allowed to diffuse into the gel containing the antiserum.

PLATE 23

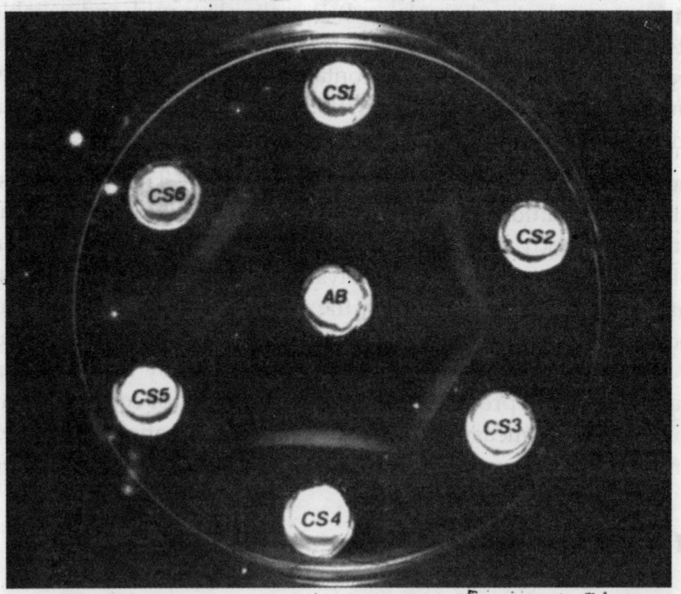

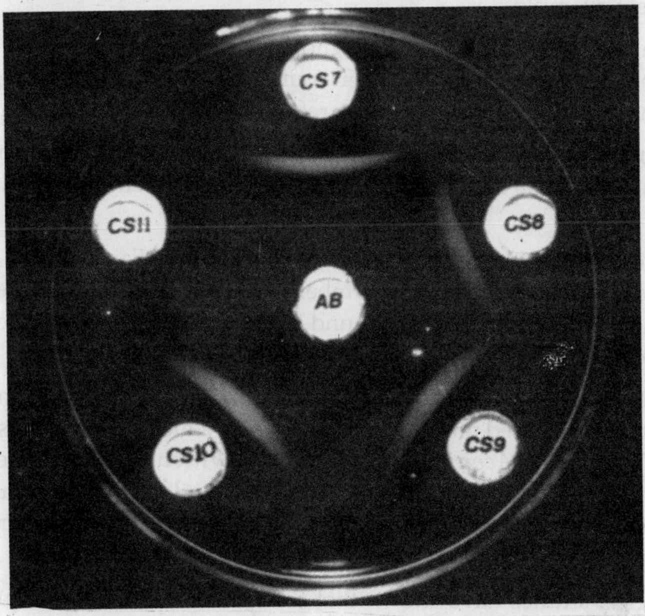

Ouchterlony — double immunodiffusion between the various antigenic proteins of the pollen of *Cassia siamea* (CS1-CS11) in the outer wells) and the antibody (AB in the inner well) against the total protein from the same pollen

➢ **Ouchterlony - double immunodiffusion** — In this method, both the antigen and antibody are allowed to diffuse into the gel and meet each other.

(b) **Rocket immunoelectrophoresis** — It is a simple, rapid and reliable method in which, rocket-like immunoprecipitate is formed when the desired protein (antigen) is electrophoresed in an agarose gel containing its monospecific antiserum. A comparison of the height of the peaks of the unknown and standard samples also allows the unknown protein concentration to be determined.

(c) **Enzyme-linked immunosorbant assay (ELISA)** — This method is generally used for quantitative estimation of a particular protein in a mixture, but can also be used to study the antigen-antibody reaction. The antibodies against a particular antigen are adsorbed to a solid support, in most cases a polystyrene microtiter plate. The support after coating with antibody is washed and then the antigen is added, which binds to the adsorbed antibodies. An enzyme-linked antibody molecule called the conjugate is then added, which also binds to the antigen, which is followed by a chromogenic substrate for the enzyme. The colored product generated is observed for confirmation of antigen-antibody reaction as well as measured for quantitative estimation. The intensity of the color is proportional to the bound enzyme and thus to the amount of the bound antigen.

8.7.3 Role of serology in taxonomy

Serological studies using crude plant protein extracts have been widely used in elucidating the taxonomy of a wide variety of higher-level taxa and in estimating phylogenetic relationships. Following are a few examples:

- A close relationship among the Magnoliidae, Hamamelididae and Comiflorae (Dahlgren, 1980) of the angiosperms has been found, based on comparative serological studies of their major seed proteins. This has refuted the idea of their independent evolution.
- The homogeneity of the iridoid-producing Comiflorae has been confirmed by serological studies, which has supported the inclusion of the Gentianaceae in it (Jensen & Greven, 1984).
- Based on phytoserological studies, Pickering and Fairbrothers (1970) have proposed the classification of the family Umbelliferae into Hydrocotyloideae, Saniculoideae and Apioideae, and Apioideae was found to be more closely related to Saniculoideae than to Hydrocotyloideae.
- Jensen (1968) discussed the relationships and classification of 20 genera of the Ranunculaceae based on serological evidence, which supports close relationship between *Delphinium* and *Aconitum*. *Hydrastatis,* which is sometime classified with the Berberidaceae has more close serological similarity to the Ranunculaceae than the Berberidaceae. *Trollis,* which is regarded as a link between follicular and achenial fruit bearing genera in the family, resembles closely with *Adonis.*
- Serological studies have shown that the genus *Linodendron* is quite distinct from other members of the family Magnoliaceae, and the genera *Magnolia* and *Michelia* are most closely related within the family (Johnston & Fairbrothers, 1965).
- Serological analysis has been useful in estimating the generic kinship in the Caprifoliaceae (Hillebrand & Fairbrothers, 1969, 1970) and the relationship between the Nymphaeaceae and Nelumbonaceae (Simon, 1971).

- Lee and Fairbrothers (1978) used serological techniques to study the systematics of Rubiaceae and related families. The quantitative data emphasized the similarity of Rubiaceae to Cornaceae and Caprifoliaceae and the pre-saturation tests revealed similarity with Apocynaceae, Asclepiadaceae and Gentianaceae. Similarly, two genera under Rubiaceae viz. *Asperula* and *Galium* were found to be serologically similar to each other while being the most dissimilar from all other genera of the family.
- In an attempt to estimate generic interrelationships of the tribe Genisteae of the Leguminoseae, Cristofolini and Chiapella (1977) made a serological survey and found that the genus *Cytisus* (*sensu lato*), including *Chamaecytisus*, *Sarothamnus* and *Carothamnus*, is a single homogenous unit apart from *C. sessilifolius*, which in turn is more closely related to *Laburnum* and *Genista* than to other species of *Cytisus*. Their studies also supported the elevation of the sect. *Asterospartum* of *Genista* into a new genus *Cytisanthus*.
- Similar studies as in *Cytisus* was also made in *Ulex* by Chiapella and Cristofolini (1981) and they allied genera of the same tribe.
- A cross-reactivity serological reaction (i.e., the reaction between an antiserum and any antigenic material other than the antigenic material used in its formation) of the seed proteins of *Phaseolus vulgaris* with those of other species of the same genus by Kloz (1971), showed a decreasing reaction in the series *P. coccineus, P. polyanthus, P. acutifolius, P. lanatus* and *P. aureus*, showing that, *P. vulgaris* is most closely related to *P. coccineus* and most distant to *P. aureus*.
- The assignment of *Phaseolus aureus* and *P. mungo* to the genus *Vigna* is strongly supported by serological evidence by Chrispeels and Gartner (1978).
- The prevailing idea of reticulate evolution in the genus *Phlox* has been corroborated by serological studies of Levin and Schaal (1972).
- On the basis of double diffusion tests, Gell, Hawkes and Wight (1960) and later Hawkes and Lester (1966, 1968) compared potato species and showed that Mexican species fell in three groups and *Solanum morelliforme* is only distantly related to other tuber-bearing solanums.

8.8 Ecology in Relation to Taxonomy

Ecological criteria are comparatively of little direct importance in taxonomy (Valentine, 1978). Although morphological data is no doubt the primary criterion in classical taxonomy, ecological criteria at the infraspecific level cannot be neglected. Habitat and environmental factors play a vital role from an evolutionary point of view and evolution is the fundamental process, which shapes taxonomic differentiation.

Every individual has a genotype, which refers to its innate genetical constitution, and a phenotype, which refers to the sum total of attributes, which it exhibits during its life span, and is not merely a manifestation of the genotype but the reflection of its interaction with the environment. Thus the ability of the genotype to produce different phenotypes in response to varying environmental conditions is referred to as **phenotypic plasticity**. According to Stebbins (1974), the evolutionary success and the potentiality for further diversification of a species or a taxon, depends upon its adaptation to :

- Survival of the seedling and adult plant under various environmental conditions — The individual may undergo environmental modifications of its phenotype as an adaptation for survival. This is what is known as the **ecotypic differentiation**. The local populations, which are made up of individuals with similar environmental tolerances, are specialized to meet local environmental conditions and may be termed **ecological races or ecotypes**. The different variations in plants, are the result of the plastic response of the individual to different factors of the environment, including:

➢ Climatic factors
➢ Edaphic factors
➢ Biotic factors

These ecological factors have an indirect effect on the genotype, but a direct effect upon the phenotype of an individual. Thus, an individual is the result of its genetic constitution and of environmental influences during its development. The environmental modifications become permanent, when incorporated into the genetic make-up, thus bringing about further differentiation within the populations and their ecotypes. However, some species appear to be genetically conservative and cannot adapt to new environments.

- Cross-pollination — Adaptation to cross-pollination involves adaptive modification of characters, which favor pollination by the most effective pollinator. These modifications are however quickly incorporated into the genotypes.
- Development and dispersal of seeds and establishment of seedlings — This involves adaptations for proper seed development, seed dispersal and seedling establishment, and the modifications to meet these diverse demands have largely lead to the evolution of major categories of flowering plants.

Generally characters that are greatly modified by environmental conditions are not considered to be taxonomically significant. As a result, environmentally variable characters have largely been rejected in taxonomic studies, and greater attention has been given to more stable floral characters. However, according to Dobzhansky (1970), any distinction between genetic and environmental influences upon phenotype is artificial and rather meaningless, as the capacity of phenotypes, to respond to environmental influences is itself under genetic control (Bradshaw, 1965), and there are evidences that environmental modifications become genetically fixed (Waddington, 1953). Even results of physiological studies of environmental modification (Bjorkman & Holmgren, 1963; Whitehead, 1963) show that, it is not random or biologically insignificant, but is frequently adaptive and of evolutionary and ecological importance (Bradshaw, 1965; Cook, 1968; Heslop-Harrison, 1964). Thus, studies in genetics, ecology, and physiology indicate that, phenotypic variation is of considerable importance from the ecological and evolutionary point of view. Although the vegetative characters are much variable in response to environmental conditions, they should also be adequately studied and used in keys (Snaydon, 1973). The recently developed techniques in molecular biology has opened a new vista in correlating the data obtained from the two diverse fields of taxonomy and ecology, with an enormous potentiality towards a new general evolutionary theory. Molecular markers are now being used not only to check the validity of taxonomic hypotheses and evolutionary relationships, but also to help tackle evolutionary processes unapproachable otherwise.

Some of the basic ecological evidentiary characters include:

- Habitat — sand, mud, rock, marsh, desert, etc.
- Abiotic relations — climate, light, moisture, soil, etc.
- Biotic relations — nutritional, symbiotic, reproductive, etc.
- Spatial relations — distributional, aerial, etc.
- Temporal relations — phenology, succession, periodicity, etc.

Ecological evidences have proved helpful in understanding the distribution of taxa, speciation within taxa, and adaptations of plants. However, many plant families are ecologically quite homogeneous, e.g. Chenopodiaceae, Ericaceae, Lentibulariaceae, Orobanchaceae, Plumbaginaceae, Potamogetonaceae and Salicaceae. They have also provided an explanation about the origin of the flowering plants, and also about the development of classification of several taxa, mainly below the level of the genus. They have particularly helped determine the taxonomic status of a species, as species are the highest taxonomic level that is of appreciable ecological value. As for example:

- According to Misra (1967), *Euphorbia hirta* bears three ecotypes:
  - ➢ Upright form growing in protected habitats.
  - ➢ Prostrate form appearing as a compact form of the footpath.
  - ➢ Prostrate form appearing as diffuse one of the grazed lands.
- *Euphorbia thymifolia* was found to exhibit ecotypic differentiation by Ramakrishnan (1960), in response to the calcium content of the soil. The red ecotype was found to be a facultative calcicole and the green one an obligate calcifuge. The red plant was further found to show three physiologically distinct ecotypes.
- *Boerhaavia diffusa*, *Gomphrena celosioides* and *Mecaraonia dianthera* have been found to give rise to appropriate calcicolous and calcifugous ecotypes, depending on the soil exchangeable calcium.
- Two species of *Lindenbergia*, *L. polyantha* and *L. urticaefolia*, have been demonstrated to be ecotypes of the same species, on the basis of developmental morphology, culture experiments and other analytical data. *L. polyantha* is a calcicolous or a miniature form of *L. urticaefolia* (Misra & Siva Rao, 1948).
- Depending on the environment, the fodder grasses, *Dichanlfnum annulatum* and *D. cancosum* (Pandeya, 1953), though similar in appearance, show variation in shape and size of branches, leaf and spike. However, *D. cancosum*, which thrives in dry sandy places, with one spike per raceme, may grow into a form resembling *D. cancosum* var. *mullicomus* if the water supply is more favorable.
- Pandeya (1967) has reported:
  - ➢ Edaphic ecotypes in *Heteropogon conforms* and *Selaria glauca*;
  - ➢ Photoperiodic ecotypes in *Ageratum conyzoides* and *Xanthium strumarium*;
  - ➢ Latitudinal ecotypes in *Anagallis arvensis* and *Cassia tora*; and
  - ➢ Moisture and shade ecotypes in *Leucas aspera*.

8.9 Palaeobotany in Relation to Taxonomy

Fossils have a potentially important role to play in taxonomy and phylogenetics. However, their use and interpretation has its own pitfalls and contentious issues. Under proper conditions, animals, plants and even microorganisms can leave remarkably good traces in rocks. In certain cases the actual substance of the fossilized material is replaced by minerals or become greatly changed chemically. But exceptions do occur, and fossils are sometimes formed under conditions that permit preservation of antigenic materials (Westbroek et al., 1979) or even DNA, that can be isolated and put to systematic use (Golenberg et al., 1990). Some well-preserved taxa provide a historic record of life on earth and thus provide a window into the past.

PLATE 24

Fossil showing impression of *Vertiveria* (leaf) — collected from lower Gondwana

Fossil showing impression of *Vertiveria* (root) — collected from lower Gondwana

PLATE 25

↳ Fossil showing impression of *Glossopteris*(Leaf) — collected from lower Gondwana

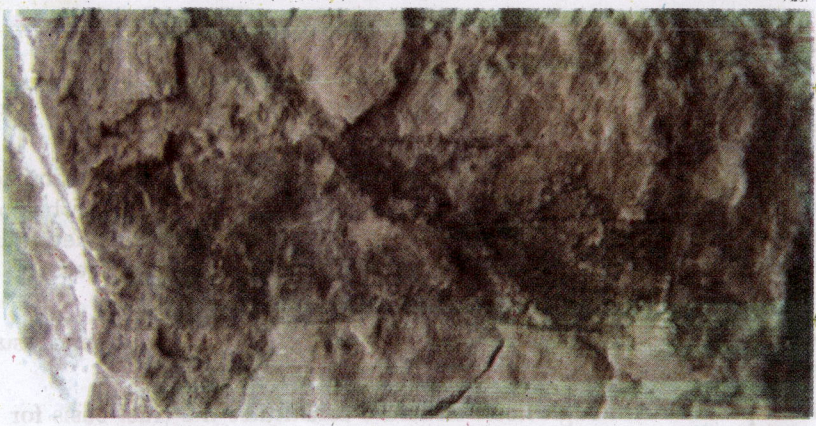

↳ Fossil showing impression of *Glossopteris* (Leaf) — collected from lower Gondwana

8.9.1 Uses of Fossil Pollen

Any theory of evolution is incomplete without any palaeobotanic evidence. According to Lam (1959) no evolutionary doctrine would be perfect without palaeobotanic evidence. Palaeobotanic studies have proved helpful in:

- The study of comparative morphology - Fossils not only provide details of the structure and sometimes the biology of extinct organisms, they can also provide estimates of the ages of the groups concerned.
- Providing phylogenetic evidence:
  - ➢ They provide additional taxa that can be scored for character states and included in a phylogenetic analysis.
  - ➢ They give a minimum estimate of the age of origin of particular taxa and their character states.
- Determining the evolution of the floras of the past.
- Determining the ecological conditions of the past.

The palaeobotanic studies have proved helpful in the phylogeny of angiosperms (Beck, 1976; Taylor, 1981). According to Dilcher (1979), palaeobotany provides important information about several aspects of the origin and diversification, as well as the early history of angiosperms. On the basis of palaeobotanic studies a number of new concepts have been developed about the primitive flow types by Dilcher (1979).

The evolutionary history of present-day plants, are often furnished by fossil records. Since the fossils of land-plants are scattered and poor, difficulties often arise in the interpretation of these data. The entire basis of fossil plant classification was changed by the appearance of angiosperms. A large number of fossil angiosperms, particularly leaves, seem to belong to extant families and to be related to recent genera. However, in some cases such affinity is unknown, either due to a lack of knowledge of living plant structure or due to poor state of preservation resulting in the absence of diagnostic features.

8.9.2 Types of plant fossils

Both, macrofossils (e.g. stems, leaves, etc.) and microfossils (e.g. pollen) are used as the soul of systematic data.

- **Stem** — Petrified wood is the best-known plant fossil. But due to the absence of proper knowledge about living plant structure, this sort of investigation has been greatly hampered for a long time. Spurred on by Metcalfe and Chalk's key to living wood anatomy, the proper study of fossil angiosperm wood has started and progressed considerably.
- **Leaves** — Leaves are only preserved under rather special circumstances such as in the anoxic bogs that ultimately produced present day coal measures.
- **Floral parts** — Although floral structure constitutes the chief basis for finding out relationships and evolutionary trends, flowers are hardly observed as fossils due to their short life span and delicate nature. Even if fossilized, they are often crushed due to which, important taxonomic details are obscured. However, if they are at all well preserved, the chances of correlation with parent plants are few and conclusions regarding affinity remains uncertain. For example, a number of well preserved petrified flowers with associated fruits from the Eocene Deccan Intertrappean Series of India was referred to the Lythraceae by Shukla (1944), which had relationship with *Decodon*.

- **Fruits and seeds** — Fossil fruits and seeds have been found, although not as numerous as leaves. Some of these have been found to retain anatomical structure in detail, while others, have, been found, to be preserved merely as frustrating casts or impressions. For example, as classification of grasses is based largely on seed characters, Elias (1946) classified his well-preserved Miocene prairie grass seeds from the high plains of northern mid-America in a similar way.
- **Pollen** — Fossil pollen provides important data for reconstructing past vegetation. Pollen have evolved a very tough and resistant outer layer of sporopollenin, a durable organic polymer, which preserves well in the fossil record. This has enabled the scientists to concentrate on the fossil pollen for study. They are preserved best if the sedimentary environment lacks oxygen or is acidic, conditions unfavorable for the organisms that decompose pollen. Pollen falling into lakes or peatlands accumulates with other sediments, layer after layer, year after year. Some lake sediments typically lack oxygen necessary for decomposing organisms, and if the sediment remains wet, the pollen will be preserved for thousands or even millions of years. Fossil pollen can be used for the following:
  - ➢ Fossil pollen is a very important kind of proxy data for reconstructing past climates, as vegetation is sensitive to climate. They can help reconstructing past environments for a number of reasons:
    - ♦ Pollen, which is released into the air, is transported some distance. The assemblage of pollen in the sediment of a particular region is representative of the vegetation of that region.
    - ♦ Abundant fossil pollen will be present in the sediment of vir´ `ally any permanent, natural lake. A cubic centimeter of lake sediment may contain tens or hundreds of thousands of pollen grains.
    - ♦ Scientists can reconstruct vegetation change over thousands of years, as pollen accumulates continuously, year after year. They can determine when and how rapid changes occurred. Like the annual ring of trees, some lakes even have annual layers that make it possible to reconstruct short-term climate variability over hundreds or thousands of years.
  - ➢ Paleopalynology (study of fossil pollen) helps in relative age dating (biostratigraphy).
  - ➢ The colouration and type of palynomorphs also provide information on the thermal maturity and type of organic material in the rock - important features for assessing hydrocarbon potential.
- **Fossil DNA** — DNA sequence information from fossils can also help to resolve many phylogenetic questions relating to both recently and not so recently demised species. The potential of the recently developed technique of polymerase chain reaction (PCR) not only helps in the analysis of DNA samples from living organisms, but also has the potential to make gene sequences available from at least some recently fossilized organisms. This is possible because of the surprisingly robust nature of DNA. DNA, like all other biological macromolecules, is very unstable, and spontaneously breaks down. In living cells, DNA is maintained by repair mechanisms, but after death DNA self-destructs at a rather rapid rate. Thus, although fossilized material usually contains very little of its original DNA and whatever is present may be largely denatured, PCR allows to amplify many million-fold whatever small amounts of DNA is present, which can be further sequenced as if it were new. Further, most fossil DNA is difficult to sequence because of oxidation of the bases, especially the pyrimidines and the sequenceable

fragments tend to be rather short, only a hundred or so base pairs long. Sykes (1991) has commented that *in vitro* estimates of the rate of spontaneous hydrolysis imply that no DNA would remain intact much beyond 10,000 years. Hence, it seems feasible that useful DNA sequences tens of thousands of years old could be recovered, particularly if the fossil has been retained at low temperature (Lindahl, 1993). Clearly therefore, the success of obtaining usable DNA sequences from fossils largely depends on the circumstances of preservation. Nevertheless, even with these and with carefully selected probes, it may be possible to obtain phylogenetically usable data. Recently, for example, PCR has made it possible to obtain gene sequences from 17-20 million years old *Magnolia* leaves from lake bottom sediments of Miocene age deposits in Idaho (Goienberg et al., 1990), which had become buried in the lake under volcanic clay deposited rapidly creating a virtually anoxic environment. Most of the DNA, which has been recovered is from plants preserved in amber, as for example from a *Hymenaea* leaf thought to be 25-40 million years old (Poinar et al. 1993).

8.9.3 Limitations of paleobotany

- Fossilization almost invariably involves distortion, chemical degradation and replacement in addition to the normal processes of death and decay. Thus, fossils are seldom if ever perfect. Unfortunately the palaeontologist has to work on with whatever material is available and this often means interpreting less than ideally preserved fragmentary or distorted material.
- As fossils are by their very nature incomplete representations of past organisms and often poorly preserved, interpreting what is a species from fossils is not easy. Further, it is highly unlikely that the fossils will display adequately the sorts of small differences that are so often used today to separate extant taxa. Thus, fossil 'species' will only correspond to biological species notions if the latter happen to be well correlated with fairly obvious morphological differences. Hence, if fossils are to be classified up to species level, they may require a modified species concept.
- Another problem, which a palaeontologist face is the fact that, the full details of how species arise over time, is not well understood from fossil data.
  - ➢ According to Hennigian philosophy and modern cladistic techniques, all speciation events involve dichotomous (or polychotomous) splits. This type of speciation, which, results from a lineage splitting into two (or more) separate species is referred to as **cladogenesis** (Fig. 8.30). In contrast to this, many fossil records suggest that speciation can occur over time without any necessity for splits to occur. This type of speciation, which, results from accumulation of change over time in a single population is termed **anagenesis.**
  - ➢ Secondly, according to Darwinian concept, the evolutionary change is likely to be a gradual process, with new species often forming over geologically active periods of time, which is referred to as **phyletic gradualism.** If it is so, then any near complete fossil sequence would be expected to show a gradually evolving sequence of forms, in which case, it would be very difficult for any taxonomist to decide where to draw the line between species. Unfortunately, very few fossil series are complete enough to permit assessment of phyletic gradualism.
- The incompleteness of the fossil record often poses as a drawback in including fossil data in cladistic analyses. Many data sets are incomplete and most extant species are still unknown to science.

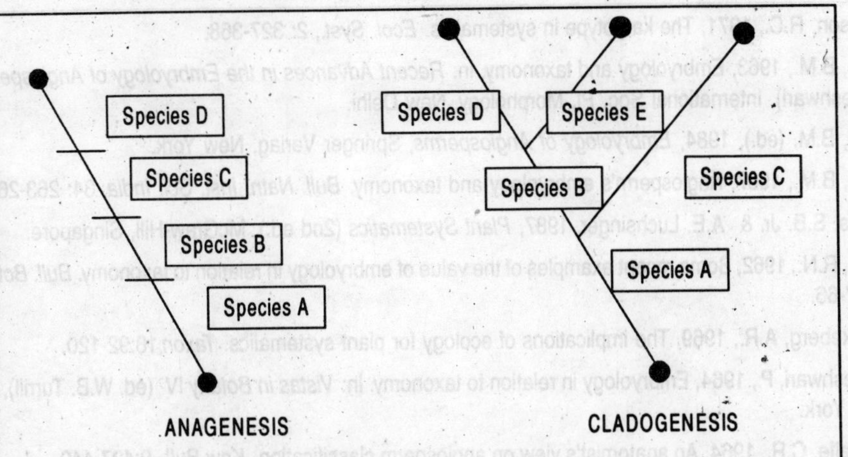

Fig. 8.30 : Types of speciation

SUGGESTED READING

- Bailey, I.W., 1957, Wood anatomy in the study of phylogeny and classification of angiosperms. *J. Arnold Arbor. Harv. Univ.*, 38: 243-254.
- Behnke, H.D., 1972, Sieve tube plastids in relation to angiosperm systematics: An attempt towards a classification by ultrastructural analysis. *Bot. Rev.*, 38: 155-197.
- Behnke, H.D., 1977, Transmission electron microscopy and systematics of flowering plants. *Pl. Syst. Evol. Suppl.*, 1: 155-178.
- Boyden, A., 1964, Perspectives in systematic serology. In: *Taxonomic Biochemistry and Serology* (ed. C.A. Leone), Ronald Press, New York, pp. 75-99.
- Cole, G.T. & H.D. Behnke, 1975, Electron microscopy and plant syslematics. *Taxon*, 24:3-15.
- Davis, G.L, 1966, *Systematic Embryology of the Angiosperms*, Wiley, New York.
- Davis, P.H. & V.H. Heywood, 1965, *Principles of Angiosperm Taxonomy*, Van Nostrand, Princeton, N.J.
- Dickison, W.C., 1975, The bases of angiosperm phylogeny: Vegetative anatomy, *Ann. Mo. Bot. Gard.*, 62: 59-62.
- Erdtman, G., 1952, *Pollen Morphology and Plant Taxonomy of Angiosperms*, Almqvist & Wicksell, Stockholm.
- Erdtman, G., 1966, *Pollen Morphology and Plant Taxonomy*, Vol. I. *Angiosperms*, Hafner, New York.
- Eyde, R.H., 1975, The bases of angiosperm phylogeny: Floral anatomy, *Ann. Mo. Bot. Gard.* 62: 521-537.
- Fairbrothers, D.E., 1977, Perspectives in plant serotaxonomy. *Ann. Mo. Bot. Gard.*, 64: 147-160.
- Fairbrothers, D.E., 1983, Evidence from nucleic acid and protein chemistry, in particular serology, in angiosperm classification. *Nord. J. Bot.*, 3: 35-41.
- Gibbs, R.D., 1974, *Chemotaxonomy of Flowering Plants*, 4 Vols. McGill-Queens Univ. Press, Montreal.
- Harborne, J.B. & B.L. Turner, 1984, *Plant Chemosystematics*, Academic Press, New York.
- Hawkes, J.G. (ed.), 1968, *Chemotaxonomy and Serotaxonomy*, Academic Press, London.
- Howard, R.A., 1963, The vascular strûcture of the petiole as a taxonomic character. *Adv. Hort. Sci.*, 3: 7-13.

- Jackson, R.C., 1971, The karyotype in systematics. *Ecol. Syst.*, 2: 327-368.
- Johri, B.M., 1963, Embryology and taxonomy. In: *Recent Advances in the Embryology of Angiosperms*, (ed. P. Maheshwari), International Soc. Pl. Morphology, New Delhi.
- Johri, B.M. (ed.), 1984, *Embryology of Angiosperms*, Springer Verlag, New York.
- Johri, B.M., 1967, Angiosperm's embryology and taxonomy. *Bull. Natn. Inst. Sci. India*, 34: 263-268.
- Jones, S.B. Jr. & A.E. Luchsinger, 1987, *Plant Systematics* (2nd ed.), McGraw-Hill, Singapore.
- Kapil, R.N., 1962, Some recent examples of the value of embryology in relation to taxonomy. *Bull. Bot. Surv. Ind.*, 4: 57-66.
- Kruckeberg, A.R., 1969, The implications of ecology for plant systemaiics. *Taxon*,18:92-120.
- Maheshwari, P., 1964, Embryology in relation to taxonomy. In: *Vistas in Botany* IV (ed. W.B. Turrill), Macmillan, New York.
- Metcalfe, C.R., 1964, An anatomist's view on angiosperm classification. *Kew Bull.*, 9:427-440.
- Metcalfe, C.R. & L. Chalk, 1950, *Anatomy of the Dicotyledons*, Vols. I, & II, Oxford Univ. Press, Oxford.
- Mondal, A.K., Parui, S. and Mandal, S. (1998). Biochemical analysis of four species of *Cassia* L. pollen. *Aerobiologia*, 14: 45-50.
- Mondal, A.K., Parui, S. and Mandal, S. (1998). Analysis of the free amino acid content in pollen of nine *Asteraceae* species of known allergenic activity. *Annals of Agricultural and Environmental Medicine*, 5: 17-20.
- Mondal, A.K., Parui, S. and Mandal, S. (1998). Phylogenetic interrelationships between four species of *Cassia* L. based on pollen morphology, chemotaxonomy and phytoserology. *Bangladesh Journal of Plant Taxonomy*, 5(1): 63-75.
- Mondal, A.K., Mondal (nee Parui), S. and Mandal, S. (2000). Molecular taxonomy of the allergenic proteins of *Cassia* L. based on seed protein and mitochondrial DNA RFLP. *Phytomorphology*, 50(1): 15-25.
- Murty,Y.S. & V. Puri, 1980, Floral anatomy in relation to taxonomy. *Glimp. Pl. Res.*, 5: 222-243.
- Nair, P.K.K., 1980, *Glimpses in Plant Research* V. *Modem Trends in Plant Taxonomy*, Vikas Publ., New Delhi.
- Paliwal, G.S. & S.K. Anand, 1978, Anatomy in relation to taxonomy. *Acta Bot. Indica*, 6:1-20.
- Parui, S., Mondal, A.K. and Mandal, S. (1998). Peroxidase isozyme profiles of immature and mature pollen of seven tropical plants from eastern India – A Comparative study. *Grana*, 38: 228-232.
- Philipson, W.R., 1974, Ovular morphology and the major classification of dicotyledons. *Bot. J. Linn. Soc.*, 68: 89-108.
- Puri, V., 1952, Floral anatomy in relation to taxonomy. *Agra Univ..J. Res.* (*Sci.*), : 15—35.
- Puri, V., 1958, Floral anatomy and taxonomy. *Ind. Bot. Ser. Mem.*, 1: 15-18.
- Puri, V., 1962, Floral anatomy in relation to taxonomy. *Bull. Bot. Surv. India*, 4:161-165.
- Radford, A.E., 1986, *Fundamentals of Plant Systematics*, Harper and Row, New York.
- Sharma, A.K., 1954, Cytology as an aid in taxonomy. *Bull. Bot. Soc. Bengal*, 18: 1-4.
- Sharma, O.P., 1987, Pollen morphology of *Cyperus*. *Bangladesh J. Bot.*, l6(2): 141-150.
- Sharma, O.P. & R. Shiam, 1981a, Occurrence of cuticular papillae in *Cyperus*. *Curr. Sci.*, 50(5): 236.
- Sharma, O.P. & R. Shiam, 1984, Epidermal structures of culm in *Cyperus* with a discussion of silica bodies in Cyperaceae. *Bangladesh J. Bot.*, 13(1): 16-24.
- Smith, P.M., 1976, *The Chemotoxonomy of Plants*, Edward Arnold, London.
- Stace, C.A., 1965, Cuticular studies as an aid to plant taxonomy. *Bull. Br. Mus. (Nat. Hist.) Bot.*, 4: 1-78.

9 NUMERICAL TAXONOMY

AT A GLANCE

Numerical taxonomy or taximetrics, nowadays frequently and perhaps more appropriately referred to as **phenetics**, refers to the application of various mathematical procedures to numerically encoded character state data for organisms under study. Thus, it is the analysis of various types of taxonomic data by mathematical or computerized methods and numerical evaluation of the similarities or affinities between taxonomic units, which are then arranged into taxa on the basis of their affinities. According to Heywood (1967) the **numerical taxonomy may be defined as the numerical evaluation of the similarity between groups of organisms and the ordering of these groups into higher ranking taxa on the basis of these similarities.**

The period from 1957 to 1961 saw the development of first methods and of theory of numerical taxonomy. Plants as we all know are classified based on their characters. It was Michel Adanson (1763), a French botanist, who for the first time put forward a plan for assigning numerical values to the similarity between organisms and proposed that equal weightage should be given to all the characters while classifying plants. He used as many characters as possible for the classification, and such classifications came to be known as **Adansonian classifications**. Numerical taxonomy was however largely developed and popularized by Sneath and Sokal (1973). The application of Adansonian principles and use of modern methods and electronic data-processing techniques, have helped in the evolution of several new classifications of plants during the past few decades.

9.1 Principles of Numerical Taxonomy

Numerical taxonomy involves two aspects :

(a) **Construction of taxonomic groups —**

- In numerical taxonomy, first, individuals are selected and their characters spotted out. There is no limitation to the number of characters to be considered. However, the larger the number of characters, better is the approach for generalization of the taxa.
- The resemblances among the individuals are then established on the basis of character analysis, which can often be worked out with the help of computers, the accuracy of which depends on the appropriateness in character. The best way to delimitate taxa is, to utilize maximum number of characters, with similar weightage given to all of them.

(b) **Discrimination of the taxonomic groups** — When the taxonomic groups chosen for the study show overlapping of characters, discrimination should be used to select them. Discrimination analysis can be done by various techniques, specially devised for such purposes.

Numerical taxonomy is thus, based on certain principles, also called **neo Adansonian principles**. Following seven principles of numerical taxonomy have been enumerated by Sneath and Sokal (1973):

(i) The greater the content of information in the taxa, and more the characters taken into consideration, the better a given classification system will be.

(ii) Every character should be given equal weightage in creating new taxa.

(iii) The overall similarity between any two entities is a function of the individual similarities in each of the many characters, which are considered for comparison.

(iv) Correlation of characters differ in the groups of organisms under study. Thus distinct taxa can be recognized.

(v) Phylogenetic conclusions can be drawn from the taxonomic structure of a group and from character correlations, assuming some evolutionary mechanisms and pathways.

(vi) The science of taxonomy is viewed and practiced as an empirical science.

(vii) Phenetic similarity is the base of classifications.

9.2 Steps to the Construction of Taxonomic Groups

Two major bases have been recommended for designing a natural system of classification. These include :

- **Phenetics** — It involves the similarity between individuals under consideration, based on a set of phenotypic characters. Thus, phenetic classification is based on overall affinity as judged by using as many characters as available. It does not embody any relationship through ancestry. The typical tree diagrams or **dendograms,** with taxa and groups of taxa, being connected at different levels according to their overall similarity, will not reflect evolutionary relationships accurately and such dendograms are referred to as **phenerograms.**

- **Cladistics** — It is based on the relationship between individuals with regard to their evolutionary history. The relationship means common ancestry. The goal is to produce objective and repeatable branching diagrams (evolutionary trees) depicting hypothetical evolutionary histories. The evolutionary trees with anatomizing branching leading to various taxa are called **cladograms.**

Following steps are used in the construction of taxonomic groups :

(a) **Operational Taxonomic Units (OTUs) :** Operational taxonomic unit is the basic unit in numerical taxonomy. It can be an individual, species, genus, family, order or class. Since the taxonomic units employed in numerical methods are not always comparable to formal taxonomic units, they are termed as operational taxonomic units. Comparison of OTUs of equal rank is always made in numerical taxonomy. In case of OTUs above the level of individual, adequate representation of various polymorphic forms is essential. For example, when genera are compared, they should be represented by different species. Similarly, when families are compared, they should be represented by different genera and so on.

(b) **Unit Characters :** Unit characters are the characters used in numerical taxonomy. According to Sokal and Sneath (1963), unit character is defined as a taxonomic character of two or more states, which within the study at hand cannot be subdivided logically, except subdivision brought about by changes in the method of coding. Only the phenotypic characters are used as unit characters, e.g. presence or absence of an awn in a grass spikelet.

(i) **Types of unit characters :** Unit characters are of two types :

- **Binary characters** — Those unit characters, which exist in two states are called binary characters, e.g. presence or absence of trichomes. They can be represented by the simplest form of coding, where characters are divided into + and – or as 1 and 0. In such cases, the positive characters are recorded as + or 1 and negative characters are recorded as – or 0. In case the organ possessing a given character is missing in an organism, the character is scored NC, which means "no comparison".

- **Multistate characters** — Those unit characters, which exist in more than two states are called multistate characters. Such characters can be coded into number of states (1,2,3...) corresponding to their range of variation. They may be further of two types :

➢ **Qualitative multistate characters** — These characters contain three or more contrasting forms and each form is ranked on equal footing, e.g., flower color, which can be in any number of states – red, white, purple, yellow, etc.

➢ **Quantitative multistate characters** — These characters represent measures of the size on a continuous scale, e.g., length of the leaf, which can be 2cm, 3cm, 4cm, etc., or height of the plant, or amount of pubescence on a leaf, etc.

(ii) **Admissible unit characters** : The following characters can be categorized as admissible unit characters :

- It cannot be subdivided.
- It must show variation among the taxa under comparison.
- It should be inherent in the organism.
- It should not be succeptible to environmental changes or affected by experimental conditions.
- It must be of some diagnostic value.

(iii) **Inadmissible unit characters** : As in other disciplines of taxonomy, the proper selection of unit characters is a critical point in the application of numerical taxonomy. Sokal and Sneath (1963) have listed certain characters, which cannot be utilized for numerical taxonomy and called them as inadmissible characters. These include :

- Attributes, which are not a reflection of the genotypes of the organisms themselves.
- Any property, which is a logical consequence of another, either partly or wholly.
- Characters, which do not vary within the entire sample of organisms.

(iv) **Proper selection of unit characters** : Sneath and Sokal (1973) have given the following suggestions for proper selection of unit characters:

- They should come from all parts of the organism.
- They should belong to all the stages of the life cycle of the organism.
- Variable characters within the group should be used.
- Due attention should be given to characters related to morphology, physiology, ecology and distribution of the organisms.

(c) **Estimation of resemblances** : Most phenetic methods involve taxon-to-taxon distance, similarity or dissimilarity measures. Distance and dissimilarity are sometimes treated as the same thing, though a distinction can be made between them. As the name implies, distance and dissimilarity measure increase with dissimilarity between taxa, while similarity measures decrease with dissimilarity. Thus the resemblance between two OTUs is estimated or measured either:

- in terms of similarity i.e., percentage of characters in which they agree, or
- in terms of dissimilarity i.e., percentage of characters in which they do not agree.

So far three methods have been devised for estimating the resemblance between the taxonomic groups, which include :

(i) **Coefficients of association** — It is measured by the following formula:

$$S = \frac{N_S}{N_S + N_d}$$

where, S is the numerical index or Simple matching index or coefficient;

N_S = the number of positive features shared by any two OTUs;

N_d = the number of positive features in one OTU and number of negative features in the other OTU.

Binary characters can be dealt with in two ways depending on whether they can be obviously polarized:

- **In terms of presence or absence, or**
- **Whether the character simply has two states and it is undesirable or impossible to classify in terms of presence or absence.**

Data are presented in the form of a truth table of test results between the two taxa or OTUs.

Thus,

		OTU 1	
		+	—
OTU 2	+	a	b
	—	c	d

Simple matching index (S) or coefficient is defined as the number of tests giving identical results (a or d) divided by the total number of tests, thus

$$S = \frac{a + d}{a + b + c + d}$$

(range 0 to 1)

(ii) Coefficients of correlation (r) — It is measured by the following formula

$$rjk = \frac{\sum^{n}[(X_{ij} - X_i)(X_{ik} - X_k)]}{\sqrt{\{[\sum^{n}(X_{ij} - X_j)^2][\sum^{n}(X_{ik} - X_k)^2]\}}}$$

where, j and k stand for two units under comparison;

X_{ij} stands for the value of the character i in unit j ;

X_{ik} stands for the value of the character i in unit k;

X_i and X_k stand for the mean of all the characters in units j and k;

n stands for number of characters.

(iii) Measurement of taxonomic distance between OTUs — using the convention of multidimensional space with one dimension for each character. Taxonomic distance (d) can be measured by the following formula :

$$d_{jk} = \sqrt{\{\sum^{n}(X_{ij} - X_{jk})2/n\}}$$

where X_{ij} is the character state of unit j for the character i;

X_{jk} is the character state of unit k;

$\sum^{n}$ stands for the sum over n characters;

The value of the taxonomic distance (d) is the distance in a phenetic space divided by $\sqrt{n}$.

(d) **Cluster Analysis :** Cluster analysis or clustering is a type of multivariate statistical analysis. It is used to group organisms into separate clusters based on their statistical behavior. The main objective of clustering is to find similarities between organisms, and then group similar organisms together to assist in understanding relationships that might exist among them. Thus, different OTUs are grouped together on the basis of degree of similarity and these groups of OTUs are called **clusters**.

(i) **Cluster analysis characteristics :** Cluster analysis is based on a mathematical formulation of a measure of similarity. There are a number of characteristics that can distinguish different approaches to cluster analysis, which include:

- Numerical, statistical, and conceptual clusters.
- Agglomerative vs. divisive - Agglomerative methods start with individual taxa and seek to connect them into pairs, etc. in such a way that similarities between pairs or group members are maximized at each level. In contrast, divisive clustering splits are sought in the group of taxa that will in some way maximize the collective phenotype disparity between the two groups formed by the split.
- Overlapping vs. disjoint clusters.
- Incremental vs. non-incremental.
- Flat vs. hierarchical representatives.

ering methods : Clustering can be achieved in two ways :

lonothetic system — This system employs the attributes one at a time. The nothetic method obviously leads to artificial clustering.

onsidering all their attributes simultaneously. This method gives a natural grouping.

cluster or a group is separated from the other by a dividing line, which indicates a nct gap between the two. In case of intraspecific clusters, there are often few ntinuities.

- **Phenetic clustering methods** — Phenetic clustering methods have largely been overtaken by cladistic methods in order to relate data to evolutionary history, and they are now rarely employed in taxonomy as they often lead to substantially different clusters when applied to real data. The most widely used phenetic clustering methods in taxonomy include the following:

  ➢ **Nearest neighbor clustering method** — It is also called single linkage clustering. In this method, phenograms are constructed by joining OTUs and groups on the basis of their most similar members, i.e., the shortest distance. Whether an OTU will join an existing cluster depends on its maximum similarity (or minimum distance from) any member of that cluster. The phenogram is initiated by joining the two (or more) most similar OTUs. An example is given next page :

Similarity matrix

	C	B	A
A	0.40	0.63	1.0
B	0.53	1.0	
C	1.0		

Sequence of procedures involved in the phenetic clustering of the above three taxa :

Step	Action	Result	Action	Tree
1	Find greatest similarity less than unity	0.63 (A-B)	Join A to B at level 0.63	A B
2	Find next highest similarity	0.53 (B-C)	Join C to A+B cluster at level 0.53	A B C

➤ **Furthest neighbor clustering method (complete linkage)** — Construction of furthest neighbor phenogram is essentially similar to that for nearest neighbor ones, with the only difference that the OTUs are joined at a level corresponding to the greatest distance between their component members instead of the shortest distance as in case of the latter. This method tends to emphasize differences between clusters.

➤ **Unweighted pair-group method using arithmetic averages (UPGMA)** — In this method an OTU is joined to existing clusters on the basis of their average (mean) distance to the members of that cluster (Fig. 9.1). The mean distance between two clusters of OTUs is calculated by adding together the distances between all possible pair-wise combination of OTUs in the two clusters and dividing by the number of combinations. The tree construction method is similar to nearest neighbor clustering. Since this method uses mean distances, it is not so subjected to the effects of aberrant OTUs, and hence preferred over nearest neighbor methods.

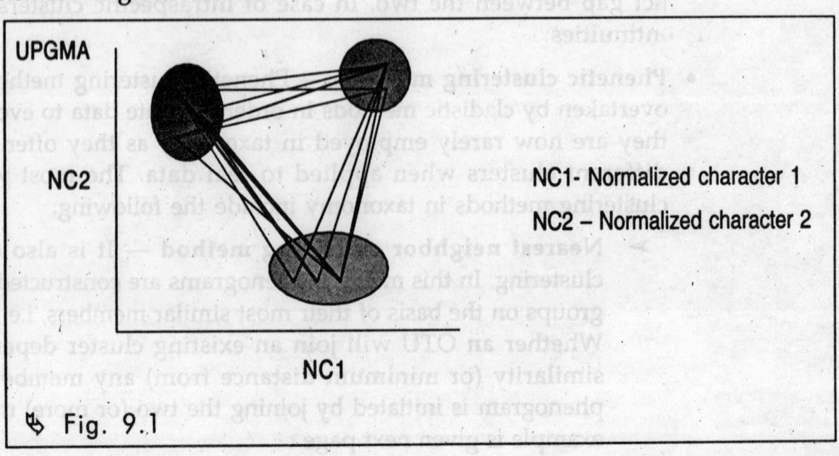

Fig. 9.1

➢ **Weighted pair-group method using arithmetic averages (WPGMA)** — This is a set of methods rather than a single one. They are essentially similar to UPGMA except that, they weight distances such that there is a tendency to emphasize separations between large clusters. They are also therefore less subjected to the effects of single or small groups of outliers and hence preferred over nearest neighbor methods. In this method, the distances are usually weighted by some function of the number of OTUs in each cluster.

➢ **Centroid clustering** — Its basic aim is similar to UPGMA and WPGMA, but it requires the actual taxa versus character raw data, or more precisely normalized data, rather than inter-taxon distances. This method makes use of the Euclidian distances within the character hyperspace between the centroids of clusters of OTUs (Fig. 9.2 & 9.3). Euclidian distance is equivalent to the real physical distance by which the taxa would be separated if they were plotted in a space in which each axis represents one normalized variable and each axis is perpendicular to all others. Calculation of Euclidian distance is shown below :

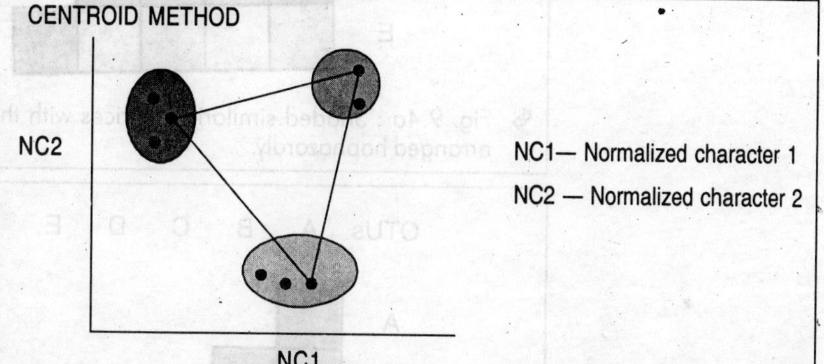

Fig. 9.2

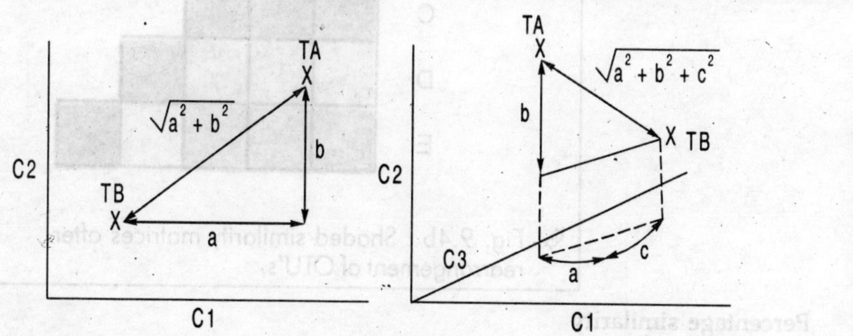

Fig. 9.3 : Illustration of the calculation of the shortest distance (Euclidean distance) between two points in 2-dimensional (left) and 3-dimensional (right) space.
[C1- Character 1; C2- Character 2; C3- Character 3; TA – taxon A; TB – taxon B]

- **Differential shading of the similarity matrix :** This is a method to describe structure in matrices of similarity coefficients, in which similarity coefficients are grouped into five to ten evenly spaced classes, each of which are represented by different degrees of shading in the squares of half matrix. Generally a darker shade is used for the highest value, and the lightest shade for the lowest value. Thus, the half matrix can be seen as a pattern to different shades, limited by a diagonal of squares with the darkest shade (Fig. 9.4a). Clusters can be more sharply defined by rearrangement of the sequence of OTUs (Fig. 9.4b).

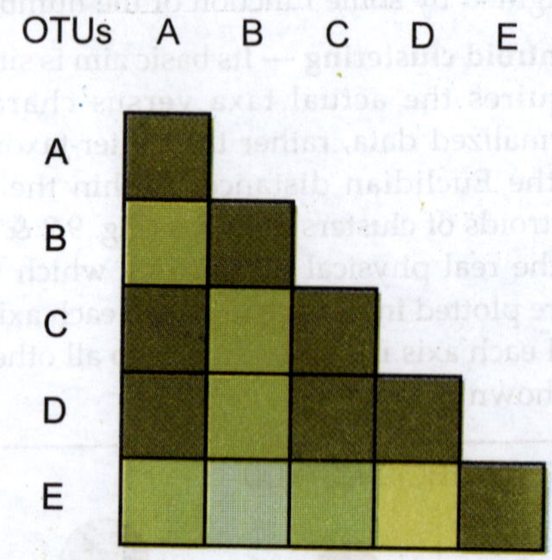

Fig. 9.4a : Shaded similarity matrices with the OTU's arranged haphazardly.

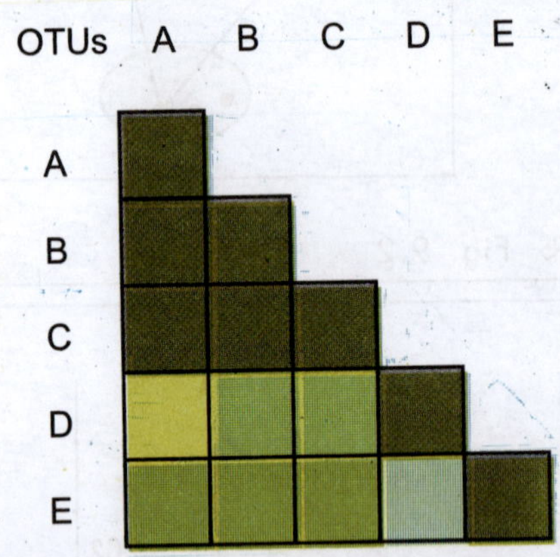

Fig. 9.4b : Shaded similarity matrices after rearrangement of OTU's.

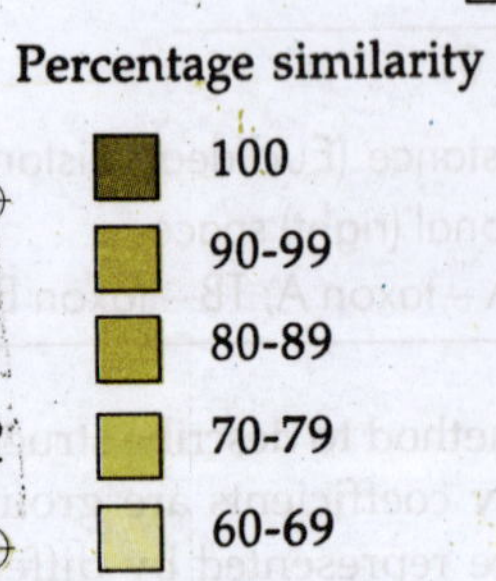

➢ **Cluster analysis :** A large number of numerical techniques can be used to analyze the groups of related OTUs based on high similarity coefficients. These techniques include elementary cluster analysis, clustering by single, complete or average linkage, central or nodal clustering, etc. The groups of similar organisms organized in this manner are termed phenons, which are arbitrary and relative groups. The clusters of phenons are then rearranged in a dendogram (Fig. 9.5) , in which the mutually most similar taxa are paired and the pairs successively joined by the average similarity. This finally results in a tree like dendogram when all units have been joined together, with the taxa at the tips of branches. The number of clusters is determined by drawing a horizontal (phenon line) intersecting the vertical lines of the dendogram.

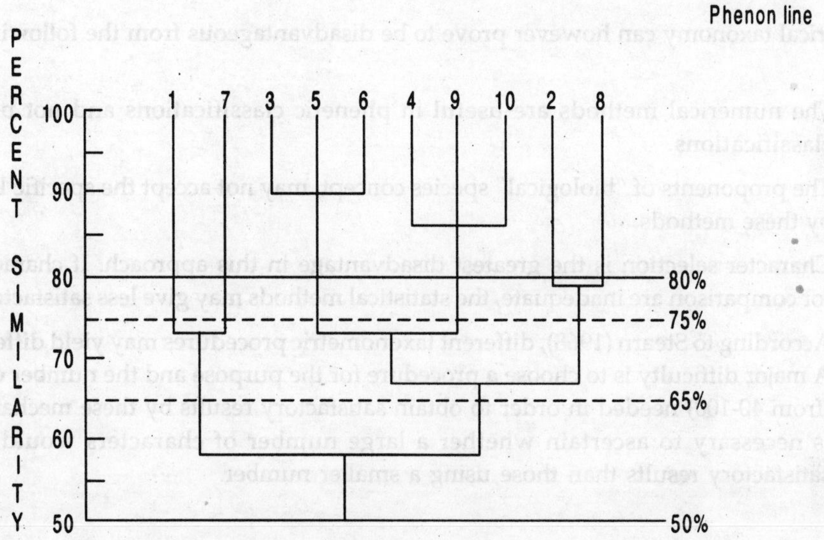

Fig. 9.5 : Dendogram to show formation of phenons

The delimitation of phenons is done by drawing a horizontal line across the dendrogram at a similarity value. For example a line at 75%, creates five 75-phenons 1; 7; 3, 5, 6; 4, 9, 10; and 2, 8., If the OTUs 1 to 10 in the above dendrogram had been species, an 80-phenon line could indicate 6 subgenera and a 65-phenon line two genera.

9.3 Merits of Numerical Taxonomy

According to Sokal and Sneath (1973), numerical taxonomy has the following advantages over conventional taxonomy:

- The data of conventional taxonomy is improved by numerical taxonomy as it utilizes better and more number of described characters. The data are collected from a variety of sources, such as morphology, chemistry, physiology, etc.

- As numerical methods are more sensitive in delimiting taxa, the data obtained can be efficiently used in the construction of better keys and classification systems, creation of maps, descriptions, catalogues, etc. with the help of electronic data processing systems. Numerical taxonomy has in fact suggested several fundamental changes in the conventional classification systems.
- The number of existing biological concepts have been reinterpreted in the light of numerical taxonomy.
- Numerical taxonomy allows more taxonomic work to be done by less highly skilled workers.

9.4 Demerits of Numerical Taxonomy

Numerical taxonomy can however prove to be disadvantageous from the following points of view :

- The numerical methods are useful in phenetic classifications and not phylogenetic classifications.
- The proponents of "biological" species concept, may not accept the specific limits bound by these methods.
- Character selection is the greatest disadvantage in this approach. If characters chosen for comparison are inadequate, the statistical methods may give less satisfactory solution.
- According to Stearn (1968), different taxonometric procedures may yield different results. A major difficulty is to choose a procedure for the purpose and the number of characters (from 40-100) needed in order to obtain satisfactory results by these mechanical aids. It is necessary to ascertain whether a large number of characters would really give satisfactory results than those using a smaller number.

9.5 Applications of Numerical Taxonomy

Numerical taxonomy has been successfully applied in the following studies :

- Study of similarities and differences in bacteria, other micro-organisms and several animal groups.
- Delimitation of several angiospermic genera like *Oryza* (Morishima & Oka, 1960), *Sarcostemma* (Johnson & Holm, 1968) *Solanum* (Soria & Heiser, 1961), and other groups including Farinosae of Engler (Hamann, 1961) and a few others.
- In the study of several other angiospermic genera including *Apocynum, Chenopodium, Crotalaria, Cucurbita, Oenothera, Salix, Zinnia,* wheat cultivars, Maize cultivars, etc. (Sneath & Sokal, 1973).
- Phytochemical data from seed protein and mitochondrial DNA RFLP studies has been numerically analyzed by Mondal et al. (2000) to study the interspecific variations among eight species of *Cassia* L. Based on the results of electrophoretic patterns, the degree of pairing affinity (PA) or similarity index was calculated by the following formula, according to the method of Sokal & Sneath (1963) and Romero Lopes et al. (1979) :

$$PA = \frac{\text{Bands common to species A and B}}{\text{Total bands in A and B}} \times 100$$

Separate dendograms expressing the average linkage were computed using the cluster method UPGMA (Sneath & Sokal, 1973), which showed that the eight species could be placed into two categories or clusters (Fig. 9.6) with *C. alata, C siamea, C. fistula and C. reginera,* all being trees or large shrubs and characterized by the absence of foliar glands on petiole or rachis and presence of dense axillary terminal racemes greater than 30 cm long, being clustered into one group, whereas the other four species, i.e., *C. occidentalis, C. sophera, C. mimosoides* and *C. tora*, forming the other cluster, all being herbs or undershrubs and characterized by the presence of short corymbose racemes less than 10 cm long and with foliar glands, either on petiole or rachis (further reading Mondal et al., 2000).

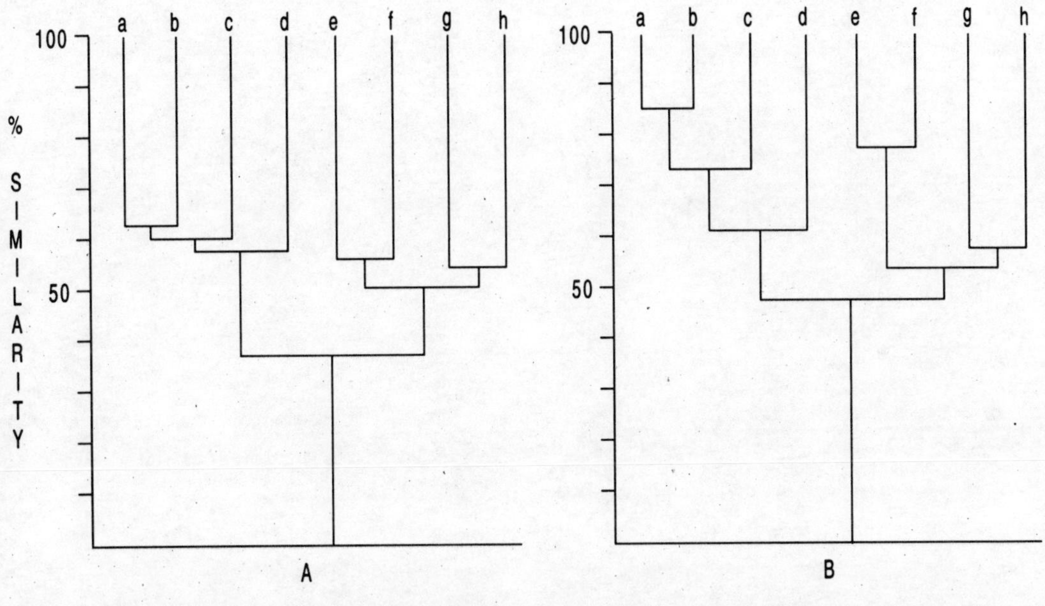

Fig. 9.6 : Dendogram representing the 'average linkage' relationship among the eight species of *Cassia* shown by (A) seed protein electrophoresis and (B) mtDNA restriction fragments (a) *C. occidentalis,* (b) *C. sophera,* (c) *C. mimosoides,* (d) *C. tora,* (e) *C. alata,* (f) *C. siamea,* (g) *C. fistula,* (h) *C. renigera.*

SUGGESTED READING

- Cole, A.J., 1969, *Numerical taxonomy,* Academic Press, London.
- Cullen, J., 1968, Botanical problems of numerical taxonomy. In: *Modern trends in Plant Taxonomy* (ed. V.H. Heywood), Academic Press, London, pp. 175-184.

Contd.

- Mondal, A.K., S. Mondal (nee Parui) & S. Mandal 2000, Molecular taxonomy of the genus *Cassia* L. based on seed protein and mitochondrial DNA RFLP. *Phytomorphology*, 50(1): 15-25.
- Rohlf, F.J. & R.R. Sokal, 1965, Coefficients of correlations and distance in numerical taxonomy. *Univ. Kansas Sci. Bull.*, 45: 3-27.
- Romero Lopes, C., E.J. De Lucca, A.M. De Andrade & C.M. Jacoia Faulin, 1979, A phylogenetic interpretation of chromosomal and electrophoretic data in Columbiformes. *Cytologia*, 44: 39-47.
- Sneath, P.H.A. & R.R. Sokal, 1962, Numerical taxonomy. *Nature*, 193: 855-860.
- Sneath, P.H.A. & R.R. Sokal, 1973, *Numerical Taxonomy*, W.H. Freeman and Company, San Francisco.
- Sokal, R.R. & P.H.A. Sneath, 1963, *Principles of Numerical Taxonomy*, W.H. Freeman and Company, San Francisco.

10 ORIGIN AND EVOLUTION OF ANGIOSPERMS (ANTHOPHYTA)

AT A GLANCE

The green kingdom : The Plantae includes an amazing range of diverse forms of land plants including the mosses, ferns, conifers, flowering plants, and so on. Plants have been around for a very long time. Although the plants first appeared in the Ordovician, they did not begin to resemble modern plants until the Late Silurian. By the close of the Devonian, about 360 million years ago, there were a wide variety of shapes and sizes of plants around, including tiny creeping plants and tall forest trees. Angiosperms are the most diverse group of the plant kingdom, comprising of about 250,000 species in 350 families (Kenrick, 1999). The distinctive feature of this group is the flower, a cluster of highly specialized leaves, which participate in reproduction. Not all families under the angiosperms are equivalent in diversity. Certain families (e.g. Asteraceae, Orchidaceae, Poaceae) have much larger numbers of species than others (Stebbins, 1981).

The origin of angiosperms still remains a mystery. More than a hundred years ago, Darwin called the origin of angiosperms an "abominable mystery". He referred to the dreadful lack of fossil evidence that would help link angiosperms to gymnosperms, seed ferns, or other extant or extinct groups. There are only a few early fossils of angiosperms, and those available are troublesome. Many of the early fossils show a mix of features, which define modern groups, making them difficult to interpret. The lack of fossils was attributed to the thought that angiosperms may have evolved in montane habitats, in dry, upland areas that were not optimal for fossilization. In the fossil record, angiosperms appear rather suddenly, with no obvious ancestors for a period of about 80 to 90 million years prior to their appearance. Not even fossil leaves or pollen are known from this earlier period. The rapid diversification of angiosperm taxa, with an almost exponential increase in diversity, began in the Albian, in the mid-Cretaceous, and has continued till date. There does not appear to have been any major extinctions of groups in between and also there is no indication of where the taxa are coming from. Although rapid diversification began in the Albian, it is not until the Cenozoic that angiosperms began to take on important ecological roles.

There has been considerable difference of opinion as to the facts about the origin of angiosperms. 'Key innovations' have been explored, however, there is no consistency between groups that may lead to a general explanation for the evolution of angiosperm diversity (Stebbins, 1981). According to Takhtajan (1976) neoteny (a delay in development causing juvenile forms of the body or of particular organs to be expressed in the adult stage) played a significant role in angiosperm evolution. Reproductive structures (Stebbins, 1981; Takhtajan, 1976) and insect pollination (Burger, 1981) are also considered to be important factors involved in angiosperm evolution. Regarding their time of origin:

- Some morphologists consider the seed plants of the Carboniferous or Permian periods, some 200-250 million years in the past, as their possible ancestors.
- Some consider the Jurassic about 150 million years ago, to be the more probable period of their origin.
- Some regard them to be more recent in origin, not older than the Cretaceous, about 100 million years ago.

Accordingly, at one time or another, all major groups of plants including Lycopodiales, Filicales, Cycadales, Pteridospennales, Bennettitales, Gnetales, etc. have been considered as ancestral to angiosperms. Even algae have been taken to be possible precursors by some (Takhtajan, 1969).

Three major issues thus currently surround the origin and evolution of the angiosperms:

- The time of their origin.
- The geographic location of their origin.
- The ancestral plant group from which they evolved.

10.1 What Makes a Plant an Angiosperm?

Angiosperm is a Greek term, meaning *vessel seed* i.e., angiosperms are plants bearing seeds, which are borne within a vessel or ovary. Bessey (1915) however treated the angiosperms in a different phylum *Anthophyta*, which is also a Greek term meaning *flowering plant*. Thus, angiosperms may be defined as a group of flowering plants, which bear seeds enclosed within the carpel, and are also characterized by the presence of vessels in the stem (Eames, 1961). However there are exceptions to these characters in certain families of angiosperms, and the angiosperms can be distinguished from the gymnosperms by the presence of the following characters:

- **Vascular tissue** — Angiosperms have vessels in secondary xylem. However, some Magnoliid angiosperms lack vessels (e.g. the Winteraceae, Tetracentraceae, Trochodendraceae, etc.). Most gymnosperms lack vessels, except the Gnetales. Hence, the presence of vessels cannot be used as evidence that a plant is an angiosperm.
- **Leaves** — Angiosperms have either parallel or reticulate venation with intercostal veins. On the other hand gymnosperms often have parallel venation with few branching (intercostal) veins. Extant gymnosperms such as *Gnetum* have strikingly angiospermous leaves. Hence, leaf morphology can only provide supportive, not conclusive evidence of a plant being an angiosperm.
- **Flowers** — Angiosperms are flowering plants. However, the presence of a flower, as a character to distinguish angiosperms from gymnosperms, is a matter of artificial definition, as it is not possible to define a flower with morphological precision, so as to exclude the cone or cluster of fertile sporophylls that characterize the gymnosperms.
- **Pollen wall structure** — Outer wall of the pollen grain is the ektexine. In angiosperms, ektexine (outer wall of the pollen grain) is either columellate or tectate (Fig. 10.1). Gymnospermous pollen are atectate i.e., lacks tectae. On the other hand endexine (the inner wall of the pollen) is spongy in gymnosperms, but not spongy in angiosperms.
- **Pollen apertures** — Advanced angiosperm pollen are triaperturate i.e., have three apertures whereas gymnosperms and Magnoliid angiosperms (most primitive group) pollen are uniaperturate i.e., have only one aperture. In the more advanced angiosperms, the pollen may also be monocolpate and tricolpate. So, pollen aperture number cannot be used to infer whether a plant is a gymnosperm or an angiosperm.
- **Reduced female gametophytes** — Angiosperms have no archegonium as is found in gymnosperms. However, some gymnosperms also lack archegonia, e.g. within Gnetales, only *Ephedra* has one.
- **Carpels** — The best evidence that a plant is an angiosperm is the presence of ovules enclosed within a carpel (megasporangium within an inrolled megasporophyll). However, according to Endress (1994), none of these carpels are truely closed because secretions from the inner lining of the carpel fill the gap.
- **Double fertilization** — In angiosperms, triploid endosperm is produced by the fusion of a sperm nucleus and a polar nucleus. A similar form of double fertilization has been found to occur in *Ephedra* (Gnetales) [Friedman, 1990], but no endosperm is formed, instead simply multiple embryos are formed. Friedman hypothesized that if this process was homologous with angiosperms, endosperm could have evolved from such "supernumerary fertilization events."

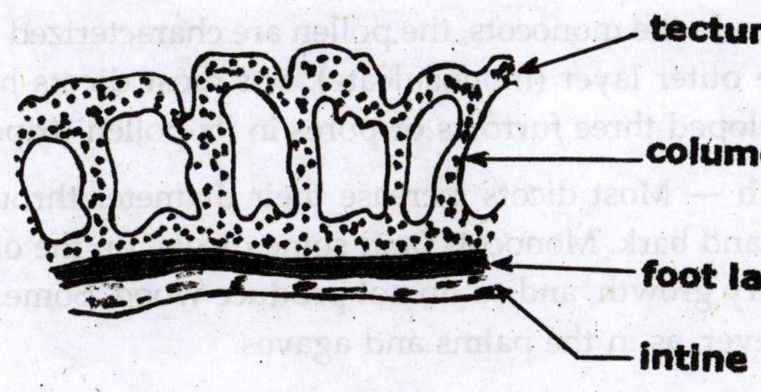

Fig. 10.1 : Tectate-collumellate wall (exine) of angiosperm pollen

10.2 The Two Classes of Flowering Plants (Monocots versus Dicots)

Traditionally, the flowering plants have been divided into two major groups, or classes: the Dicots (Magnoliopsida) and the Monocots (Liliopsida). Theophrastus (circa 370 BC) is credited for first recognizing differences between the two groups. However, until the 1600s, classification of plants was based upon overall growth form and it was in 1682 that, Dicotyledones and Monocotyledones were first given formal taxonomic standing by John Ray in his book *Methodus Plantarum Nova*, which was later popularized by Antoine Laurent de Jussieu in his *Genera Plantarum* of 1789. Following are some characters, which distinguishes between dicots and monocots, although there are many exceptions to these characters in both groups:

- **Number of cotyledons** — The actual basis for distinguishing the two classes of angiosperms is the number of cotyledons found in the embryo, and is the source of the names Monocotyledonae (one cotyledon) and Dicotyledonae (two cotyledons).
- **Venation** — In monocots, there are usually a number of major leaf veins, which run parallel the length of the leaf whereas in dicots, there are usually numerous auxillary veins, which reticulate between the major ones. This character is also not always reliable, as there are many monocots with reticulate venation, example the Aroids and Dioscoreales.
- **Vascular arrangement of stem** — Vascular bundles within the stem of dicots are arranged to form a cylinder, appearing as a ring across the stem. On the other hand in monocots, the vascular bundles appear scattered through the stem, with more of the bundles located toward the periphery of the stem than in the center. This arrangement is unique to monocots and some of their closest relatives among the dicots.
- **Root development** — In most dicots the radicle (the part of the embryo from which the root develops) gives rise to an apical meristem, which continues to produce root tissue for most part of the plant's life. However in monocots, the radicle aborts and new roots arise adventitiously from nodes in the stem.
- **Number of flower parts** — The monocot flowers tend to have a number of parts that is divisible by three, usually three or six. Dicot flowers on the other hand, tend to have

parts in multiples of four or five (four, five, ten, etc.). This character is however not always reliable, and is not easy to use in some flowers with reduced or numerous parts.

- **Pollen structure** — In the monocots, the pollen are characterized with a single furrow or pore through the outer layer (monosulcate), but most dicots have descended from a plant which developed three furrows or pores in its pollen (triporate).
- **Secondary growth** — Most dicots increase their diameter through secondary growth, producing wood and bark. Monocots (and some dicots) on the other hand have lost the ability of secondary growth, and so do not produce wood. Some monocots can produce a substitute however, as in the palms and agaves.

MONOCOTS	DICOTS
1. Embryo with single cotyledon	1. Embryo with two cotyledons
2. Major leaf veins parallel	2. Major leaf veins reticulated
3. Stem vascular bundles scattered	3. Stem vascular bundles in a ring
4. Roots are adventitious	4. Roots develop from radicle
5. Flower parts in multiples of three	5. Flower parts in multiples of four or five
6. Pollen with single furrow or pore	6. Pollen with three furrows or pores
7. Secondary growth absent	7. Secondary growth often present

10.3 Are Angiosperms Monophyletic or Polyphyletic?

A fundamental problem that must be dealt with is whether the angiosperms are monophyletic (i.e. a group consisting of all descendents derived from a single ancestor) or polyphyletic (i.e. a group that does not have a common ancestor). Due to inadequate fossil records, the question of phyly of the angiosperms still remains unsolved. However, angiosperms are a natural group and contains characters, which make them unique from all other vascular plants (Takhtajan, 1969; Stebbins, 1974). There are two opposing views regarding the origin of angiosperms :

(a) Monophyletic origin

- As a group, the angiosperms have typically been viewed as being monophyletic (e.g. Dahlgren, 1983; Doyle & Donoghue, 1993; Faegri, 1980; Takhtajan, 1969). However, no definite fossil evidences are available in favor of the monophyletic origin. This view is based on the fact that present-day angiosperms show remarkable consistency in their characters, i.e. presence of sieve tubes in all, uniform staminal structure, characteristic endothedial layer of the anther wall, double fertilization, and formation of triploid endosperm, which are considered defining features of angiosperms (Dilcher, 2000; Stebbins, 1976) and support the monophyletic grouping (Crane et al., 1995).
- The monophyletic origin of angiosperms is also supported by Hickey & Doyle (1977) on the basis of their studies of monosulcate pollen.
- Dahlgren (1963) believes that the ancestor of the present-day angiosperms was a gymnospermous member.

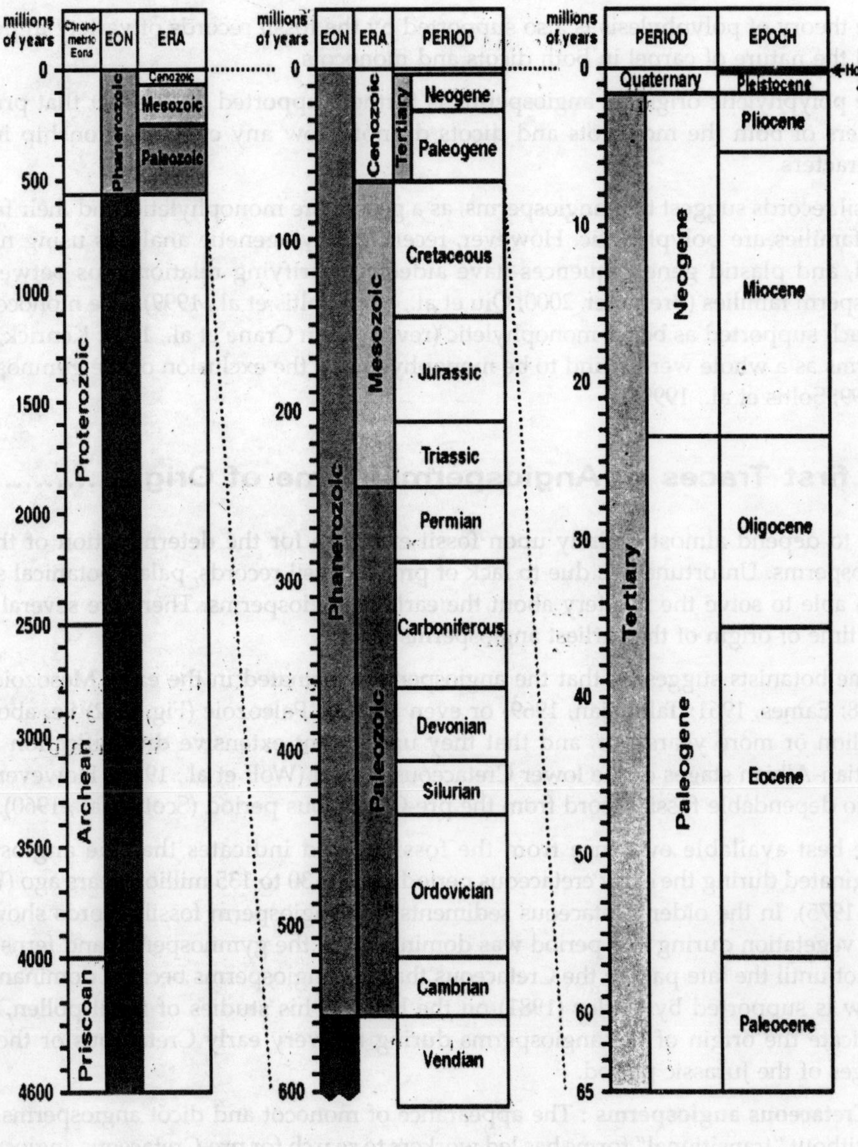

Fig. 10.2 : Geological Time Scale

(b) **Polyphyletic origin**

- Several phylogenists including Cronquist (1965), Hughes (1994), Games (1961), Krassilov (1977) and Meeuse (1970,1975) have argued that the angiosperms are polyphyletic i.e. dicots and monocots originated from different primitive stocks at different times, and attained their present status through parallel or convergent evolution.
- The theory of polyphylesis is also supported by the fossil records of variety in perianth and the nature of carpel in both dicots and monocots.
- The polyphyletic origin of angiosperms is further supported by the fact that primitive orders of both the monocots and dicots do not show any close relationship in their characters.

Thus fossil records suggest that angiosperms, as a group, are monophyletic, and their families or groups of families are polyphyletic. However, recently, phylogenetic analyses using nuclear, mitochondrial, and plastid gene sequences have aided in clarifying relationships between the various angiosperm families (Bremmer, 2000; Qiu et al., 1999; Soltis et al., 1999). The monocots and eudicots are each supported as being monophyletic (reviewed in Crane et al., 1995; Kenrick, 1999). The angiosperms as a whole were found to be monophyletic to the exclusion of the gymnosperms (Qiu et al., 1999; Soltis et al., 1999).

10.4 The first Traces of Angiosperms (Time of Origin)

We have to depend almost entirely upon fossil evidence for the determination of the first traces of angiosperms. Unfortunately, due to lack of proper fossil records, palaeobotanical studies have not been able to solve the mystery about the earliest angiosperms. There are several views regarding the time of origin of the earliest angiosperms :

- Some botanists suggested that the angiosperms originated in the early Mesozoic (Just, 1948; Eames, 1961; Takhtajan, 1969) or even the late Paleozoic (Fig. 10.2) i.e. about 250 million or more years ago, and that they underwent extensive diversification by the Aptian-Albian stages of the lower Cretaceous period (Wolf et al., 1975). However, there is no dependable fossil record from the pre-Cretaceous period (Scott et al., 1960).
- The best available evidence from the fossil record indicates that the angiosperms originated during the early cretaceous period about 130 to 135 million years ago (Wolf et al., 1975). In the older Cretaceous sediments, the angiosperm fossil records show that, the vegetation during this period was dominated by the gymnosperms and ferns and it is not until the late part of the Cretaceous that the angiosperms became dominant. This view is supported by Taylor (1981) on the basis of his studies of fossil pollen, which indicate the origin of the angiosperms during the very early Cretaceous or the latter stages of the Jurassic period.

(a) **Pre-Cretaceous angiosperms :** The appearance of monocot and dicot angiosperms in the fossil record without "transitional" forms has led workers to search for pre-Cretaceous angiosperms. Several authors have also hypothesized the possible mechanisms of the evolutionary change. Many of the purported pre-angiosperm ancestors have "angiosperm" leaf characters (net-like venation pattern), which had arisen independently in several clades.

- **Sanmiguelia** is a herbaceous plant of the Late Triassic, whose leaves although appearing "palm-like" lack the midvein and petiole features of true palms. This enigmatic plant has been allied by some workers with the cycads, whereas others have allied it with pteridosperms.

- **Furcula** is also a Late Triassic genus, which shows angiosperm-leaf characteristics along with that of other groups, which include a forked midrib (primary vein) from which secondary veins arise. Intercostal veins between these anastomose to form a reticulate pattern that delimits poorly defined areoles. Stomatas are syndetocheilic, which is also an angiosperm character.
- **Pannaulika**, a Late Triassic leaf, is the only known specimen having brochidodromous secondary veins with tertiary veins forming loops outside the secondary loops, which is also an angiospermic character.
- **Marcouia** is another Late Triassic pinnate leaf, which is palmately compound with four-five pinnae, that are basally united. Secondary veins originating from a well-defined midrib dichotomize and anastomose. This genus probably does not have any relation to angiosperms.
- Monocolpate pollen grains having an outer exine, which is differentiated into a well-defined tectum supported by columellae, characteristic of angiosperms, have been described from the Late Triassic-Early Jurassic.

The leaf impressions from Triassic and Jurassic deposits, which have been attributed to angiosperms, are disputable. Similarly, the pollen from the pre-Cretaceous deposits, are also very problematic and in most cases have proved to be pollen grains of gymnosperms. Even some deposits of this period, which were considered to be pollen grains of angiosperms at one time, have been suggested to be the unicellular green algae *Tetraedron*. Pollen sequences have clearly shown a pattern of increasing diversity in the cretaceous from primitive types in older strata to more derived types in younger sediments. The first monosulcate angiosperm pollen grains, characteristic of the primitive dicotyledons and the monocotyledons, appear in the Barremian stage of the lower cretaceous. Tricolpate pollen grains, characteristic of the more advanced dicotyledons were first reported from slightly younger Aptian rocks. The increasing diversity of pollen through the cretaceous indicates that angiosperms underwent much diversification during this period (Raven & Axelrod, 1974; Wolfe et al., 1975).

Further, recent investigations have shown that double fertilization, anastomosing leaf venation, reduction of the male gametophyte, tetrasporic megagametophyte with free nuclei serving as eggs, and a feeder in the embryo, the primary characters of angiosperms, occurs sporadically in *Ephedra*, and a welwitschian-type reproductive structure (*Archaestrobilus*) found in the Late Triassic of Texas, which has been interpreted to have been a precursor to angiosperms. Thus, there is little doubt that angiospermic features originated from a gymnosperm ancestor. Recently, however, a fossil of *Archaefructus liaoningensis*, the world's oldest flower, from a lake deposit in China, in which limestones and volcanic ashes accumulated, exhibiting the presence of closed carpels was found, revealing that angiosperms were present in the Late Jurassic (Sun et al., 1998). Pollen and pollen-bearing organs were not found in the specimen. Whether *Archaefructus* flowers were unisexual or bisexual, also remains unknown.

(b) Upper-Cretaceous angiosperms : The upper Cretaceous period is marked by the abundance of genera and families of angiosperms. The floras of the Vancouver Island (Bell, 1957), Amboy clays of New Jersey (Newberry, 1895), the basin of the Rocky Mountain system (Dorf, 1936, 1938 & 1942), Dakota sandstone near Kansas, Nebraska and Minnesota (Lesquereux, 1891), etc. were particularly rich in angiospermous genera, which altogether account for about 700 to 750 forms, of which most are referable to extant genera. Several fossil forms including abundant fruits and seeds have been described subsequently from the Tertiary floras of Europe and Western United States, the valaughnian deposits of Southern France (Chandler, 1958) and Northern California (Chandler & Axelrod, 1961).

Angiosperms steadily and progressively occupied all vegetated areas of the world from Albian to Senomanian times, none of which could however be classified in recent families. Angiosperms appear to be comparatively rare up to Albian times, apparently forming only small populations with smaller number of individuals. Occurrence of angiospermous woods, leaves, pollen, etc. became considerably more numerous from Aptian-Albian times onwards, and became widely distributed throughout the world at the close of the Albian period or towards the middle of the Cretaceous period during which, they appeared in great diversity of form and quickly became dominant. Most of these Cretaceous angiosperms belong to the extant genera, and there are representatives of both more or less primitive forms, as well as the highly evolved ones.

10.5 The Geographic Location of the Origin of Angiosperms ...

The geographic origin of the angiosperms has been under considerable debate. Just as the fossil evidence is insufficient to decide the time, as well as the nature of ancestors of angiosperms, it is also inadequate to settle the problem of the place of origin. There are three divergent views regarding the place of the origin of angiosperms, which are under mentioned :

(a) **The hypothesis of Arctic (or Antarctic) origin** : The dominant view regarding the origin of angiosperms during the late nineteenth and early twentieth centuries, had been that the angiosperms originated high in the northern latitudes and even in the polar-regions. Heer (1868) first suggested the northern polar region to be the center of origin of angiosperms and the hypothesis was strongly supported by a large number of palaeobotanists as well as phytogeographers including Saporta (1877), Hooker (1879), Gray, Engler, Seward, Berry, Chaney, Just, Arnold and many others. This hypothesis holds that angiosperms originated in the polar region of Holarctic and from the polar Arctic and Antarctic regions, the angiosperms spread in successive waves across the whole earth. It provides an easy explanation for the present distribution of many genera in widely separated continents, chiefly the northern America and the eastern Asia. This view has however been disputed for a more tropical "center of origin" (between 45° N to 45°S), as at the beginning of the Cretaceous, angiosperms were found only between 45° N and 45°S latitudes and not beyond these, from which they migrated towards the polar regions (Axelrod, 1959). The fossil evidence suggests that it was not until the Late Cretaceous that angiosperms established in the Arctic, as angiosperm fossils from the Early Cretaceous have been found only at low latitudes (Takhtajan, 1969). According to Hickey et al. (1983) the Arctic may have been an important site for some level of floral evolution in the Late Cretaceous. It is only at the end of the early Cretaceous that the angiosperms reached higher latitudes, eventually replacing the relict Jurassic-flora at higher latitudes towards the end of the Mesozoic (Axelrod, 1959). Thus, Takhtajan (1969) is of the opinion that "the hypothesis of Arctic (or Antarctic) origin for angiosperms" must be firmly rejected.

(b) **Theory of the south-east Asian center of origin** : The angiosperm center of origin has been predicted to be in south-eastern Asia (Takhtajan, 1969 & 1987). Hallier (1912) proposed that the angiosperms originated first in the basin of the Pacific Ocean, at places such as New Zealand, New Caledonia, Andes and the Hawaiian Islands. Takhtajan (1969) has suggested that the 'cradle of the angiosperms' occurs somewhere between Assam and Fiji (including Burma, Thailand, Indo-China, and Malaysia), due to the high abundance of primitive angiosperm families (e.g. Amborellaceae, Austrobaileyaceae, Calycanthaceae, Degeneriaceae, Eupomatiaceae, Gomortegaceae,

Himantendraceae, Lactoridaceae, Magnoliaceae, Trochodendraceae, Tetracentraceae, Winteraceae, etc.) found in that region of the Pacific basin. However, this does not mean that all primitive angiosperms are concentrated in the South-Eastern Asia. They occur outside this region as well. Takhtajan (1987) therefore considers south-east Asia to be the most likely region where angiosperms originated, and descendants of the first angiosperms may have dispersed far from their point of origin. The lack of fossil evidence is the serious drawback of this theory (Takhtajan, 1969). The present acceptance of the theory is largely based on the study of distribution and comparative morphology of presumably primitive angiosperms.

(c) **"Species pump" hypothesis** : According to Stebbins (1974), the tropical rain forests with equable climate have most probably served as the "museums" rather than "cradle" of angiosperms i.e., they are not the communities within which angiosperms originated and differentiated, but they have preserved the diversified plant groups because of low rate of extinction. He proposed that active and rapid speciation, leading to the origin of diversified groups of angiosperms is likely to occur in marginal or ecotonal regions of tropical forests, due to increasing drought. The museum hypothesis is supported by the fact that tropical forests do not contain any transitional forms or when present they are less abundant. However, they contain a complete spectrum of families ranging from the most primitive to the most advanced forms. According to the "species pump" hypothesis developed by Valentine (1967), the newer populations produced in the marginal areas are "pumped" into the stable communities of tropical forests from time to time. Thus, angiosperms, in all probabilities have developed in semi-arid situations, marked by seasonal drought and a short season favorable for the formation of flowers and seeds. The regions in which climatic factors have greatly promoted the diversification of modern flora are the Ethiopian highlands, the Cape region of South Africa, parts of the Indian Peninsula, the coast of Equador, south-central Mexico, northern Venezuela, and the northern provinces of Argentina.

10.6 The Nature of Probable Ancestors of Angiosperms .... ➧

The identity of the ancestors of the flowering plants has been a very difficult job due to lack of proper fossil evidence. Various plant groups have been considered as ancestral stock for angiosperms. Different workers have proposed various orders of gymnosperms (Bennettitales, Cordaitales, Cycadofilicales, Cycadales, and Gnetales), pteridophytes (Psilophytales, Lycopodiales and Equisetales), and even algae as the probable ancestors of angiosperms. But, none of these hypotheses appear to be tenable or based on sound footing.

(a) **Were the first angiosperms trees or herbs?**

Besides the origin of angiosperms, another big question about early angiosperms is the nature of their growth habit i.e., were the first angiosperms woody trees and shrubs, or were they small herbs? Two competing hypotheses have been proposed, based on recent discoveries of a herbaceous angiosperm fossil record in the Early Cretaceous, which paint very different pictures about the biology of the earliest flowering plants i.e., about the implications for angiosperm diversification. The Paleoherb Hypothesis suggests that the basal lineages were herbs with rapid lifecycles, while the Magnoliid Hypothesis suggests that the basal lineages were small trees with slower lifecycles (Fig. 10.3). Although each hypothesis has been favored at one time or another, the question however still remains unanswered. Early angiosperms were paleoherbs, rather than trees or shrubs, although scientific opinion currently favors a woody origin.

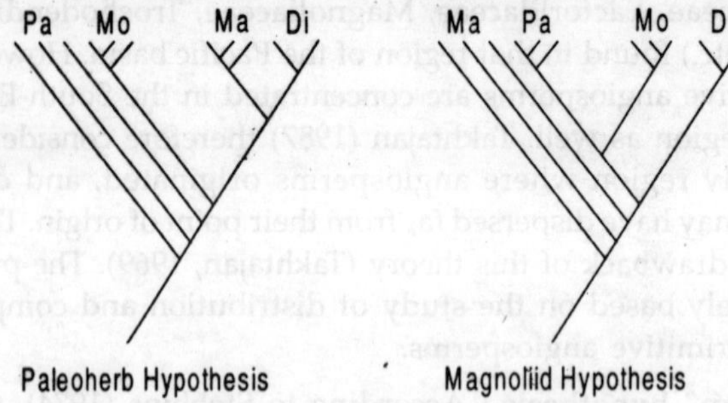

Fig. 10.3 : Paleoherb and Magnoliid Hypothesis (Di = tricolpate dicots; Ma = magnoliids; Mo = monocots; Pa = paleoherbs)

- **The Paleoherb Hypothesis** — This view is based on an herbaceous origin for the angiosperms. It has been promoted by results of cladistic analyses conducted by paleobotanists Taylor and Hickey. Their cladistic analysis of angiosperms suggests that, the basal angiosperms are tropical paleoherbs, a group of flowering plants with uncomplicated flowers and a mix of monocot and dicot featu"es. The implication of their proposal is that in the earliest angiosperms, there was presence of the key innovations of flowers and a rapid life cycle. Taylor and Hickey have suggested that changes in climate or geography provided opportunities for these early angiosperms to diversify.
- **The Woody Magnoliid Hypothesis** — This hypothesis is the result of cladistic analyses by Doyle and Donoghue, according to which, the basal angiosperm stock is an early angiosperm with morphology similar to living members of the Magnoliales and Laurales, which are comprised of small to medium-sized trees with long broad leaves and large flowers with indeterminate number of perianth parts. The carpels are imperfectly fused, and make a physical intermediate between a folded leaf and fused pistil. This hypothesis is currently favored by systematic botanists, as it is supported by molecular studies [evolutionary studies based upon DNA sequencing (plastid gene rbcL)]. It suggests that the earliest angiosperms were understory trees and shrubs, and that the flower was not the key innovation for the rapid diversification of angiosperms. In fact woody magnoliids are not particularly diverse, even today.

(b) Some theories about probable ancestors of angiosperms : Various well-known plant groups, have been deemed to be closely related to ancestral stock or stocks of flowering plants. Following are the views of various workers:

- **The Isoetes - Monocotyledon Theory :** This theory was proposed by Campbell (1928), and is essentially based on marked similarity between *Isoetes* and Monocots especially *Najas flexilis.* He advocated that monocotyledons have been derived from *Isoetes* via Ophioglossaceae. Both eusporangiate and leptosporangiate ferns, and their allies like Lycopodiales, Isoetales, etc., have been shown to possess many common features with those of the monocotyledons. *Isoetes* an aquatic or amphibious genera, is predominantly herbaceous and geophilous and like most of the Filicineae, it is found in the humid tropics, a habitat which is also characteristic of a large number of monocotyledons. The marked similarity in habit and in the embryo and anatomy of the older sporophyte of *Isoetes* and some other lower aquatic monocotyledons, have also been pointed out by Campbell. He suggests that the herbaceous character in angiosperms is primitive, inherited from filicinean ancestors.

The probable origin of monocotyledons from various groups of Pteridophytes, through an intermediate and hypothetical group, the protangiosperms has also been postulated by Engler and his associates, according to whom due to the herbaceous nature of most of the members of this group, they are not represented in the fossils. Engler also postulated the direct derivation of monocotyledons from Ophioglossaceae on the basis of similarities in their vascular bundles. However, this theory of Pteridophytes ancestry is highly unacceptable, as the monocotyledons are now considered as most advanced group of angiosperms and derived from dicotyledons.

- **The Coniferales-Amentiferae Theory :** This theory has been proposed by Eichler (1875), Engler (1882 & 1892), Engler & Prantl (1924), Rendle (1904 & 1930), Hagerup (1934 & 1936) and Doyle (1945). According to this theory, the Coniferales have been treated as the probable ancestors of angiosperms by several phylogenists including Engler (1892) and Rendle (1904). They pointed out several resemblances between angiosperms and conifers and proposed that the Coniferales might have given rise to primitive (hypothetical) group of angiosperms known as Amentiferae. These similarities include:
  - ➢ Similarities between the inflorescences of the amentifers like Casuarinaceae, Salicaceae, Fagaceae, etc., consisting of simple and naked flowers, with those of the conifers.
  - ➢ Similarities in the stamens and bisporangiate microsporophylls of many conifers.
  - ➢ Covered nature of seed in *Agathis* and *Araucaria* as found in angiosperms.
  - ➢ Similarities in the seed-scale complex of *Juniperus* with the gynoecia of Amentiferae (Hagerup, 1934 & 1936).
  - ➢ Similarity in the fertilization process of certain conifers with that of angiosperms, where the pollen grains are deposited on the scale and only the pollen tube enters the micropyle (Doyle, 1945).

In spite of these similarities, it is now very well documented that these resemblances are only superficial and due probably to parallelism. Further this theory is not acceptable because the Amentiferae group is now considered as an advanced group of angiosperms. The advanced features in Amentiferae include:

- ➢ The primitive angiosperm flower is bisexual whereas cones of conifers are uniformly unisexual.
- ➢ The inflorescence of amentiferous families exhibit specialized structures. The flowers of these families are not primitively simple but are highly reduced.
- ➢ The wood anatomical data suggest that Amentiferae are highly advanced. Some of the angiosperms like *Trochodendron*, *Tetracentron* and others, which lack vessels, have a structurally more primitive secondary xylem than conifers as well as amentifers.
- ➢ The ovuliferous scales of conifers and carpels of angiosperms follow distinctly different lines of evolutionary specialization.
- ➢ The carpels of angiosperms are modified leaves (phyllosporous), whereas the seed-bearing organs of conifers are never so and probably represent modified telome-systems.

Thus, this theory, which proposes Coniferales as the ancestors of angiosperms and Amentiferae as primitive group of angiosperms is ruled out.

- **The Gnetales-Angiosperm Theory :** This theory was proposed by Richard von Wettstein (1901) and supported by Markgraf (1930) and Fagerlind (1947). According to this theory, Gnetales, the transitional group between angiosperms and gymnosperms, are considered as the ancestors of angiosperms. Thus, among living plants, the gnetophytes (*Ephedra*, *Gnetum*, and *Welwitschia*, forming a monophyletic clade themselves), are believed to be most closely related to the flowering plants. The gnetophytes, like the other extant (non-angiosperm) seed plants, had their greatest diversity in the past. However today, *Gnetum* forms a tropical moist-habitat plant, while *Ephedra* (containing about 30 species with a global distribution) and *Welwitschia* (represented by a single species that is restricted to some of the driest portions of the Namib Desert in Southwest Africa) are both, dry-climate or desert plants.

  The Gnetales resemble angiosperms in many respects :

  - The leaves of *Gnetum*, have reticulate venation, resembling that of some dicot angiosperms. On the other hand *Ephedra*, in contrast to other conifers, whose leaves are vascularized by a single trace, produces highly reduced spike-like leaves, vascularized by a pair of traces that exit from the eustele and enter the leaf. An even more bizarre leaf arrangement is found in *Welwitschia*, which produces a single pair of large, parallel-veined leaves that continue to grow throughout the life of the plants. This is a very derived condition that might best be interpreted as progenetic.
  - The Gnetales, like the flowering plants, possess vessels in their wood. However, the vessels in the Gnetales have a subtly different developmental origin and thus may not be strictly homologous with those of angiosperms (Thompson, 1918).
  - All the three members of Gnetales have two cotyledons.
  - The stamens of all the three genera of Gnetales are apparently similar to those of angiosperms.
  - Unlike most gymnosperms, ovules of Gnetales have close resemblance with those of most angiosperms in possessing two integuments instead of only one. These envelopes have been variously homologised with perianth or carpels of angiosperms.
    - ❑ In *Gnetum* the inflorescences are unisexual, which can be easily compared with the catkins of many amentiferous angiosperms. The strobili are arranged on axis with a conspicuous node-internode organization. Microsporangiate strobili have two fused bracts that form a cupule, and surrounds fertile shoots, each being composed of two fused bracteoles, which surround the microsporophyll. In the megasporangiate strobilus, the cupule subtends a whorl of ovules, the outer integument of which are wrapped with two external envelopes, the tissue of which may be sclerified and fused to the integument to form a seed-coat-like structure, which makes the *Gnetum* seed appear like angiosperms.
    - ❑ *Ephedra* is dioecious, with mega- and microsporangiate strobili. The microsporangiate strobili consist of several pairs of bracts, from the axis of each of which arises a shoot bearing bracteoles (smaller bracts), each bracteole in turn surrounding a stalked microsporangiophyll bearing two or more

pollen sacs (microsporangia). The megasporangiate cone is also similarly arranged with pairs of bracts, the top two of which subtend an ovule, which appears to be surrounded by two integuments. However, according to most interpretations, the inner layer is the true integument, while the outer layer is a reduced bracteole similar to that surrounding the microsporangia.

- ❑ The male flowers of *Welwitschia* are truly bisexual and show that they are derived by reduction of the female parts.

- ➢ Gametophytes of *Gnetum* and *Welwitschia* are highly reduced like those of angiosperms.
- ➢ In most angiosperms, in the process of double fertilization, two sperm are involved in fertilization, one of which unites with the egg cell to form a diploid zygote, while the other fuses with the two polar nuclei to form a triploid endosperm, which provide nutrition to the developing embryo. *Ephedra* and its sister taxon, *Gnetum*, undergoes a form of double fertilization, which is very much similar to that of angiosperms. In *Ephedra*, two egg cells are produced, one of which is fertilized to become the embryo while, the other begins but does not complete development into an embryo, thus not providing nutrition to the developing embryo as does the endosperm of angiosperms (Friedman, 1994). The degenerate embryo is believed to be an intermediary step between the normal seed plant condition (where embryo is nourished by a larger gametophyte) and that of angiosperms (megagametophyte so reduced as to not be effective for nourishing the embryo). This can also be explained on the basis of reduction in the polyembryony as observed in many non-angiosperm seed plants, for example in some conifers, where several embryos form from fertilizations by several pollen grains, but ultimately, only one embryo per seed survives.

Unfortunately, in spite of these outward resemblances, the details of the vascular anatomy and developmental studies, strongly refutes the suggestion of the existence of any similarity between the two groups. Compared to that of some of the primitive angiosperms, the members of Gnetales have attained a higher evolutionary level in several respects. For example:

- ➢ They have vessels in the secondary wood, which are absent in the xylem of a few genera of angiosperms. Thompson (1918) has shown that the vessels of Gnetales, originated in an entirely different way from those of angiosperms.
- ➢ The scalariform pitted tracheids characteristic of primitive angiosperms, are eliminated from both the primary and secondary xylem in the Gnetales and replaced by circular bordered pits in the secondary xylem. This development of vessels is unique in type and is entirely different from that of vessels in angiosperms and certain vascular cryptogams (Bailey, 1944; Cheadle, 1953).
- ➢ The similarity between the strobilus of Gnetales and the aments of certain angiosperms, is now considered as superficial and due to some degree of convergence.
- ➢ The ovules of Gnetales are naked, the nature of their envelopes is uncertain and they also differ in the general structure.
- ➢ The similarity in the number of cotyledons between Gnetales and angiosperms also appears to be due to convergent evolution. Angiosperms might have been derived from poly-cotyledonous rather than dicotyledonous ancestors.

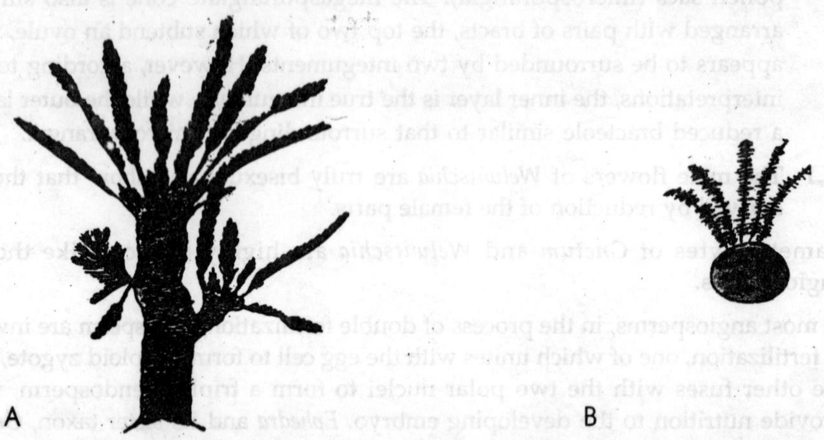

↳ Fig. 10.4 : (A) Reconstuction of *Williamsonia sewardiana* with spirally arranged leaf scars. (B) *Cycadeoidea* trunk and foliage drawn for comparison of proportions.

- **The Anthostrobilus (Bennettitalean) Theory :** The bennettitaleans, which first appeared in the Triassic and became extinct toward the end of the Cretaceous, were originally mistaken for cycads because of their cycad-like growth form. However, the reproductive structures of these plants and presence of syndetocheilic stomata, place them soundly in the anthophyte clade.

  Two major groups of bennettitaleans have been recognized (Fig. 10.4) :

  ➢ *Cycadeoidea* — They are with stout trunks and bisporangiate reproductive structures, and relatives of *Williamsonia.*

  ➢ *Williamsoniella* — They have slender, branching trunks and either bisporangiate or monosporangiate strobili.

  Both groups produced entire simple leaves or pinnate foliage and had woody stems that were heavily armored with persistent leaf bases, much like modern cycads. The groups only differ in the details of stomatal morphology.

  The Bennettitalean theory was first proposed by Saporta and Marion (1885), followed by Arber and Parkin (1907) according to which, Bennettitales have been proposed as possible ancestors of angiosperms, on the basis of the resemblance in structure between the strobili of the Mesozoic genus *Cycadeoidea* and the flower *of Magnolia.* Both these structures are:

  ➢ bisexual (Fig. 10.6-10.7), and

  ➢ contain an elongated axis having protective bracts, microsporophylls and megasporophylls, arranged successively from below upwards.

  However, in spite of these superficial resemblances, further studies have shown several differences, which are as follows:

  ➢ In *Magnolia,* the microsporophylls (stamens) are free and are spirally arranged on the axis, whereas in Bennettitales they are whorled and mostly connate (Fig. 10.5).

  ➢ In Bennettitales, the megasporophylls are greatly reduced, simplified stalk-like structures, each bearing a solitary terminal erect ovule. Between megasporophylls,

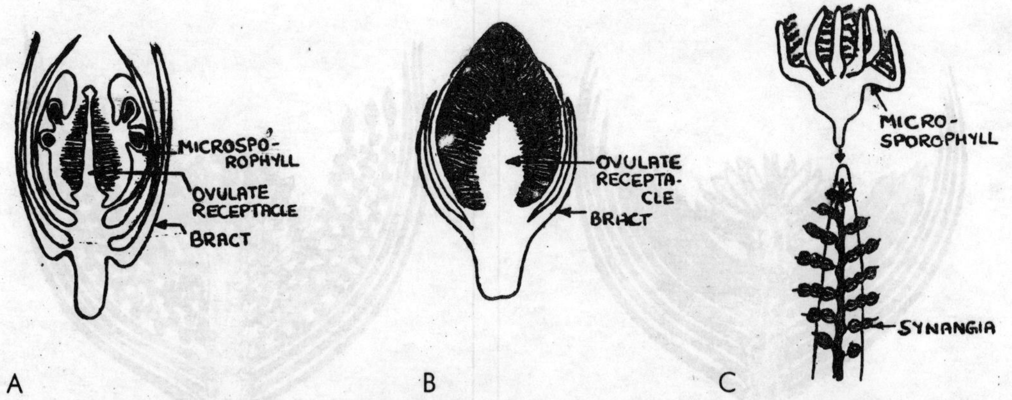

Fig. 10.5 : Examples of bennettitalean reproductive structures. (A) *Williamsoniella* bisexual reproductive structure (microsporophylls with microsporangia embedded within). (B) *Williamsonia* ovulate structure. (C) *Weltrichia* pollen-bearing structure with detail of microsporophyll showing attached synangia.

there are sterile scales (interseminal scales), which are protective in function. No such structures are present in the flowers of *Magnolia*.

- **The micropylar tube formed in the ovules of Bennettitales are absent in the angiosperms and the pollen grains are shed on the stigma of the carpel (megasporophyll).**
- **The seeds of *Magnolia* and other primitive angiosperms are with copious endosperm and small embryo while those of Bennettitales are non-endospermic with a large embryo.**

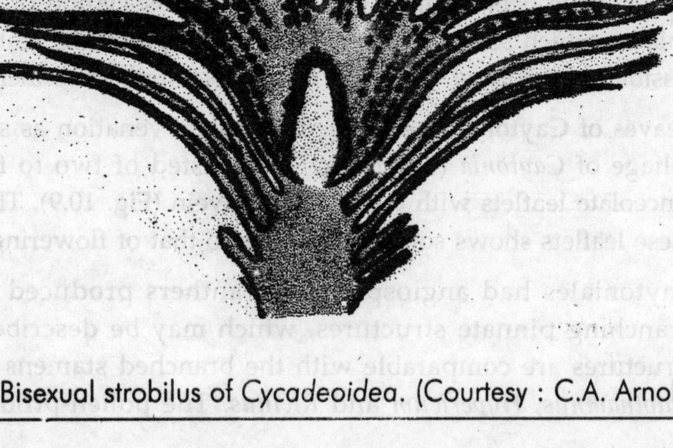

Fig. 10.6 : Bisexual strobilus of *Cycadeoidea*. (Courtesy : C.A. Arnold, 1947)

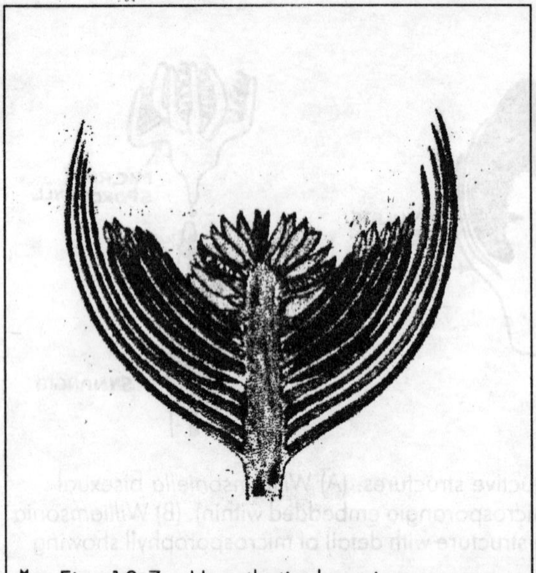

Fig. 10.7 : Hypothetical angiospermous strobilus in the primitive condition. (Courtesy : Arber & Parkin, 1907)

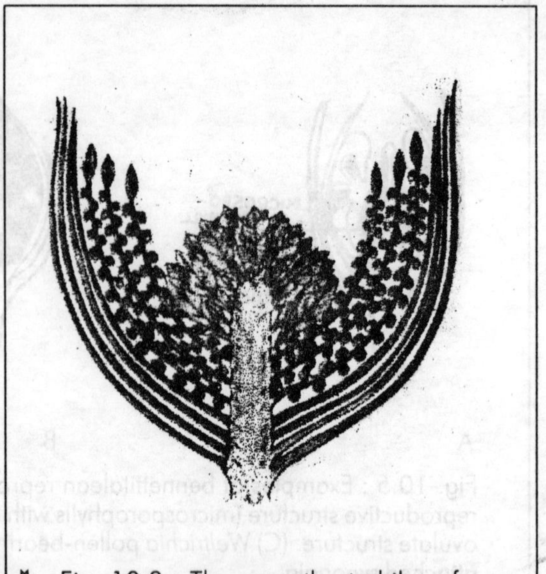

Fig. 10.8 : The proanthostrobilus of hypothetical Hemiangiospermae. (Courtesy : Arber & Parkin, 1907)

- In the bennettitalean stem there is a large pith, a thin vascular cylinder and a thick cortex, while the angiosperm stem has a small pith, a thick vascular cylinder and a thin cortex.

These differences indicate that Bennettitales cannot be considered as the ancestors of angiosperms. The similarities with angiosperms, most probably, might have resulted due to a common ancestry and parallel evolution. Arber and Parkin (1907) have postulated that the two groups did have a common origin from seed ferns and they might have diverged very early.

- **The Caytonialean Theory :** This theory was proposed by H. Hamshaw Thomas (1925 & 1936) on the basis of comparative studies of modern flowering plants, with those of Caytoniales, which are certain fossil plants of the middle Jurassic. This theory was later supported by Stebbins (1974). *Caytonia* had a wide distribution and it has been described from Triassic to Cretaceous localities in most of today's Northern Hemisphere. It is believed to have been a small tree, based on association of foliage with woody axis. Caytoniales, which seems to have a few angiosperm-like features, are looked upon as the possible ancestors of angiosperms. These similarities include:

  - Leaves of Caytoniales exhibited reticulate venation as seen in angiosperms. The foliage of *Caytonia* (*Sagenopteris*) consisted of two to four palmately arranged lanceolate leaflets with a distinct mid-vein (Fig. 10.9). The epidermal structure of these leaflets shows some similarities to that of flowering plants.

  - Caytoniales had angiosperm-like anthers produced in groups or singly on branching pinnate structures, which may be described as sporophylls. Such structures are comparable with the branched stamens found in such plants as *Calothamnus, Hypericum* and *Ricinus*. The pollen-producing organ of *Caytonia*

(*Caytonanthus*), consisted of synangial units arranged along an axis, which probably represented a modified pinnate microsporophyll (Fig. 10.10). Each synangial unit had four fused, elongate microsporangia that showed some resemblance to the angiosperm anthers.

➢ In Caytoniales, the ovules were enclosed by a curved cup-like structure called cupule. The megasporophyll of *Caytonia* consisted of an axis bearing sub-opposite pairs of stalked, rounded cupules (Fig. 10.11). Each cupule was recurved, had a lip-like projection near the point of attachment, and contained several ovules arranged in a curved row with their micropyles facing the cupule opening. The micropyles of the ovules were connected with the outer lip of the cupule via a canal. Proponents of caytoniales as the ancestor of the angiosperms point to the almost sealed cupule as suggestive of the way a carpel might have evolved. According to Thomas, the carpel wall of angiosperms may represent, a pair of concrescent cupules and he suggested that the possible origin of stigma should be considered in the light of these ancient forms.

➢ According to Gaussen (1946) and Stebbins (1974), the ovule of angiosperms is homologous to the cupule of advanced pteridosperms, particularly Caytoniales, including both the Corystospermaceae (with uniovulate and indehiscent cupules) and the Caytoniaceae (with multiovulate and dehiscent cupules), where the position of the cupules relative to their stalks was strikingly similar to that of an anatropous ovule having a well developed funiculus, rather than to the unitegmic, orthotropous ovule of other seed plants (Thomas, 1933 & 1955; Harris, 1951).

However, the Caytoniales differed from the angiosperms in the following respects:

➢ The caytonialian order was strongly pinnately veined, but the angiosperm carpel is palmate.

➢ Presence of pollination drop and direct pollination.

➢ Highly cuticularized winged pollen.

➢ Absence of connective like structure in tetralocular synangia.

➢ Presence of fleshy canals in the inner side of the curved cupules.

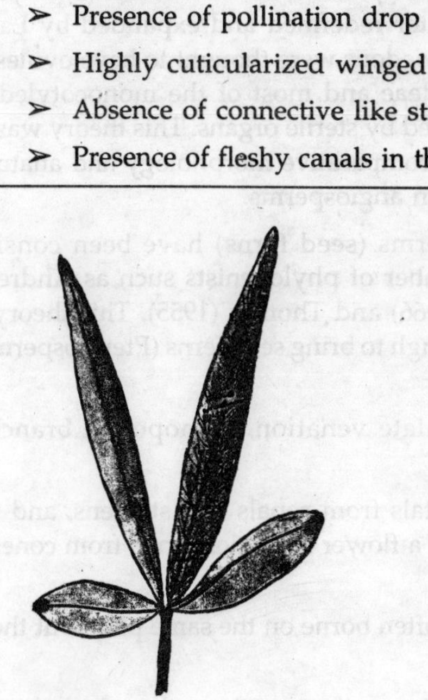

Fig. 10.9 : *Caytonia* — *Sagenopteris* foliage

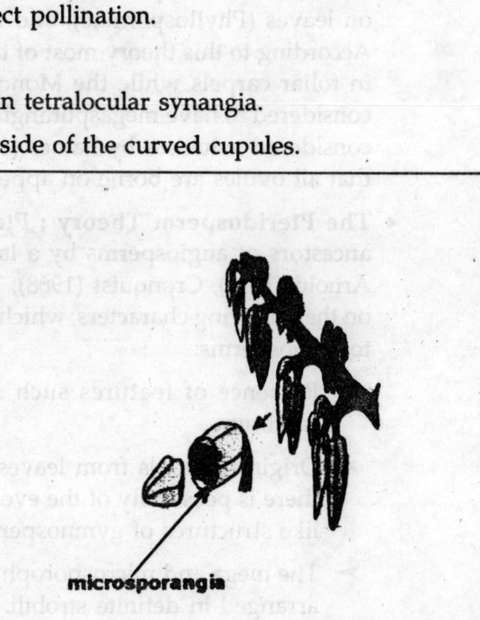

Fig. 10.10 : *Caytonanthus* microsporophyll with synangia - branched stamens

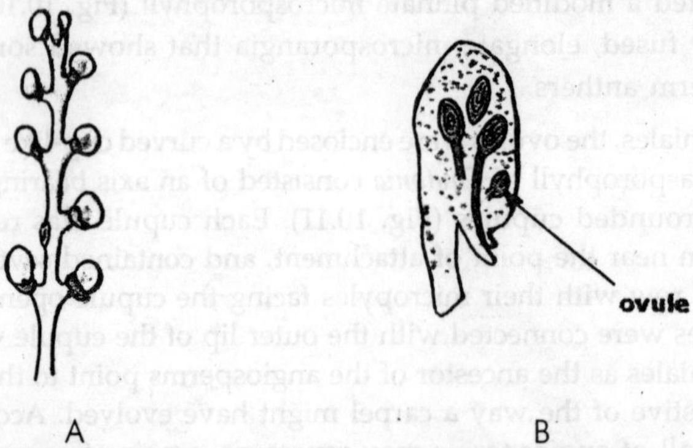

Fig. 10.11 : *Caytonia* – (A) megasporophyll with pinnately arranged cupules; (B) longitudinal section of cupule showing ovule position and orientation.

- Caytoniale ovules had single integument while most angiospermic ovules have two integuments.

These dissimilarities strongly negate the theory of Caytonialean ancestry of angiosperms. Caytoniales, are now classified as Mesozoic remnants of the Pteridosperms as they have been shown to have relationships with the latter group. The morphological studies on the nature of the carpel of angiosperms suggest that the ancestors of angiosperms must have had open megasporophylls with exposed ovules. Therefore, the ovary-like pouches of caytoniales cannot be taken to be the fore-runners of angiosperm carpel.

- **The Stachyospory-Phyllospermae Theory** : This theory was proposed by Sahni (1920) on the basis of supposed differences in ovule position borne on stem (Stachyospory) or on leaves (Phyllospermae), and it was later redefined and expanded by Lam (1948). According to this theory most of the dicotyledons were thought to have ovules enclosed in foliar carpels while the Monochlamydeae and most of the monocotyledons were considered to have megasporangia protected by sterile organs. This theory was however considered valueless by Eames (1961), as comparative morphology and anatomy show that all ovules are borne on appendages in angiosperms.

- **The Pteridosperm Theory** : Pteridosperms (seed ferns) have been considered as ancestors of angiosperms by a large number of phylogenists such as Andrew (1947), Arnold (1949), Cronquist (1968), Long (1966) and Thomas (1955). This theory is based on the following characters, which are enough to bring seed ferns (Pteridosperms) closest to angiosperms:

  - Presence of features such as reticulate venation, monopodial branching and cambium.
  - Origin of sepals from leaves and petals from sepals and stamens, and therefore, there is possibility of the evolution of a flower in angiosperms from cones or cone-like structures of gymnosperms.
  - The mega and microsporophylls are often borne on the same plant but they are not arranged in definite strobili.
  - Similarities of seed structure and existence of one or several ovules subtended by a cupule. The angiosperm carpel is regarded as derived from a dorsoventral bivalved pteridosperm cupule by Long (1966).

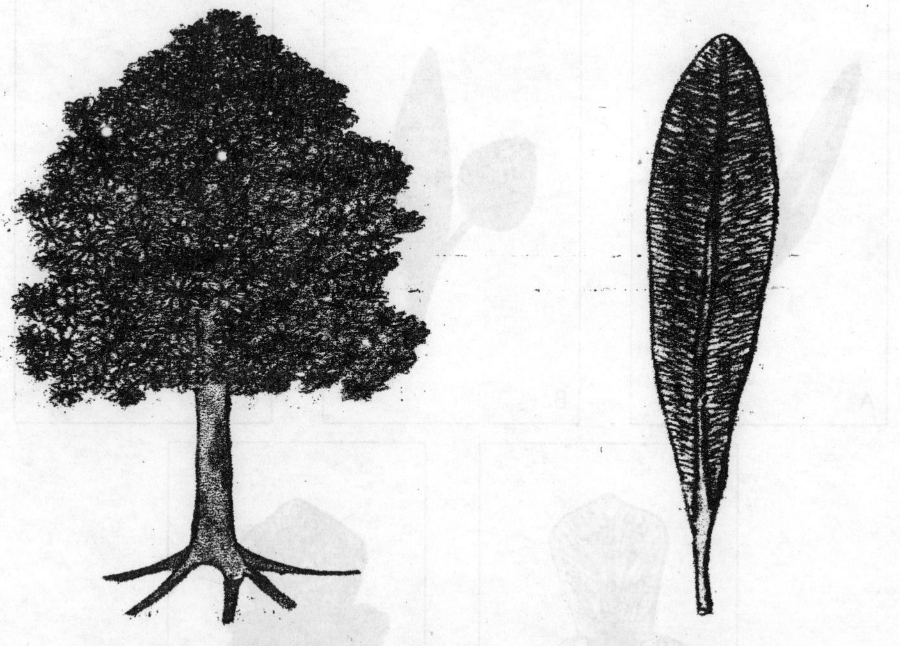

Fig. 10.12 : Reconstruction of the whole plant and a single leaf, *Glossopteris*

- Similarity in the morphological nature of the megasporangium or nucellus, as established by Hofmeister (1862).
- Development of triploid endosperm and extreme reduction of female gametophyte.
- Some primitive angiosperms lack vessels in the secondary wood, a feature also found in seed ferns.
- The reproductive branch in *Glossopteris* (a pteridosperm) is somewhat comparable with that of the present-day angiosperm *Dichapetalum* (Melville, 1960). Glossopterid leaves were abundantly and ubiquitously present in Pennsylvanian to Triassic rocks of Australia, Africa, South America, Antarctica, and the Indian peninsula, which lead geologists to the conclusion that these continents had once formed a continuous land-mass – the Gondwanaland. The glossopterids are reconstructed (Fig. 10.12) as shrubs or large trees and the diversity of their reproductive structures suggests that they constituted a species-rich group, with a great number of species (over 200 described). *Glossopteris* leaves are tongue-shaped, commonly with an entire (smooth) margin. They have a distinct mid-rib and reticulate venation adjacent to the mid-rib. Leaves are borne in spirals or whorls, probably on short shoots. Basically, both pollen and ovulate structures (Fig. 10.13) are borne separately on the adaxial side of a more or less modified *Glossopteris* leaf. In the case of ovulate fructifications, ovules are arranged on a dorsiventral structure that has been variously called capitulum, megasporophyll or cupule. On the basis of this and other evidences, Melville has proposed a new theory regarding the origin of angiospermic flower known as the **"Gonophyll theory"**. Advocates of the *Glossopteris*-origin theory propose the glossopterid vegetative leaf to be homologous with the angiosperms carpel, and the megasporophyll with the outer integument of an angiosperm seed, based on some peculiar specimens with abaxially borne seeds.

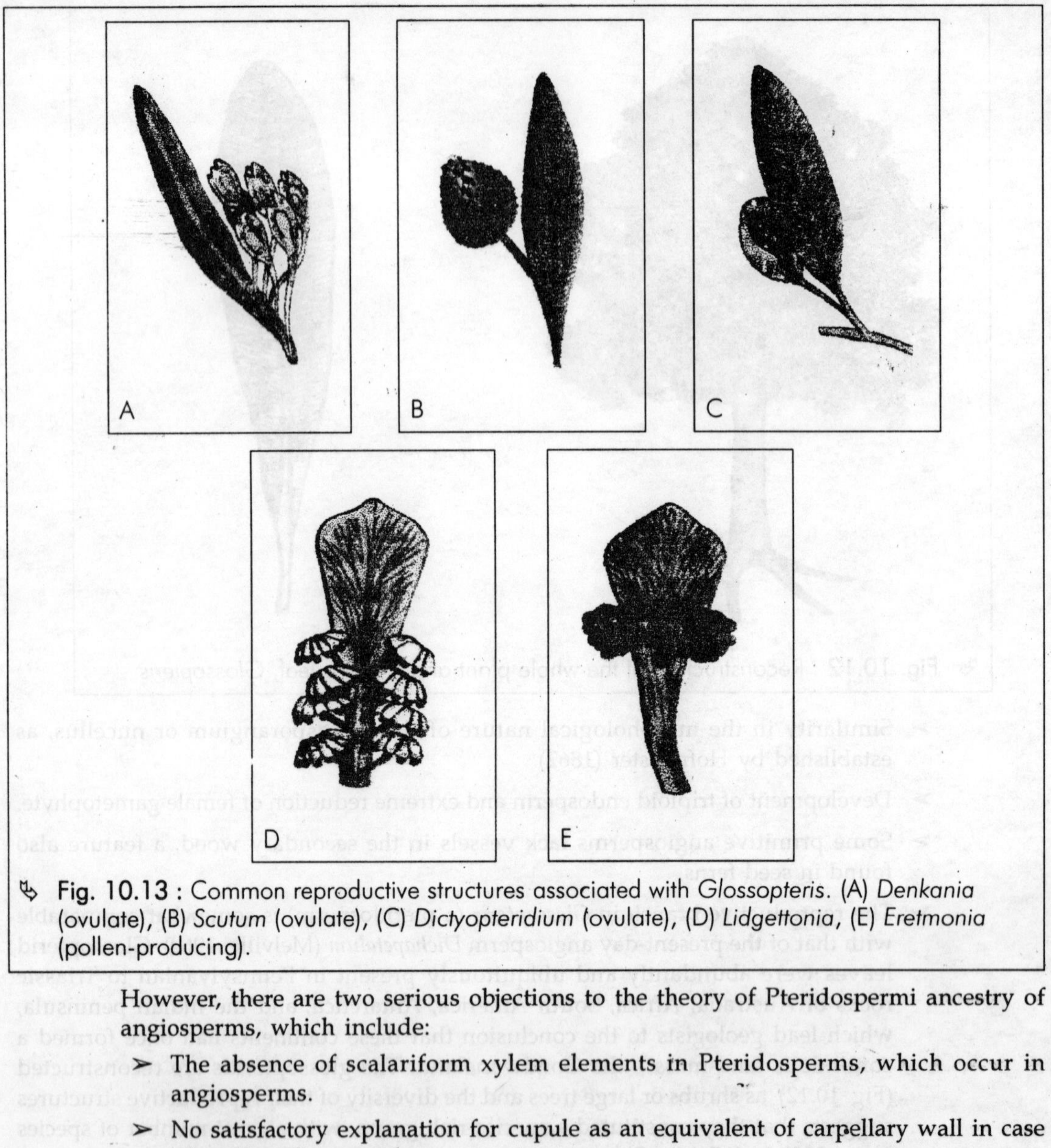

Fig. 10.13 : Common reproductive structures associated with *Glossopteris*. (A) *Denkania* (ovulate), (B) *Scutum* (ovulate), (C) *Dictyopteridium* (ovulate), (D) *Lidgettonia*, (E) *Eretmonia* (pollen-producing).

However, there are two serious objections to the theory of Pteridospermi ancestry of angiosperms, which include:

- The absence of scalariform xylem elements in Pteridosperms, which occur in angiosperms.
- No satisfactory explanation for cupule as an equivalent of carpellary wall in case of multiovulate cupules.

In spite of these objections, the similarities found between the pteridosperms and angiosperms suggest that angiosperms have some very close connection with pteridosperms. Due to the absence of any direct evidence, there is a growing tendency among phylogenists to accept the pteridospermic ancestry of angiosperms, at least tentatively.

- **The Pentoxylales Theory** : Meeuse (1961) compared *Pandanus*, a monocot, with Pentoxylales, a group of fossil pteridophytes. Like *Glossopteris* foliages, *Pentoxylon* is also a Gondwanan taxon, which has been found in India, Australia and New Zealand. It takes its name from the five wedges of wood that characterize its stem. *Pentoxylon* first appears in the later part of the Paleozoic, but its greatest diversity and abundance appears to be in the Jurassic, continuing into the Early Cretaceous •

Pentoxylon has a distinct short-shoot-long-shoot morphology. Stems are eustelic with secondary wood developing in five or six wedges around a mesarch primary xylem strand. Foliage arises from short shoots, where leaf scars are commonly visible (Fig. 10.14A). Leaves are long and narrow with a prominent mid-rib, and secondary veins arise at right angles to it and continues, unbranched till the leaf margin. *Pentoxylon* does not have bisexual reproductive structures like many other members of the anthophyte clade. Ovule-bearing structures are aggregated into strobili, which are attached to short shoots (Fig. 10.14B). Individual cupules almost completely enclose a single ovule, the integument of which is composed of an inner sclerenchymitized layer and a fleshy outer layer (Fig. 10.14C). Pollen organs are also born in clusters on short shoots.

Meeuse (1961) listed a number of common features between *Pandanus* and *Pentoxylon*, which include :

- Both the members are erect or sub-erect dioecious plants with slender cylindrical stems.
- Both bear terminal tufts of strap-shaped leaves in spiral arrangement.
- The inflorescences of both members are borne among the leaves or on short lateral branches in the axils of the scaly bracts.
- In both cases, the peduncles of inflorescences contain tracheids with spiral thickenings and bordered pitting.
- The seeds are enclosed in a fleshy layer called sarcotesta, and a seed coat consisting of two layers.
- Both bear seeds with copius endosperm and with minute embryo.

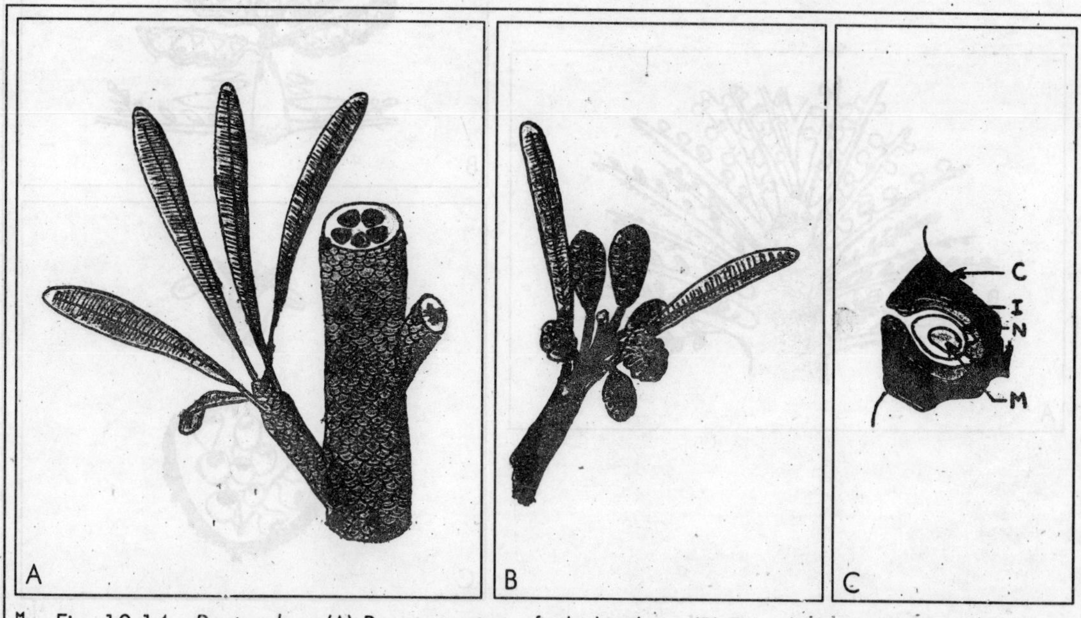

Fig. 10.14 : *Pentoxylon*. (A) Reconstuction of whole plant. (B) Female branch with ovule-bearing structures. (C) Section of ovulate cone in longitudinal section [C-cupule, I-integument, N–nucellus, M–megagametophyte]

Based on these similarities, Meeuse (1961) considered Pentoxylales as ancestors of angiosperms. He proposed that *Pandanus* "is a direct descendent of a group of plants practically identical" with Pentoxylales. However, in spite of these resemblances, authors like Pant and Kidwai (1971) hold that they could equally be the result of parallel evolution.

- **The Durian Theory :** *Durio zibethinus*, a member of Bombacaceae from Burmese and Malayan forests was considered by Corner (1949), as the only "surviving member of primitive angiosperms". This theory is named after the name of the genus *D. zibethinus* is a cauliflorous tree with large, colored, loculicidal, spiny capsules and fleshy arillate seeds (Fig. 10.15). According to Corner (1949), arillate genera are found in about 45 families of angiosperms, mostly distributed in the tropics like *Durio*, and within these arillate families, a reduction series can be traced from genera with arillate fruits to genera having fruits of other types. According to this theory, the primitive angiosperms are :

  - ➢ Mesophytic.
  - ➢ Tropical in distribution.
  - ➢ With cycad-like tree habit.
  - ➢ Have compound leaves.
  - ➢ Are probably monocarpic.
  - ➢ Have a large terminal cluster of arillate follicles.

  The theory has however been strongly criticized by a number of morphologists including Fiji (1952), Parkin (1953), Metcalfe (1954) and Eames (1961).

Fig. 10.15 : Corner's reconstructions of ancestral angiosperms according to Durian theory : (A) Ancestral pachycaulous spinous flowering plant with terminal flowers and fruits; (B) Primitive fruit with spinous arillate pods; (C) Fruit and flower of *Durio*.

10.7 Lines of Evolution in Angiosperms

Among the angiosperms, Ranales have been considered as the most primitive due to the following reasons:

- Absence of vessels in a large number of families under this order e.g., Alismataceae, Magnoliaceae, Winteraceae, etc.
- Peculiar floral morphology and anatomy.
- Pollen grains of the genus *Schizandra* of Magnoliaceae having a triradiate ridge.
- Binucleate pollen at the time of release from the microsporangium in several families such as Magnoliaceae, Ranunculaceae, Winteraceae, etc.

According to most phylogenists there are atleast three main lines of general evolution observed from the Ranalian plexus.

- **The first line** — It consists of Rhoeadales, Sarraceniales, Parietales, Malvales, Geraniales, Sapindales, Rhamnales and some Tubiferales. In this case the hypogynous condition remained unchanged.
  - ➢ Labiatae is the most advanced family of this line. It is characterized by fusion of sepals, petals and carpels, and epipetalous condition and the number of floral parts are reduced to four or two.
  - ➢ Solanaceae is also characterized by fusion of sepals, petals and carpels, and epipetalous condition.
  - ➢ Rhoeadales and Sarraceniales evolved parallely from Ranales.
  - ➢ A definite relation persists between Parietales and Rhoeadales due to the development of the gynophore in Capparidaceae.
  - ➢ Malvales evolved through Bombacaceae and Malvales are related to Geraniales.
  - ➢ Geraniales are related to Sapindales and Sapindales in turn are related to Rhamnales.
- **The second line** – It consists of Campanulales, Rosales, Myrtales, Umbellales and Rubiales. In this case, there is a change from hypogyny to epigyny through perigyny, which is seen clearly in Rosales.
  - ➢ Campanulales top this line with the presence of pappus, irregular flowers, and fusion of floral parts.
  - ➢ In Myrtales, there is a change of perigyny to epigyny due to the development of hypanthodium.
  - ➢ In Umbellales, epigyny and syncarpy show an advancement over Myrtales, and the former are related to Rubiales due to presence of characters such as suppression of calyx, epigyny, the presence of umbel-like inflorescence, etc.
- **The third line** – It includes Liliales and Orchidales. It shows a few resemblances with monocotyledons (e.g. Alismataceae) and dicotyledons (e.g. Ranunculaceae).
  - ➢ Liliales differ from Ranales in the number of floral parts and fusion of carpels.
  - ➢ Iridaceae and Amaryllidaceae are more advanced than Liliales due to presence of epigynous condition.

> ➢ Poaceae seems to have originated from Liliaceae. However the reduction in the number of stamens and carpels and also the loss of perianth parts clearly indicate that Poaceae is more advanced than Liliaceae.
>
> ➢ Orchidaceae is characterized by insect-pollinated, irregular and showy flowers, and is thus the most advanced family of monocotyledons.

10.8 The Origin of Monocotyledons

Several hypotheses have been proposed by different phylogenists to explain the origin of monocots, but none appears to be wholly satisfactory.

- According to Engler and his associates, Isoetales, Ophioglossales and a few other pteridophytes are ancestors to monocots. The herbaceous monocots have been considered to be more primitive than woody dicotyledons by them. However, this view is contradicted by the fact that, the woody habit is now generally accepted to be more primitive, and monocots most probably have diverged from some primitive dicotyledons in the early history of evolution, which also implies the monophyletic origin of angiosperms as a group.
- Several workers have proposed the dicot families such as Lardizabalaceae (Hallier, 1912), Menispermaceae, Ranunculaceae (Hutchinson, 1959) and Berberidaceae as possible ancestors of monocots. Hutchinson's view is generally not accepted as majority of monocots are characterized by monocolpate pollen, while the pollen in Ranunculaceae is tricolpate, pantocolpate or pantoporate and is never monocolpate. Takhtajan (1954 & 1961) and Eames (1961) are of the view that ancestors of monocots must therefore be the dicots with primitive monocolpate pollen. In dicots, monocolpate pollen is found in Magnoliales, some Laurales, Nymphaeales (excluding *Nelumbo*), majority of the Piperales and in the genus *Saruma of* Aristolochiaceae.
- Takhtajan (1943, 1959 & 1961), Cronquist (1968a), Arber (1925), Hallier (1905) and Parkin (1923) are of the opinion that some extinct vesselless herbaceous plants, with apocarpous gynoecia and monocolpate pollen, resembling the modern Nymphaeales, were most likely the immediate ancestors of monocots. This view is based on the fact that vessels originated independently in dicots and monocots, which has been proved by comparative anatomical studies (Cheadle, 1953). Thus, the ancestors of monocots must have been the vesselless dicots such as Magnoliales and Nymphaeales. There are several common features between the primitive monocots and Nymphaeales, which include features of leaf structure and development, stem with scattered, closed vascular bundles, more or less reduced primary root and their root structure and apocarpous gynoecia with "diffuse" placentation (Cronquist, 1968a; Takhtajan, 1969). The primitive monocots and Nymphaeales only differ is the presence of two cotyledons in the latter.
- According to various workers, monocots have an aquatic ancestry as well (Cronquist 1968a; Henslow, 1911). Takhtajan (1969) has considered the order Nymphaeales as a hygrophilous derivative of some ancient Magnoliales, the vegetative organs of which have more or less degenerated in an aquatic environment. The monocot line was most probably initiated from such ancient herbaceous dicots of Nymphaealean type, but more primitive and less reduced than their present-day representatives.

- Sargant (1903, 1904 & 1908) proposed that monocots have arisen as a result of adaptation to a geophilous (marsh loving) habit and called them "aquatic geophytes". The common ancestors of both Nymphaeales and monocots were apparently the amphibious geophytes, in which geophily at first arose under terrestrial conditions, most probably under the forest canopy or in the forest margin.
- On the basis of comparative study of various dicot and monocot embryos, Hegelmaier (1874 & 1878) suggested that, the monocot embryo (consisting of a single cotyledon) arose as a result of the failure of one of the two cotyledons of a typical dicot embryo (consisting of two cotyledons) to develop. This abortion-hypothesis has also been agreed upon by Eames (1961), Henslow (1893 & 1911), Metcalfe (1936), Winkler (1931), and many others. However, Stebbins (1974) believes that the single cotyledon of monocots is derived through intercalary concrescence of the petioles of two original cotyledons by way of syncotyly and subsequent reduction of lobes.
- Burger (1977) proposed the dicots, very similar to Chloranthaceae of the order Piperales, which are more closely related to monocots than other living dicots, as probable ancestors of monocots. He developed a hypothesis suggesting that the simple flowers, similar to those of Chloranthaceae, came together by the loss of internodes to form three-parted flowers of Piperales and many monocots. He also postulated that the ancestral simple flowers had a single small bract, two stamens and a single adaxial pistil. Evidence from Lactoridaceae, Saururaceae, Potamogetonaceae, Alismatales and others support his hypothesis.

10.9 Molecular Evidence to Angiosperm Origin

Until recently the "anthophyte hypothesis" whereby Gnetales are sister to angiosperms was the most popular concept. However, with the advent of molecular data, it has become quite apparent that there is a conflict between this idea and one in which gymnosperms were monophyletic and in which this entire clade was sister to angiosperms. This conflict between the morphology and molecular data has been attributed to the "inability of molecular data to resolve relationships at this level" by Doyle (1998). The large number of morphological characters have been found to contradict the molecular tree topologies. The problem is that there are now many molecular data sets that all point to the same tree topology, i.e. the gymnosperms are monophyletic and sister to angiosperms and this relationship is borne out by genes derived from nuclear, plastid, and mitochondrial genes.

SUGGESTED READING

- Axelrod, D.I., 1952, A theory of angiosperm evolution. *Evolution*, 6: 29-60.
- Bessey, C.E., 1915, Phylogenetic taxonomy of flowering plants. *Ann. Mo. Bot. Gard.*, 1:109-164.
- Burger, W.C, 1981, The monocot theory of angiosperm evolution. *Evol. Theory*, 5:189-225.
- Cheadle, V.I., 1953, Independent origin of vessels in monocotyledons and dicotyledons. *Phytomorphology*, 3: 23-44.

Contd.

- Daghlian, C.P., 1981, A review of the fossil record of monocotyledons. *Bot. Rev.*, 47:517-555.
- Doyle, J.A., 1978, Origin of angiosperms. *Ann. Rev. Ecol. Syst.*, 9: 365-392.
- Krassilov, V.A., 1977, The origin of angiosperms. *Bot. Rev.*, 43: 143-176.
- Meeuse, A.D.J., 1975, Origin of the Angiosperms—Problem of inaptitude. *Phytomorphology*, 25: 373-379.
- Scott, R.A., E.S. Berghoom & E.B. Leopold, I960, How old are the angiosperms ? *Am. J. Sci.*, 258A: 284-299.
- Stebbins, G.L., 1974, *Flowering Plants: Evolution Above the Species Level*, The Belknap Press, Cambridge.
- Takhtajan, A., 1969, *Flowering Plants: Origin & Evolution*, Smithsonian Inst. Press, Washington.
- Turin, T.G., 1952, Phylogeny of flowering plants: Facts or Fiction? *Nature* (London) 169:126-127.
- Wolfe, J.A., J.A. Doyle & V.M, Page, 1975, The basis of angiosperm phylogeny: Palaeobotany. *Ann. Mo. Bot. Gard.*, 62: 801-824.

11 REMOTE SENSING TECHNOLOGY AND VEGETATION CLASSIFICATION

AT A GLANCE

Remote sensing is the observation of an object from a distance, as for example Aerial Photography and Satellite remote sensing. The term "remote sensing" was first used by Ms. Evelyn Pruitt of the U.S. Office of Naval Research in the 1950s. It is now commonly used to describe the science and art of identifying, observing, and measuring an object without coming into direct contact with it, by detecting and measuring the radiation of different wavelengths reflected or emitted from distant objects or materials, by which they may be identified and categorized by class/type, substance, and spatial distribution.

Satellite remote sensing involves gathering information about features on the Earth's surface from orbiting satellites. The digital data acquired by the satellites is transmitted to ground stations, which can be used to reconstitute an image of the Earth's surface not too dissimilar to an aerial photograph. Data can be obtained as hard copy photographic data products, or as digital data products, which can be viewed and manipulated on a variety of software systems.

Remotely-sensed imagery from aircraft and satellites represent one of the fastest-growing sources for raster GIS data. Remote sensing provides important coverage, mapping and classification of land cover features, such as vegetation, soil, water and forests. Although remote-sensing technology has been around for decades, recent technological advances have led to a dramatic increase in the types of imagery available. Remote sensing is thus, an interesting and exploratory science, as it provides faster and efficient images of areas of interest and demonstrates what is happening in the study area at any particular moment. Although air photos and fieldwork still remain critical as sources of information, sometimes the cost and time to carry out these methods may not be feasible for the study. In such cases, satellite and digital imagery, provide more overall detail to assist the researcher in the classification process. It can also be used to access hard to reach areas for fieldwork, and provides a permanent, more detailed and objective survey, that offers a different perspective.

11.1 Geographic Information System or GIS

GIS or geographic information system combines layers of information about a place or area to give a better understanding of it. For a full GIS, one requires hardware (computers and peripherals), software, data and sound analysis methods for interpreting the results generated by the GIS.

GIS may be used for the following :

(i) Mapping where things are, which allows one to find places that have the features one is looking for and to see where to take action. Mapping where things are lets one :

- Find a feature — One uses maps to see where or what an individual feature is.
- Finding patterns — By looking at the distribution of features on the map instead of just an individual feature, one can see patterns emerge.

(ii) Mapping quantities — One map quantities to find out the dominance of a feature of interest in a particular area i.e., where the most and least are, which allows one to find places that meet the required criteria and take action, or to see the relationships between places. Instead of simply mapping the locations of features, mapping quantities give an additional level of information.

(iii) Mapping densities — While one can see concentrations by simply mapping the locations of features, it may be difficult to see which areas have a higher concentration than others in areas with many features. A density map lets one measure the number of features using an uniform aerial unit, such as acres or square miles, so that one can clearly see the distribution.

(iv) Finding what's inside — GIS can be used to monitor what is happening and to take specific action by mapping what is inside a specific area.

(v) Finding what's nearby — GIS can be used to find out what is occurring within a set distance of a feature by mapping what is nearby.

(vi) Mapping changes — Future conditions can be anticipated by mapping the change in an area, which allows to decide on a course of action, or to evaluate the results of an action.

For GIS analysis, one must proceed in the following way :

(i) Frame the question — GIS analysis should be started by figuring out what information one needs.

(ii) Selection of data — The type of data and features one works with helps determine the method one uses.

(iii) Selection of an analysis method — One must decide on which analysis method to use based on the original question and how the results of the analysis will be used.

(iv) Processing of data — Once the analysis method has been selected, one will need to process the data in a way that makes sense for one's goal.

(v) Looking at the results — The final step is to look at the results of one's analysis and take action based on those results. The results can be displayed as a digital map, printed as a paper map, combined with spreadsheet-like tables or charts, or displayed as such. Though in GIS, a lot of emphasis is given on making maps, the software is flexible enough to allow one to display the results in the format that best suites ones needs.

Thus, the benefits of GIS lie in the fact that it allows :

(i) Improved organizational integration i.e., improved management of one's organization and resources.

(ii) Make better decisions i.e., better information leads to better decisions. A GIS is not just an automated decision making system but a tool to query, analyze, and map data in support of the decision making process.

(iii) Make maps i.e., GIS creates maps from data pulled from databases and GIS is flexible enough to map any kind of terrain, even the human body.

11.2 History of Remote Sensing

The technology of modern remote sensing has a very long history, dating back to the end of the 19th century with the invention of the camera. Initially cameras were used to take photographs on the ground, which provided (and still does) a fascinating and exciting way to capture moments in time and keep a record of something that happened, which looked more realistic than a drawing or painting, and which could be captured much quicker than by drawing or painting. The idea and practice of remote sensing first developed in the 1840s, when it was realized that a different and perhaps more revealing view of a particular landscape could be obtained by taking a photograph from a vantage point, such as an incline or building, and efforts were made to look down at the Earth's surface by taking pictures with the aid of cameras secured to tethered balloons, for purposes of topographic mapping. It was realized that the airborne perspective gave a completely different view than to what was available from the ground. The most novel platform at the end of the last century is perhaps the famed pigeon fleet that operated as a novelty in Europe. By the first World War, cameras were mounted on airplanes, which provided aerial views of fairly large surface areas and was used as a method of data and information acquisition, that proved invaluable in military reconnaissance. Thus, aerial photography remained the single standard tool for depicting the surface from a vertical or oblique perspective till the early 1960s.

The history of Satellite remote sensing can be traced back to the early days of the space age of both Russian and American programs. It actually began as a dual approach of imaging surfaces, from spacecraft, using several types of sensors. After World War II, in 1946, V-2 rockets acquired from Germany, containing automated still or movie camera, were launched to high altitudes from White Sands, New Mexico. These rockets, however never attained orbit, but took pictures of the earth's surface as the vehicle ascended. In the 1960s, with the emergence of the space program, cosmonauts and astronauts started taking photographs out of the window of their spacecraft in which they were orbiting the earth. Today, remote sensing is carried out using airborne and satellite technology, not only utilizing film photography, but also digital camera, scanner and video, as well as radar and thermal sensors (Fig. 11.1). Unlike in the past, when remote sensing was restricted to only the visual part of the electromagnetic spectrum i.e., what could be seen with naked eye, today through the use of special filters, photographic films and other types of sensors, the parts of the spectrum which can not be seen with the naked human eye can also be utilized. Thus, today remote sensing is largely utilized in environmental management, which frequently requires rapid, accurate and up-to-date data collection.

11.3 Principles of Remote Sensing

Remote sensing involves the detection and measurement of the radiations of different wavelengths reflected or emitted from distant objects or materials, which helps in their identification and categorization. It offers four basic components to measure, which include:

(i) The energy source

(ii) The transmission path

(iii) The target

(iv) The satellite sensor

Among these, the energy source or electromagnetic energy, is very important, as it serves as the crucial medium for transmitting information from the target to the sensor. It is described as an electromagnetic spectrum, on which, many forms exist that describe energy in a specific region of the spectrum (Fig. 11.2). These are visible light, radio waves, microwaves, heat, UV rays, X-rays and gamma rays. This spectrum is an overview of the continuum of electromagnetic energy from extremely short wavelengths (cosmic gamma rays) to extremely long wavelengths (radio and television waves). These divisions are not absolute and definite as overlapping may occur. The basic unit of wavelength, is measured in meters (m). Most energy in the visible (including U.V., visible, and near infrared) and infrared portions of the electromagnetic spectrum are measured in micrometers (10-6m) [1 micrometer = .10,000 angstroms]. Millimeters may be used for longer wavelengths (blue – 0.4 – 0.5 mm).

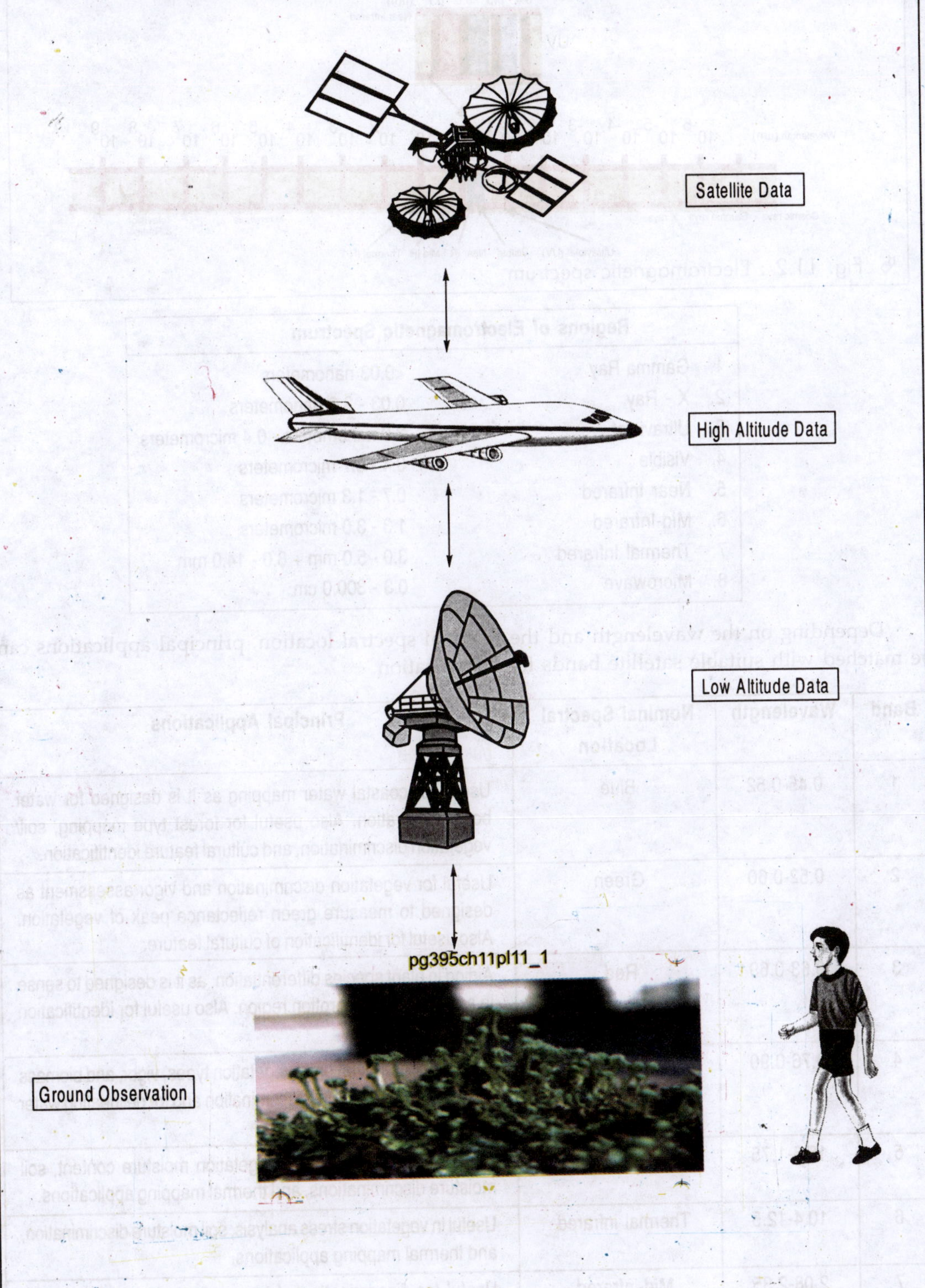

Fig. 11.1 : An overview of remote sensing

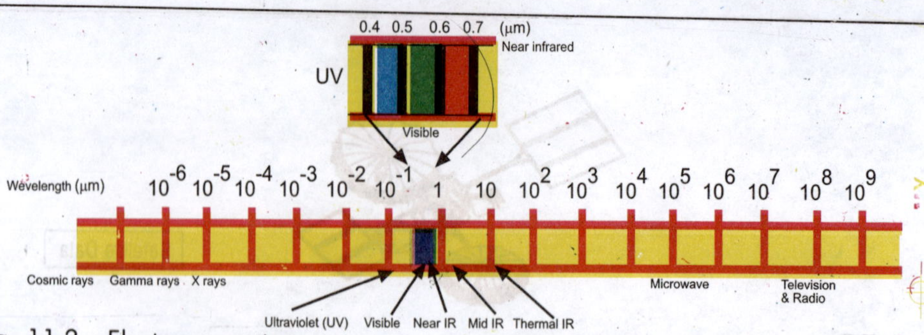

Fig. 11.2 : Electromagnetic spectrum

Regions of Electromagnetic Spectrum	
1. Gamma Ray	<0.03 nanometers
2. X - Ray	0.03 - 3.0 nanometers
3. Ultraviolet	3.0 nanometers - 0.4 micrometers
4. Visible	0.4 - 0.7 micrometers
5. Near Infrared	0.7 - 1.3 micrometers
6. Mid-Infrared	1.3 - 3.0 micrometers
7. Thermal Infrared	3.0 - 5.0 mm + 8.0 - 14.0 mm
8. Microwave	0.3 - 300.0 cm

Depending on the wavelength and the nominal spectral location, principal applications can be matched with suitable satellite bands for classification.

Band	Wavelength	Nominal Spectral Location	Principal Applications
1	0.45-0.52	Blue	Useful for coastal water mapping as it is designed for water body penetration. Also useful for forest type mapping, soil/ vegetation discrimination, and cultural feature identification.
2	0.52-0.60	Green	Useful for vegetation discrimination and vigor assessment as designed to measure green reflectance peak of vegetation. Also useful for identification of cultural feature.
3	0.63-0.69	Red	Aiding in plant species differentiation, as it is designed to sense in a chlorophyll absorption region. Also useful for identification of cultural feature.
4	0.76-0.90	Near infrared	Useful for determination of vegetation types, vigor, and biomass content, for soil moisture discrimination and for delineating water bodies.
5	15.5-1.75	Mid-infrared	Useful for determination of vegetation moisture content, soil moisture discriminations, and thermal mapping applications.
6	10.4-12.5	Thermal infrared	Useful in vegetation stress analysis, soil moisture discrimination, and thermal mapping applications.
7	2.08-2.35	Mid-infrared	Useful for discrimination of types of mineral and rock and determination of vegetation moisture content.

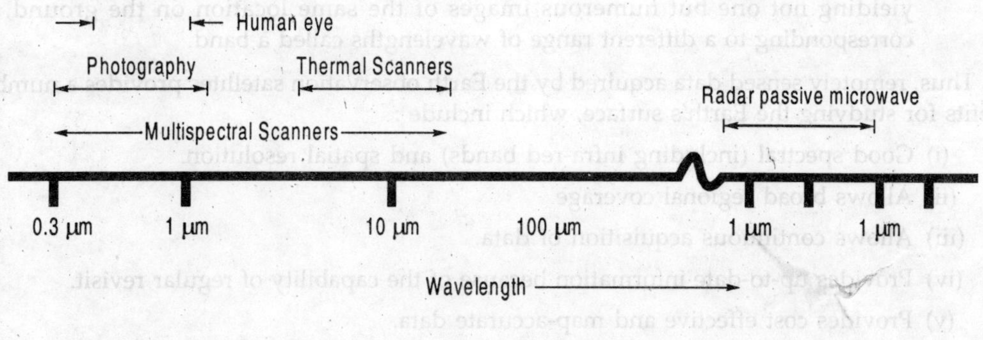

Fig. 11.3 : Common remote sensing systems

11.4 Types of Remote Sensing Technologies

Satellite remote sensing involves gathering information about features on the Earth's surface from orbiting satellites, which may carry either of the following two types of sensor systems:

(i) **Passive system** — It generally consists of an array of small sensors or detectors, which records the amount of electro-magnetic radiation reflected and/or emitted from the Earth's surface. Thus, passive remote sensing relies on naturally reflected or emitted energy of the imaged surface. Most remote sensing instruments fall into this category, obtaining pictures of visible, near-infrared and thermal infrared energy. A multi-spectral scanner is an example of a passive system (Fig. 11.3). Passive visible and near-infrared data are used in a variety of GIS applications, for example in the classification of vegetation and land-use, and may be performed at a variety of temporal and spatial scales.

(ii) **Active system** — This type of a system propagates its own electro-magnetic radiation and measures the intensity of the return signal. Thus, active remote sensing means that the sensor provides its own illumination and measures what comes back. Remote sensing technologies that use this type of system include lidar (laser) and radar. Synthetic Aperture Radar (SAR) is an example of an active system. Active remote sensing (radar and lidar) systems are rapidly increasing in use since the launch of the ERS-1 Synthetic Aperture Radar (SAR) satellite in 1991. In comparison to visible/near-infrared imagery, radars are sensitive to very different surface properties. As for example, radar images are sensitive to the shape, orientation and size of leaves and their moisture content, rather than the vegetation color. Similarly, airborne lidars have been largely used for mapping surface topography in three dimensions. Existing and planned radar and lidar altimeters will also help in monitoring closely the elevation of the world's ice caps and sea level with centimeter precision.

11.5 Advantages of Remote Sensing Technology

Remotely sensed imaging systems have several advantages over camera photography, from which it differs significantly in the following two ways :

(i) It is not just restricted to the visible part of the electromagnetic spectrum (from about 0.4 to 0.7 micrometers in wavelength), but can also measure energy at wavelengths invisible to the eye, such as near-infrared, thermal infrared and radio wavelengths.

(ii) Most remote sensing instruments can record different wavelengths at the same time, yielding not one but numerous images of the same location on the ground, each corresponding to a different range of wavelengths called a band.

Thus, remotely sensed data acquired by the Earth observation satellites provides a number of benefits for studying the Earth's surface, which include :

(i) Good spectral (including infra-red bands) and spatial resolution.

(ii) Allows broad regional coverage.

(iii) Allows continuous acquisition of data.

(iv) Provides up-to-date information because of the capability of regular revisit.

(v) Provides cost effective and map-accurate data.

(vi) Provides large archive of historical data.

(vii) Enables to combine satellite digital data with other digital data.

(viii) Has the ability to manipulate or enhance digital data.

(ix) Allows the possibility of stereo viewing.

11.6 Limitations of Remote Sensing Technology

Although remote sensing has many advantages over ground-based survey, yet remote sensing has not totally replaced ground-based survey methods, largely because of some limitations with this technology, which still exist. These include :

(i) Cost of data collection and data purchase.

(ii) Problems with data analysis and interpretation.

(iii) Potential limitations with spatial, spectral and temporal resolutions of the various sensors.

(iv) Problems with all weather capability as some sensors cannot 'see' through cloud.

Inspite of these limitations, remote sensing has however many advantages over ground-based survey in that large tracts of land can be surveyed at any one time, and areas of land (or sea) that are otherwise inaccessible can be monitored. The advent of satellite technology, and multi-spectral sensors has further enhanced this capability, with the ability to capture images of very large areas of land in one pass, and by collecting data about an environment that would normally not be visible to the human eye.

11.7 Applications of Remote Sensing Technologies

Satellite data allows the proper management of our renewable and non-renewable resources as it provides timely and detailed information about the Earth's surface. Following are a few examples of some of the important uses of satellite data :

(i) Assessment and monitoring of vegetation types and their status.

(ii) Agricultural property management planning and crop yield assessment.

(iii) Soil surveys including mineral and petroleum exploration.

(iv) Litho logic mapping.

(v) Monitoring and planning of water resources and groundwater exploration.

(vi) Geographic information

(vii) Map making and revision and production of thematic maps.

(viii) Weather and agricultural forecasts and assessment of environment and natural disasters.

(ix) Urban planning.

(x) Image processing.

(xi) Precision georeferencing.

(xii) Laser film writing and printing.

11.8 Application of Remote Sensing in Vegetation Classification

Remote sensing has been largely used in classifying vegetation, as it can determine vegetation distribution and occurrence, and how the vegetation of a particular area is influenced by factors such as soil type and drainage conditions, topographic aspect and slope, latitude, elevation above sea level, moisture, solar radiation, temperature regimes, length of the growing season, prevailing winds, salt spray, air pollutants, etc. It has also been successfully utilized to detect and prevent the spread of damaged and stressed plants.

The reflectance quality of vegetation on satellite and remote sensing images are influenced by several factors. These include:

(i) **Brightness** — It is calculated as a weighted sum of all the bands and is defined in the direction of principal variation in soil reflectance.

(ii) **Greenness** — It is related to the amount of green vegetation in the scene. It is orthogonal to brightness and is a contrast between the near-infrared and visible bands.

(iii) **Moisture** — Moisture in vegetation reflects more energy than dry vegetation.

(iv) **Leaf optical properties** — It is influenced by leaf properties such as internal or external structure, age, the health of the leaf, water status, and mineral stresses. The reflectance of the optical properties of leaves remains the same, regardless of the species. It is only the typical spectral features recorded for the three main optical spectral domains viz. leaf pigments, cell structure and water content, that may differ for each leaf.

(v) **Electromagnetic wavelengths** — It affects different parts of the plant and trees, which include their leaves, stem, stalks and limbs. The amount of reflection that occurs also depends on the length of the wavelengths. It has been found that the tree leaves and crop canopies reflect more in the shorter radar wavelengths, while tree trunks and limbs reflect more in the longer wavelengths. Further, the scattering of the wavelengths is affected by the density of the tree or plant canopy. Within the electromagnetic spectrum, bands will produce different levels of reflectance rates. For example, a lower reflectance will occur in the visible bands (400 - 700 nm) as more light will be absorbed by the leaf pigments than reflected. The blue (450 nm) and red (670 nm) wavelengths include two main absorption bands that absorb two main leaf pigments.

(vi) Seasonal differences, plant transitional zones, quality, scale and season of photography, film type and background will influence the remote sensing image. It is affected by the time of the day, sun angle, clouds, atmospheric haze, processing errors of transparencies/prints and errors in interpreting the images.

Photographic texture (smoothness and coarseness of images), total contrast or color, relative sizes of crown images at a given photo scale and topographic location, help to determine the cover types of vegetation. If the researcher has a sound knowledge of what plants and tree species are native to a particular area, and what influences their growth and distribution, it becomes easier for him to interpret the satellite images of vegetation. Remote sensing images can be further supplemented by aerial photographs, color infrared and black and white infrared photographs, which help to identify species. Details of branching characteristics, crown shapes, spatial distribution and patterns of species can be interpreted from such photographs, which may provide useful data for the interpreter.

Satellites that successfully identify vegetation types include the AVHRR (Advance Very High Resolution Radiometer), Land sat MSS, Land sat TM, SPOT HRV and RADARSAT. Satellite images can be combined with topographic data (ancillary data), to identify plant species with relation to slope direction, sun angles and other spectral characteristics. This is known as Multitemporal Image Classification. The technique of combining multi-spectral and ancillary information into a classification algorithm is referred to as Multidimensional Image Classification.

11.9 Application of Remote Sensing in Forest Resource Management

Remote sensing also allows for efficient management of forest resources, which include wood, forage, water, wildlife, recreation, timber management, maintenance and improvement of existing forest stands and fire control. Along with air photos and forest classification maps, satellite and remote sensing images help in the detection of signs of damage of forests by fire, insects or disease and in the interpretation and determination of the volume of timber harvested from an individual tree or stand of trees. When viewing air photo/satellite images, seasonal differences should be taken into account. Individual tree species will be identifiable through shape, size, pattern, shadow, tone and texture. Measurements include tree height/stand height, tree crown diameter, density of stocking and stand area. Thus, classifying vegetation is an important component in the management and planning of natural resources.

11.10 Remote Sensing and Soil

Soil studies generally require large scale surveys involving airborne methods (air photos) and field work, which provide information on soil orders and their relation to vegetation/rock cover, color of topsoil, conditions of soil formation, landforms, surface materials, regional zones, drainage and other topographic data. However small to medium scale surveys depend on the use of satellite data and air photos. There are several different areas where remote sensing can be applied for the study of soil. Some important applications are as follows:

(i) Satellite imagery can provide the visible boundaries of soil types, while remote sensing provide for a shallow penetration of soils. Additional physical data can be obtained from spectral signatures for the soil surfaces.

(ii) Remote sensing allows for classification of soils, which can be interpreted from the remote sensing images and the spectral signatures.

(iii) Remote sensing can provide information on the productivity of forests, meadows, wildlife habitat conditions, land-use and recreational suitability, which allows for future protection of the environment.

(iv) Multitemporal techniques can be used to map dynamical features, erosion and soil moisture.

(v) Remote sensing can also be used in combination with ground radar, to detect changes of diagnostic soil horizons such as albic, spodic and argillic horizons or soil/rock boundaries.

There are several factors, which influence soil reflectance in remote sensing images. These include :

(i) **Mineral composition** — The reflectance spectrum is affected by the mineral composition of soil. Increasing reflectance of soil occurs from the visible to the short-wave infrared - with absorption bands around 1.4 μm and 1.9 μm related to the amount of moisture in the soil.

(ii) **Soil moisture** — Soil moisture affects the spectrum and the ration between the spectral bands. Spectral bands of red and near-infrared bands are independent from the soil moisture. If the soil is moist, radar waves may not be able to penetrate the soil and in such cases soil moisture will develop parallel curves on the soil reflectance spectra. The soil line of the soil reflectance spectra, characterizes the soil type, defines vegetation indices, and corrects the plant canopy reflectance's from the optical soil property effects.

(iii) **Organic matter content** — It may indirectly affect the spectral influence, based on the soil structure and water retention capacity. High organic matter in soil may result in spectral interferences for band characteristics of minerals like Mn and Fe.

(iv) **Soil texture (surface)** — Soil optical properties are affected by the rough soil texture, as light is trapped in the rough surfaces of the coarse soil particles. For example, a stronger reflectance is received if iron and lime are present, than if the soil material is fine textured and dry. Variation in soil reflectance also occurs where there is a change in distribution of light and shadow areas with surface roughness areas.

(v) **Size and shape of the soil aggregate** — The reflectance properties are also influenced by soil size and shape. A decrease in reflection results when the size of a soil aggregate expands in diameter. The surface texture depends on the shape of the soil aggregate. Instead of a jagged soil aggregate, a more spherical soil aggregate results in a smooth and even surface.

Satellites that provide images to help achieve soil classification include :

(i) Land sat MSS — It allows for discrimination between the moist and dry soils.

(ii) Land sat TM high-resolution imagery — It provides useful information on soil mineralogy and arid areas.

11.11 Water and Remote Sensing

Remote sensing can be applied to water bodies as water absorbs, as well as reflects energy. The most distinctive characteristic of water, with reference to spectral reflectance, is the energy absorption at the near-infrared wavelengths. Water absorbs energy in wavelengths, which interact in a complex manner with the water and material in the water (organic and/or inorganic material). Wavelengths may be less than 0.6 μm and wavelengths in the blue-green portion of the spectrum have a high rate of transmittance. The reflectance properties of water are a function of the water and matter and is influenced by the following factors :

(i) **Amount of suspended sediment** — Higher visible reflectance results, if the water has a large amount of suspended sediment present, when compared to clearer waters.

(ii) **Turbulence of water** — The amount of energy absorbed and transmitted is also affected by turbulent water.

(iii) **Amount of chlorophyll** — The amount of water reflectance is also affected by the amount of chlorophyll. An increase in green wavelengths and a decrease of blue wavelengths results due to an increase in the amount of chlorophyll. The concentration of algae present in a water body can be detected and estimated by remote sensing.

(iv) **Temperature** — Reflectance of water is affected by temperature variations throughout the day.

Reflectance data for water bodies has been used to detect a number of water pollutants e.g. oil and industrial wastes, and to determine the presence or absence of tannin dyes from the bog vegetation in lowland areas. Certain water characteristics such as salt concentration, pH, dissolved oxygen concentration, etc. cannot be observed directly with water reflectance changes but can be correlated with observed reflectance.

SUGGESTED READING

- F. F. Sabins, *Remote Sensing.*
- J. R. Schott, *Remote Sensing.*
- G. Rees, *The Remote Sensing Data Book.*
- D. S. Wilkie and J. T. Finn, *Remote Sensing Imagery for Natural Resources Monitoring.*
- M.S. Rajan, *Remote Sensing and Geographic Information System for Natural Resource Management.*
- D.L. Verbyla, *Satellite Remote Sensing of Natural Resources.*
- R. Alexander and A. Millington (Edited), *Vegetation Mapping.*
- I.D. Yurkevich et. al., *Vegetation of Byelorussia, its Cartography, Conservation and Use.*
- J. Bolton, *NASA's Remote Sensing Tutorial.*

12 FUTURE OF PLANT TAXONOMY

AT A GLANCE

Plant taxonomy has largely been a neglected branch of science, and teaching and research in plant taxonomy has suffered a serious setback particularly in this country. It has largely been considered as a service science, particularly providing:

- names for organisms, so that one can refer to them,
- keys, so that one can identify what he or she encounters,
- classifications, which facilitate data retrieval, and
- phylogenies that holds considerable importance for other areas of biological sciences in which, the comparative method is central to experimental design and data interpretation.

12.1 Reasons for the Lacunae in the Field of Plant Taxonomy

The main reasons for the lacunae in the field of plant taxonomy, that has a bearing on its future, can be attributed to the following factors:

- As Gaston has pointed out, earlier taxonomical work has been largely size dependent i.e., earlier workers had to concentrate their attentions on larger taxa, as their optical microscopes were not up to present day standards and they did not have scanning electron microscope, which is now so important in many fields of taxonomy.
- Our predecessors also had less efficient collecting apparatus and techniques than are employed today.
- Another aspect is how much of taxonomy is taught to new generations of students (Bramwell, 1989). A recent survey has highlighted not only the relative decline in taxonomy as a research topic but also as a subject taught in higher education (Claridge & Ingrouille, 1992).
- With the recent advances in computing and phylogenetic theory and other related techniques, there has been increasing demand for systematics. However, due to the lack of sufficient number of adequately trained new taxonomists, workers in this field are unable to meet the present demands. If such trends are not stemmed soon, the decline in taxonomic training might even become irreversible because there would not be enough competent taxonomists to carry on training the next generation.
- Another important aspect relates to the groups that taxonomists work on at present, compared with the amount of taxonomic work that particular groups require. This issue has been recently examined by Gaston and May (1992), who compared the number of taxonomists working on particular groups with estimated numbers of described and undescribed species in each group, and found that the number of taxonomists working on a particular group may not be sufficient when compared to the size of the group. For example the number of taxonomists working on the microorganisms may be sufficiently low when compared to the staggeringly large number of species of microorganisms.

12.2 Importance of Continuing Taxonomic Research

In recent years however, following factors in particular have highlighted the importance of continuing taxonomic research and producing accurate phylogenies :

- The biological diversity is so vast that at present we really have no idea exactly how many species of plants exist in the world. Compared to the vast number of species of organisms inhabiting the earth, with estimates frequently in the range of 5 -15 million (May, 1988 & 1992), a small percentage of living species have as yet been described (Erwin, 1982). At the same time, human activities such as deforestation particularly of the species-rich tropical forests, are leading to increasing rate of extinction of many known and unknown species. As a result, vast number of species are going and will continue to go extinct well before any taxonomist would ever be likely to discover or describe them. For some groups, the number of extant species is known to a large extent, and these may be used to estimate the magnitude of the current crisis. It is therefore important that conservation efforts be directed more strongly to those areas that most deserve preservation, and the major problem that arises here is to decide which areas are more important than others. Systematics plays a very important role in this regard, as for many areas we know very little about the species occurring. It is necessary to decide whether it is the total number of species that holds more importance or the genetic diversity of the included species in deciding conservation priorities. Stace (1980) has pointed out that taxonomy has at least as important a role in deciding conservation priorities as have ecological considerations.
- Taxonomical research should be assigned sets of priority objectives amongst which, might be counted medicinally important groups, economically important ones and so on (Soule, 1990). The conservation of tropical plants ought to be an area of high priority from the point of view of the pharmaceutical industry. For example in the USA alone, although 25% of prescription medicines contain active ingredients of plant origin (Farnsworth, 1988), only handful of North American companies are actively screening the potential wealth of new compounds in tropical plants, and most of these are interested more from the pesticide point of view. Another excellent example in this regard is that the conservative estimates indicate that the number of extant species of fungi may be in the region of 1.5 million compared to the only 69,000 or so described till date (Hawksworth, 1991). Only some 7% of the latter described species have been studied from the point of view of their secondary metabolites. Those discovered so far include penicillin, cyclosporin and the relatively new avermectin insecticides. The potential of others to yield immensely profitable new products is obvious and remains an important area for research.
- Another reasonable objective will be to sample, preserve, describe, document, etc. a representative sample of the world's flora before too much has been lost. Such a survey would at least provide a valuable insight into the ranges of taxa that have evolved on our planet and probably a great deal more besides in terms of our understanding of the

complexity and stratification of various communities. For this purpose, employment of teams of parataxonomists can be quite beneficial. **Parataxonomists** are people with no previous higher education in biology or taxonomy but who receive a basic training in collection techniques, specimen preparation and identification and are then charged with the task of assembling collections of their particular groups of organisms which can then be submitted to relevant experts for identification.

With the advent of the **Red Data Books** and national and international conservation legislation such as the **CITES** agreement, taxonomy is gaining a whole new perspective, that of forming the basis for legal decisions on taxonomic matters, from details of nomenclature and the validity of subspecies and species to the justification for conservation of particular taxa.

GLOSSARY

A

Active Remote Sensing : Form of remote sensing, where the sensor provides its own source of electromagnetic radiation to illuminate the object understudy. Radar is an example of an active remote sensing device.

Aerial Photography : Form of remote sensing, that captures images of objects using photographic cameras, and film from platforms in the atmosphere.

Afforestation : Forest crops established by purposeful planting on land mass previously not used for tree crops.

Algae : A very simple photosynthetic plant that usually lives in moist or aquatic environments. The bodies of algae can be unicellular or multi-cellular is design.

Alkaloids : Structurally diverse, physiologically active secondary compounds derived from various amino acids or from mevalonic acid by various biosynthetic pathways.

Allelopathy : A particular form of amensalism found in plants. In this interaction, one species produces and releases chemical substances that inhibit the growth of another species.

Allogenic Succession : A succession caused by a change in environmental conditions that is unrelated to the activities of the developing plant community.

Anagenesis : Evolutionary process involving the gradual accumulation of changes in ancestor-to-descendant lineages (= phyletic evolution).

Analogy : Similarity of form between two taxa not shared by their nearest common ancestor - due to convergent evolution.

Angiosperms : Group of vascular plants, which encase their seeds in a mature ovary or fruit.

Anthocyanins : Violet, blue, red and yellow flavonoid pigments, not containing nitrogen, widespread among angiosperms.

Antibody : One of a diverse array of antigen binding receptors.

Antigen : Molecules that white blood cells recognize as foreign and that triggers an immune response.

Apetalous : Lacking of petals.

Apical placentation : Ovule or ovules attached at the apex of the ovary.

Apomyxis : Process in which, seeds are produced without fertilization. Example : Dandelion seeds.

Areole : Spiny short-shoot characteristic of the Cactaceae.

Aril : A hard to juicy often brightly colored outgrowth of the seed, usually developed from the funiculus or the outer integument.

Asexual Reproduction : Any process of reproduction that does not involve the fusion of gametes.

Autogenic Succession : Succession where the plant community causes the environment to change and this modification drives the succession.

Autotroph : An organism that produces food molecules inorganically by using a light or chemical based sources of external energy. This organism does not require outside sources of organic food energy for survival.

Axile placentation : Ovules attached to the central axis of an ovary with two or more locules.

B

Basal placentation : Ovules attached at the base of the ovary.

Base number : Inferred ancestral haploid chromosome number of a taxon usually prefixed by the letter "X".

Betalains : Nitrogenous red and yellow pigments restricted to the order Caryophyllales.

Binomial nomenclature : The system of naming developed by Linnaeus in which, each species name is composed of two words: the genus name, a noun and the specific epithet, an adjective or noun in apposition.

Biodiversity : The relative abundance and variety of plant and animal species and ecosystems within particular habitats, the diversity of life in all its forms and organization that comprise the natural world. Diversity includes genetic diversity, species diversity and ecosystem diversity.

Biological nomenclature : Rules for naming living organisms as outlined in the various nomenclatural codes.

Biological systematics : It is the classification of the immense diversity of life into a general reference system, which enables communication about particular groups of organisms and the storage and retrieval of information about these groups.

Biome : Largest recognizable assemblage of animals and plants on the Earth. The distribution of the biomes is controlled mainly by climate.

Biotechnology : The application of biological knowledge and techniques to develop products.

Bisexual flower : Flower with both androecium (stamens) and gynoecium (carpels).

Botanical nomenclature : Rules for naming plants as outlined in the International Code of Botanical Nomenclature.

Bracketed key : Key in which the two leads of a couplet (pair of contrasting choices) are placed on adjacent lines so that the leads of a couplet are not separated by intervening lines.

Bract : A reduced leaf in the axil of which arises a flower or an inflorescence branch.

Bracteole : Prophyll of a flower, usually very small borne below the flower on a pedicel, also called a bractlet.

C

C_3 : Plants having the photosynthetic pathway in which, the first intermediate formed after fixation of carbon dioxide is the three carbon compound phosphoglycerate.

C_4 : Plants having the photosynthetic pathway in which the first intermediate formed after fixation of carbon dioxide is the four carbon compound oxaloacetate.

Calcium oxalate : Oxalic acid ($HO_2C.CO_2H$) associated with calcium often occurring in plant tissues as raphide or druse crystals.

Callose : Glucose polymer, deposited during development in sieve areas of sieve tube elements.

Calyx : Collective term for all sepals of a flower.

Campylotropous ovule : Ovule that is curved so that the micropyle is positioned near the funiculus.

Carpel : Ovule bearing unit that make up the gynoecium.

Catkin : Inflorescence consisting of a dense, elongated mass of inconspicuous, usually wind pollinated flowers also called ament.

Clade : A cluster of taxa derived from a single common ancestor.

Cladistics : A mode of classification based principally on grouping taxa by their shared possession of similar (derived) characters that differ from the ancestral condition.

Cladogram : A cladogram is a "phylogenetic argumentation scheme" in shape of a tree-like graphic (dendrogram), which only specifies the relative degrees of phylogenetic relationship (sistergroup relationships) of the analyzed taxa, as well as their monophyly, based on the evidence of the recognized synapomorphies and autapomorphies.

Class : A taxonomic rank that stands between phylum and order; a phylum may include one or more classes and a class may include one or more orders.

Classification : It is the arrangement of groups of plants with particular characteristics by rank or position according to certain criteria and placing them within the taxonomic hierarchy. It includes the determination of position or rank for new taxa, as well as old taxa, which have been remodeled, divided, united, transferred or altered in rank.

Climax Community : Plant community that no longer undergoes changes in species composition due to succession.

Co-evolution : The coordinated evolution of two or more species that interact and exert selective pressures on each other that can cause each species to undergo associated adaptations.

Colpate : Pollen grain having elongate apertures called colpi.

Colporate : Pollen grain having apertures with pores within colpi.

Conservation : The preservation and protection of the environment and the wise use of natural resources.

Conservation Biology : The multidisciplinary science that deals with the conservation of genes, species, subspecies, communities, and ecosystems that make up earth's biodiversity.

Continental Drift : The theory that suggests that the earth's crust is composed of continental plates that have the ability to move. First proposed by A. Snider in 1858 and developed by F.B. Taylor (1908) and Alfred Wegener (1915). It is the movement over time of large landmasses-tectonic plates – other earth's surface relative to each other.

Convergence (Homoplasy) : The evolution of similar characters in genetically unrelated or distantly related species, mostly because they have been subjected to similar environmental selective pressures.

Convergent evolution : Where an analogous trait exists in two taxon because of an adaptation to a similar problem.

Cretaceous : Geologic period that occurred roughly 65 to 144 million years ago. During this period, the first flowering plant species appear and dinosaurs are at their greatest diversity. Dinosaurs die out at the end of this period.

D

Deforestation : The process of clearing of forests.

Dendrogram : Treelike diagram showing relationships, whether phenetic, cladistic, or evolutionary taxonomic.

Deoxyribonucleic Acid (DNA) : DNA from all organisms is made up of the same chemical and physical components. The DNA sequence is the particular side-by-side arrangement of bases along the DNA strand (such as **ATTCCGGA**). This order spells out the exact instructions required to create a particular organism with its own unique traits; first described by Francis Crick and James Watson in 1953, DNA looks like a twisted extension ladder. It is found in the nucleus and controls everything inside the cell.

Diacytic : Stomata enclosed by a pair of subsidiary cells whose common wall is at right angles to the guard cells.

Dichotomous key : Key composed of pairs of contrasting statements (couplets). They are keys in which the choices allow only two (mutually exclusive) alternative couplets.

Division : It is the major taxonomic rank within the Plant Kingdom. Alternative name for phylum.

E

Ecological Diversity : The variety of unique biological communities found on the earth.

Ecological Niche : It is all of the physical, chemical and biological conditions required by a species for survival, growth and reproduction.

Ecology : The study of the factors that influence the distribution and abundance of species.

Ecosystem : An ecosystem is a system where populations of species group together into communities and interact with each other and the abiotic environment.

Ecosystem Diversity : The variety of unique biological communities found on the earth. A component of biodiversity.

Ecotype : A phenotypic and genotypic variant of a species associated with a particular environmental habitat.

Electrophoresis : Process using an electrical current to separate molecules (usually proteins) on the basis of their size and electrical charge.

Endangered Species : A species found in nature that has so few surviving individuals that it could soon become extinct in all or most of its natural ranges.

Endemic : A species or population that is specific to a particular geographic region.

Epiphyte : Type of vegetation that gets its physical support from the branches of other plants.

Eukaryote : Organisms whose cells have a membrane-bound nucleus and many specialized structures located within their cell boundary. In these organisms, genetic material is organized into chromosomes that reside in the nucleus.

Evolution : A process by which species come to possess genetic adaptations to their environment. Its mechanism is natural selection.

Evolutionary adaptation : A genetically based characteristic expressed by a living organism.

F

Fern : A group of about 11,000 species of vascular seedless plants that belong to the division Pterophyta.

Flora : All plants of a particular region or time period.

Forest : A biological community dominated by trees and other woody plants.

Forestation : Establishment of a forest, naturally or artificially on an area, either previously forested or may not.

Forestry : The science, art and practice of managing and using trees, forests and their associated resources.

Fossils : The geological remains, impressions, or traces of organisms that existed in the past.

G

Gene Pool : Sum total of all the genes found in the individuals of the population of a particular species.

Genetic Diversity : The genetic variability found in a population of a species or all of the populations of a species.

Genetic Engineering : The process by which scientists use molecular tools to move genes from one organism to another, changing one or more traits of the receiving organism; similar to recombinant DNA.

Genome : An organism's complete set of DNA. Genomes vary widely in size; the smallest known genome for a free-living organism (a bacterium) contains about 600,000 DNA base pairs, while human and mouse genomes have some 3 billion.

Genus : A taxonomic category that stands between family and species; a family may comprise a number of genera, each of which contains a number of species that are presumably related to each other by descent from a common ancestor.

Geographical Coordinate System (GCS) : System that uses the measures of latitude and longitude to locate points on the spherical surface of the earth.

Geographic Information System (GIS) : A geographic information system merges information in a computer database with spatial coordinates on a digital map.

Geological Time Scale : The correlation between rocks and time periods of the past.

Germplasm : The total genetic variability, represented by germ cells or seeds, available to a particular population of organisms.

Global Positioning System (GPS) : System used to determine latitude, longitude, and elevation anywhere on or above the earth's surface. This system involves the transmission of radio signals from a number of specialized satellites to a hand held receiving unit. The receiving unit uses triangulation to calculate altitude and spatial position on the earth's surface.

Global Warming : Warming of the earth's average global temperature because of an increase in the concentration of greenhouse gases.

Gondwana : The super continent in the Sourthern Hemisphere formed from the breakup of the larger Pangaea landmass about 180 million years ago. Gondwana was composed of what is now South Africa, Africa, Antarctica, Australia and India.

Greenhouse Effect : The greenhouse effect causes the atmosphere to trap more heat energy at the Earth's surface and within the atmosphere by absorbing and re-emitting long wave energy. Of the long wave energy emitted back to space, 90% is intercepted and absorbed by greenhouse gases. Without the greenhouse effect the Earth's average global temperature would be -18° Celsius, rather than the present 15° Celsius.

Greenhouse Gases : Gases responsible for the greenhouse effect. These gases include : carbon dioxide (CO_2); methane (CH_4); nitrous oxide (N_2O); chlorofluorocarbons (CFxClx); and tropospheric ozone (O_3).

Gymnosperm : Plant that bears naked seeds. Representatives of this group include the conifers.

H

Hemiparasite : An organism, which lives on and derives part of its nourishment from a different organism, and is partially self-supporting.

Herb : A non-woody angiosperm whose above ground vegetation dies off seasonally.

Herbarium : It is a store-house of dried and pressed plant specimens collected from different places, which are mounted on appropriate sheets and arranged according to some known system of classification, and kept in pigeon-holes of steel or wooden cupboards specially prepared for the purpose.

Hierarchy : A term used in some evolutionary and developmental studies to designate increasing levels of complexity or organization.

Holotype : The one specimen or other element designated by the author as the nomenclatural type in describing a new species.

Hybridization : Process using one strand of DNA to create the paired strand, so that the proper double helix structure of DNA can be created; by having one strand, the use of DNA probes can be used and the discovery of the exact makeup of DNA is much easier to uncover.

I

Identification : Determination of the name of an unknown plant. Identification is the assignment of additional unidentified plants to a correct rank once a classification has been established.

Indented key : Key with each pair of contrasting choices (leads) that are equally indented and followed by the appropriate couplets and thus the two leads are often widely separated in the key.

Indicator Species : Species that can be used as an early indicator of environmental degradation of a community or an ecosystem.

Iridoids : Nine or ten carbon derivatives of a simple terpene unit.

K

Karyotype : Preparation of metaphase chromosomes sorted by length, centromere location and other features.

Keys : They are devices consisting of a series of contrasting or contradictory statements or propositions requiring the identifier to make comparisons and decisions based on statements in the key as related to the material to be identified. A single pair of contradictory statements is called a *couplet* and each statement of a couplet is termed a *lead*.

Kingdom : It is the most inclusive taxonomic rank recognized by the *International Code of Botanical Nomenclature*.

Kranz anatomy : Foliar veins surrounded by conspicuously enlarged bundle sheath cells containing chloroplasts; associated with C_4 photosynthesis.

L

Lectotype : A specimen as the standard bearer of a species or subspecies and selected from the syntype one.

Living Fossil : An existing species whose similarity to ancient ancestral species indicates that very few morphological changes have occurred over a long period of geological time.

M

Macroevolution : Evolution of taxa higher than the species level (for example, genera, families, orders, classes), commonly entailing major morphological changes.

Mangrove : Trees of wetlands located on the coastlines in warm tropical climates.

Mesophyte : Plants that have moderate water requirements.

Mesozoic era : The middle era of the Phanerozoic eon, covering the approximately 180 million years interval between the Paleozoic (ending about 245 million years ago) and Cenozoic (beginning about 65 million years ago).

Microevolution : Evolutionary changes of the kind usually responsible for causing differences between populations of a species.

Microspecies : Minimally differentiated series of populations derived from uniparental reproduction.

Molecular systematics : Molecular systematics is that part of systematics, which utilizes data from nuclear, chloroplast or mitochondrial DNA, to elucidate phylogenetic relationships between plants.

Monad : Generally, a single individual free from others, specifically a single pollen grain.

Monocolpate : Pollen grain with a single long grooved aperture.

Monograph : Comprehensive systematic study of a particular taxonomic group, including keys, descriptions, illustrations, distribution maps, etc.

Monophyletic : A group of organisms, which includes the most recent common ancestor of all of its members and all of the descendants of that most recent common ancestor. A monophyletic group is called a clade.

Monoporate : Pollen grain with a single pore like aperture.

Monosulcate : Pollen grain with a single, long and grooved aperture located at the pole.

Monotypic : Containing only one taxon of the next lower rank, e.g. applied to a family containing only one genus.

Morphological System : This system describes the process relationships or correlations between the elements of the system in terms of measured features.

Morphometry : The measurement of shape. Measurements are then manipulated statistically or mathematically to discover inherent properties.

Moss : About 9,500 species of plants that belong to the division bryophyta. These low growing plants are common in moist habitats.

N

Nomenclature : It is the naming of a taxon correctly. It is a precise and universal system of rules used by all botanists of the world for naming newly discovered plants.

Nomen nudum : It is a name published without a diagnosis or description of the entity to which it was applied, and without reference to either.

Nomen rejiciendum : It is a name rejected in favour of a 'nomen conservandum'.

Numerical Taxonomy : A statistical method for classifying organisms by comparing them on the basis of measurable phenotypic characters and giving each character equal weight. The degree of overall similarity between individuals or groups is then calculated, and a decision is made as to their classification.

O

Ontogeny : The developmental history of an individual from egg to adult.

Operational Taxonomic Unit (OTU) : Terminal taxon used in an analysis of relationships, especially in phenetic studies.

P

Palynology : Study of the form and structure of pollen and spores.

Paracytic : Stomata surrounded by two subsidiary cells parallel to the guard cells.

Paraphyletic group : Group containing a common anscestor and some but not all of its descendents diagnosed by symplesiomorphies.

Paratetracytic : Stomata surrounded by four subsidiary cells, two of them being parallel to the guard cells and the other pair being polar and often smaller.

Pericytic : Stomata surrounded by a single subsidiary cell.

pH : Scale used to measure the alkalinity or acidity of a substance through the determination of the concentration of hydrogen ions in solution. A pH of 7.0 is neutral. Values below 7.0, to a minimum of 0.0, indicate increasing acidity. Values above 7.0, to a maximum of 14.0, indicate increasing alkalinity.

Photosynthesis : Is the chemical process where plants and some bacteria can capture and organically fix the energy of the sun. This chemical reaction can be described by the following simple equation:

$$6CO_2 + 6H_2O + \text{light energy} = C_6H_{12}O_6 + 6O_2$$

Phylectic evolution : Evolutionary changes that produce two or more lineages that diverge from a single ancestral lineage.

Phyllotaxy : It is the arrangement of leaves on a stem.

Phylogenetic Classification : Classification of organisms based on genetic connections between other species.

Phylogenetic Systematics : A methodology, for the reconstruction of phylogenetic trees and the discovery of monophyletic groups by the exclusive use of shared (homologous) derived character states (synapomorphies), as well as the reconstruction of the ground plans of the discovered monophyletic groups.

Phylogeny : The evolutionary history of a species or groups of species in terms of their derivations and connections.

Physiological adaptation : The change in an organism's physiology as a result of exposure to some environmental conditions.

Plant : Organisms belonging to the kingdom Plantae. These organisms have the following general characteristics: lack of locomotion, lack of a nervous system, and cellulose cell walls. Most plants can photosynthesize.

Plantae : Group, at the kingdom level, in the classification of life. Multi-cellular organisms that have a eukaryotic cell type, chloroplasts, mitochondria and a cell wall composed of cellulose.

Point mutation : Change of a single nucleotide in a strand of DNA.

Pollen : Microspore containing a male gametophyte in angiosperms germinating to form a pollen tube, which rapidly transport the sperm to the ovule.

Polocytic : Stomata with guard cells partially surrounded by a single subsidiary cell.

Polyad : Small cluster of pollen grains.

Polyclave key : Multi-access key, as in a punched card key or computer interactive identification.

Polycolporate : Pollen grain with several long, grooved apertures each with a central pore.

Polygamous : With both bisexual and unisexual flowers on the same plant.

Polymerase chain reaction (PCR) : A laboratory technique that can replicate a sequence of DNA nucleotide into millions of copies in a very short time.

Polymorphism : The presence of two or more genetic or phenotypic variants in a population. Usually refers to genetic variations where the frequency of the rarest type is not maintained by mutation alone.

Polynomial : Before Linnaeus devised the simple system for naming plants and animals that we use today, natural historians used a phrase of several words to name a particular species.

Polyphyletic : The presumed derivation of a single taxonomic group from two or more different ancestral lineages through convergent or parallel evolution.

Protista : One of the eukaryotic kingdoms; includes protozoa, algae, slime molds and some other groups.

P-type plastids : Plastids in sieve tube elements that accumulate proteins (or proteins and starch) in the form of crystalloids or filaments.

R

Rare : Not widely known.

Recombinant DNA technology : Procedure used to join together DNA, outside of a cell or organism. The recombinant DNA can then be inserted into a cell and changes — such as more nutrients in a carrot or a chemical to block bugs from potatoes — will be produced in that plant. Differs from genetically engineered bacterium by not using a living organism to transfer information.

Reforestation : Replanting of trees on previously forested lands.

Relationship (in cladistics) **:** Meaning either ancestor-descendant or collateral (sister/cousin relationships).

Remote Sensing : The gathering of information from an object or surface without direct contact.

Remote Sensor : Mechanical devices used to remotely sense an object or phenomenon.

Respiration : The process where mitochondria of cells of organisms release chemical energy from sugar and other organic molecules through chemical oxidation. This process occurs in both plants and animals. In most organisms, respiration releases the energy required for all metabolic processes. This chemical reaction can be described by the following simple equation :

$$C_6H_{12}O_6 + 6O_2 = 6CO_2 + 6H_2O + \text{released energy}$$

Restriction Fragment Length Polymorphisms (RFLPs) : Differences between individuals in the size of DNA fragments for a particular DNA section cut by restriction enzymes. These are inherited in mendelian fashion and furnish a basis for estimating genetic variation. They also provide linkage markers used to track mutant genes between generations.

S

Sclerophyllous : Literally meaning sclerophyll hard leaves – referring to the small, tough evergreen leaves.

Sensu lato : In the broad sense; with a wide or general interpretation. Abbreviated s.i.

Sensu stricto : In the strict or narrow sense; with a restricted interpretation. Abbreviated s.s.

Serology : Technique that assesses evolutionary relationships using an animal's immunological antibody-antigen reaction when exposed to a variety of foreign substances including plant proteins.

Sexine : Outer portion of the exine of a pollen grain.

Silviculture : Silviculture is a science of forest nature, method for forest growing and forest improvement.

Southern blot : Method of determining lengths of DNA fragments by transferring denatured fragments from an electrophoresis gel to a nylon membrane.

Species : A group of interbreeding organisms that do not ordinarily breed with members of other groups.

Specific epithet : The second word of a binomial species name; usually an adjective that modifies the generic name.

Stoma : Controllable opening between two guard cells in the epidermis of the vascular plants, also called a stomata.

S-type plastids : Plastid in the sieve tube elements that accumulates starch.

Subsidiary cell : Epidermal cell lying next to a guard cell and differing in size and shape from other epidermal cells.

Subspecies : An infraspecific rank, usually applied to geographical races within a morphologically variable species.

Sulcate : Pollen grain with a sulcus.

Sulcus : Long, grooved aperture positioned at the pole of a pollen grain.

Syncolpate : Condition resulting from the fusion of the furrows (colpi) of a pollen grain.

Synonym : One of two or more names applied to the same taxon.

Systematics : It is the study of phenetic, genetic and phylogenetic relationship among taxa.

Systematist : A person trained in the discipline of systematics.

T

Tautonym : Name of a species in which the specific epithet exactly repeats the generic name, not permissible in the botanical code.

Taxon : A classification category for a group of organisms, the unit is that of a species, genus, family, order and so on.

Taxonomic Classification : Classification of organisms based on structural and physiological connections between other species.

Taxonomist : Person trained in the discipline of taxonomy.

Taxonomy : Taxonomy is the science dealing with description, identification, nomenclature and classification of living things, Taxonomy means classification following certain rules or, principles.

Tetracytic : Stomata surrounded by four subsidiary cells.

Tetrad : Group of four, usually applied to pollen grains united in a group of four that have not separated after meiosis.

Threatened : Likely in the near future to become endangered.

Tricolpate : Pollen grain with three long, grooved apertures.

Tricolporate : Pollen grain with three long, grooved apertures, each with a central pore.

Triporate : Pollen grain with three equatorial, pore-like apertures.

Type specimens : When a scientist describes and gives a scientific name to a plant or animal for the first time, the actual specimen that he was working from is known as the 'type'.

Typification : A representational or typifying form.

Z

Zonate apertures : Ring-shaped or band shaped apertures of a pollen grain.

INDEX

A

B

C

D

E

F

G

H

I

J

K

L

M

N

O

P

S

T